D1415800

Maryland Geography

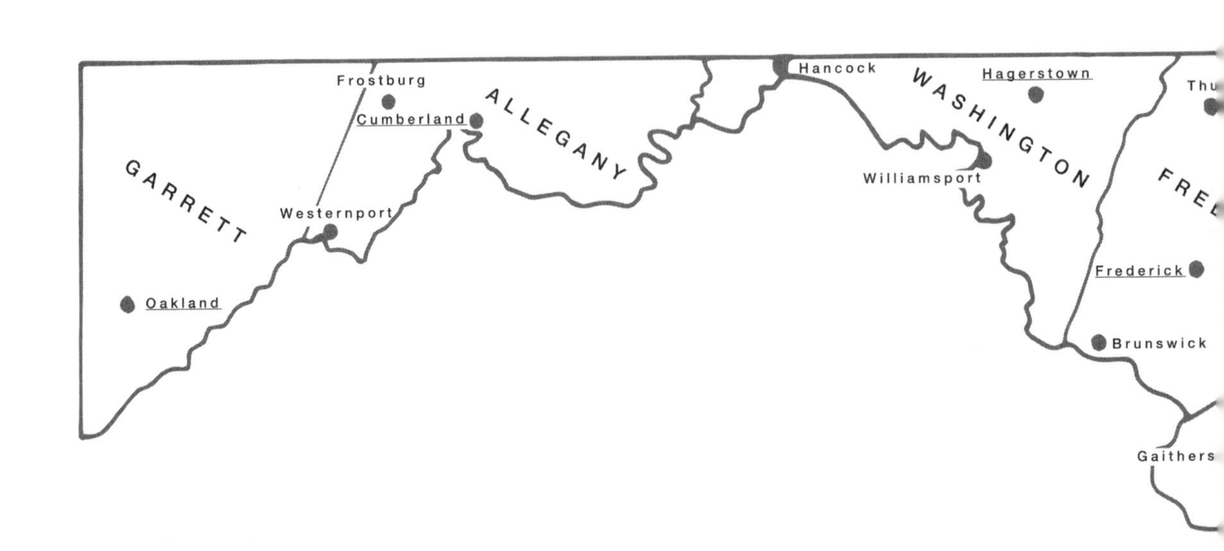

Frostburg
Cumberland
ALLEGANY
Hancock
WASHINGTON
Hagerstown
GARRETT
Westernport
Oakland
Williamsport
Frederick
Brunswick

Maryland Geography

An Introduction

James DiLisio

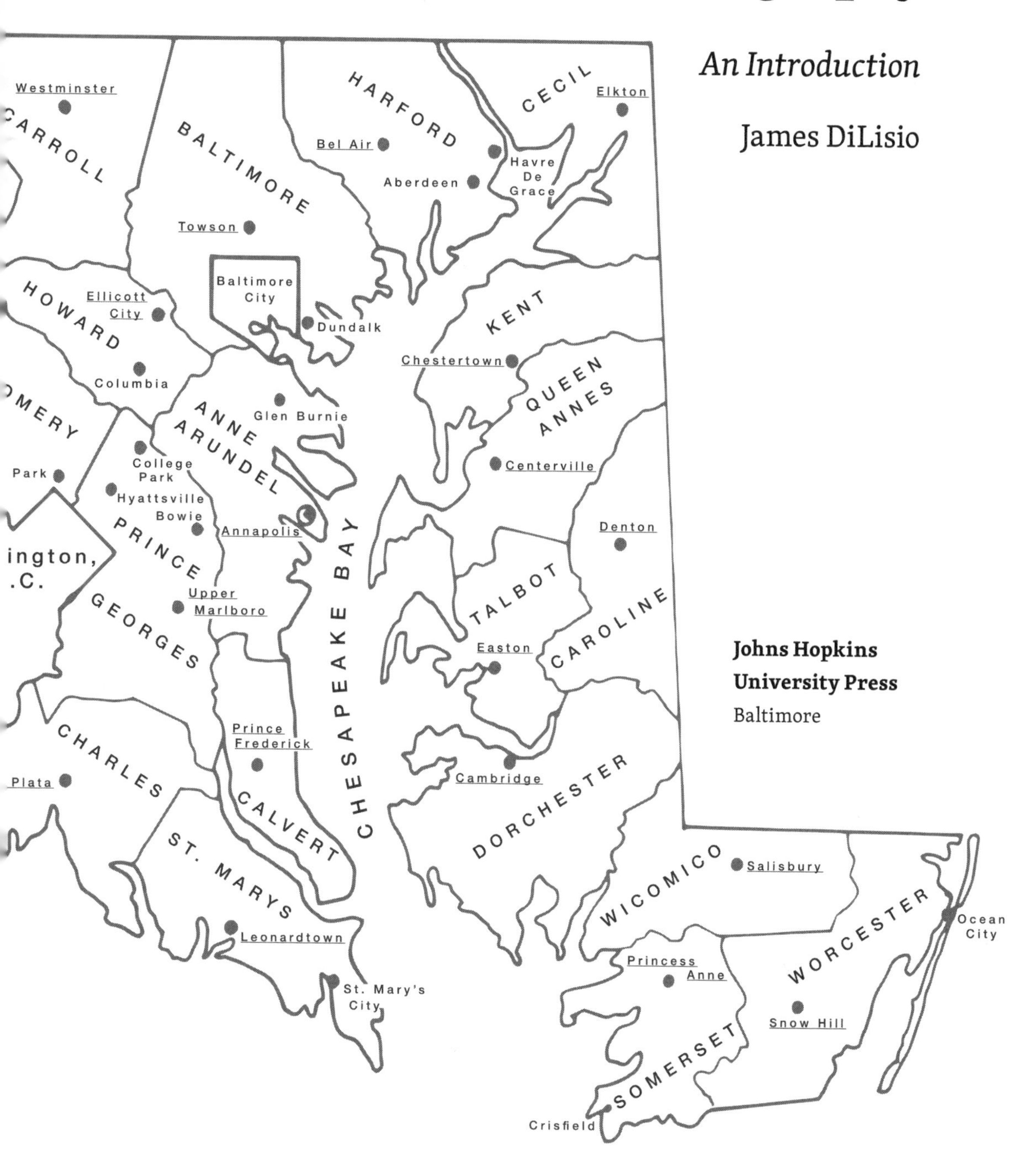

Johns Hopkins University Press
Baltimore

© 2014 Johns Hopkins University Press
All rights reserved. Published 2014
Printed in the United States of America on acid-free paper
9 8 7 6 5 4 3 2 1

Johns Hopkins University Press
2715 North Charles Street
Baltimore, Maryland 21218-4363
www.press.jhu.edu

Library of Congress Cataloging-in-Publication Data

DiLisio, James E.
Maryland geography : an introduction / James DiLisio.
pages cm
Includes bibliographical references and index.
ISBN-13: 978-1-4214-1482-9 (paperback : acid-free paper)
ISBN-10: 1-4214-1482-1 (paperback : acid-free paper)
ISBN-13: 978-1-4214-1483-6 (electronic)
ISBN-10: 1-4214-1483-X (electronic)
1. Maryland—Geography. 2. Maryland—Social conditions. 3. Maryland—Economic conditions. 4. Maryland—History. 5. Landscapes—Maryland. I. Title.
F181.8.D55 2014
917.52—dc23
2013050141

A catalog record for this book is available from the British Library.

Special discounts are available for bulk purchases of this book. For more information, please contact Special Sales at 410-516-6936 or specialsales@press.jhu.edu.

Johns Hopkins University Press uses environmentally friendly book materials, including recycled text paper that is composed of at least 30 percent post-consumer waste, whenever possible.

CONTENTS

ACKNOWLEDGMENTS

Like most works of regional geography, this book has evolved over many years. Throughout its evolution, I have been fortunate to receive help and inspiration from numerous people. Over six years ago, I shared my initial concept for the book with Bob Brugger, senior editor at Johns Hopkins University Press. Bob's insightful recommendations on how to restructure the project proved invaluable in the final draft. My thanks to him for this initial help and for coming back to the project during its final stages. During a critical period of writing in 2012, former JHUP editor Greg Nicholl stepped in to offer his expertise in editing, formatting, and graphics, and helped me cut a massive volume down to a more concise and manageable size.

More than in some other disciplines, geography books are driven to a large degree by meaningful graphics. Special thanks are owed to my son, Jay DiLisio, for taking on the job of producing a set of well-designed graphics that are consistent in style and deliver the message in a clear manner. Jay's mother and I are pleased that our investment in his degree from the Maryland Institute College of Art was well spent. Finally, I never would have completed this book without the constant support of my wife, Kay. Several times, when I was ready to put the project on the shelf, she offered the kind of encouragement that can only come from love and togetherness after nearly forty years. I give my thanks also to the many students and colleagues at Towson University who over the years have helped me learn about the geography of this great state.

Maryland Place Names

- County Seat
- Other Places
- State Capital

Maryland Geography

Introduction The Itinerary

i·tin·er·ary, *n., pl.* -ar·ies.

1. a detailed plan for a journey, esp. a list of places to visit.
2. a line of travel; route.
3. an account of a journey.
4. a guide for travelers.

—Random House Webster's College Dictionary, 1997

In November 1634, two small ships set sail from England with the goal of reaching the shores of the New World. The *Ark* and the *Dove* had an eventful Atlantic crossing. After being separated for a time, they rejoined in Barbados. With high hopes and expectations, tempered by a natural anxiety of the unknown, the ships' passengers, estimated at 122, sailed up the East Coast of North America, stopping in Virginia and eventually landing in what is now St. Mary's County, Maryland.

Among the passengers on the *Ark* and the *Dove* were three Jesuit priests and the governor of the new colony, Leonard Calvert, brother of Cecil Calvert, Baron of Baltimore and proprietor of the colony granted to him by King Charles I on June 30, 1632, in the Charter of Maryland. The most prominent of the Jesuits was the Reverend Andrew White, whose account of the voyage tells the story of the Atlantic crossing and landing in Maryland:

> On the twenty second of the month of November in the year 1633, being St. Cecilia's Day, we set sail from Cowes, in the Isle of Wight, with a gentle east wind blowing. The winds increasing and the sea growing more boisterous, we could see the pinnace [the *Dove*] in the distance, showing two lights of her masthead. Then indeed we thought it was all over with her, and that she had been swallowed up in the deep whirlpools, for in a moment she had passed out of sight, and no news of her reached us for six months afterwards. But after returning to England and made a fresh start from thence . . . and overtook us months later in the Antilles.
>
> Twenty seventh of February reached Point Comfort in Virginia. We were kindly treated by the English. Set sail on March 3, 1634 into Chesapeake Bay turning north into the mouth of the Potomac River.

> Never had I ever beheld a larger or more beautiful river than the Potomac. The Thames seems a mere rivulet in comparison with it. The first island we came to we called St. Clements Island on the day of the Annunciation of the most Holy Virgin Mary in the year 1634 . . . we celebrated mass for the first time on this island. We erected a great cross hewn from a tree as a tribute to Christ the Savior. Since the island contains only 400 acres, it would not afford room for a new settlement. We went 9 leagues from St. Clements Island up the north side of the Potomac to the St. Mary's River. About a mile from the shore we laid out the plan of a city, naming it after Saint Mary. (Maryland Historical Society, Fund Publication No. 1, Baltimore, Maryland, 1874; http://www.thearkandthedove.com/narrative.html)

Although the new Maryland colony had a modest and inauspicious beginning, it soon turned into a grand experiment. Today, more than 375 years later, the fruits of those early adventurous explorer-settlers are evident on the landscape of Maryland, one of the most prosperous, diverse, and progressive states in the United States.

This book takes the reader on an itinerary through time and across the physical and human landscape of Maryland. Part I, Shaping the Landscape, addresses its historical geography, regional setting, and physical environment. Part II, Old Economy, New Economy, describes how Marylanders have shaped the economic landscape and how they make a living. Part III, Human Footprints on the Marylandscape, focuses on the people of Maryland by looking at their many characteristics.

Throughout the book, I discuss the challenging issues that Marylanders today face as they struggle to sustain the uniqueness of their state. Some of the concerns related to maintaining Maryland's quality of life are listed in table I.1. Although Maryland's shifting employment patterns, declining industries, environmental degradation, deforestation, urban sprawl, and social and political demands are challenges that other states face, they combine on the Maryland landscape in a geographically unique pattern. It is how Marylanders have responded creatively that makes Maryland a national leader in addressing many of these issues.

One cannot overlook what the great diversity of physical environments has meant to Maryland through the years. Maryland's central location in a huge population cluster and market, known as a megalopolis, ensures its economic success and influence. Each region of Maryland has its own character that has contributed in significant ways to the state's historical and contemporary story. As a comprehensive geographical survey, this book synthesizes the complex and varied relations between Maryland's physical diversity and its human geography. It describes how its people have integrated and maintained their regional cultures, politics, and outlooks.

Although much of central Maryland is part of the old American Manufacturing Belt, sometimes called the Rust Belt, suffering a de-

Table I.1. Major challenges to sustaining Maryland's quality of life

Chapter	*Major challenges*	*Major assets*
2	• part of the old American Manufacturing Belt (Rust Belt) in western Maryland's Appalachian region	• central location in megalopolis
3, 4	• deforestation • degradation of Chesapeake Bay • air and water pollution	• varied physical geography • mineral resources • Chesapeake Bay
12	• older, decaying urban centers	• rich cultural heritage
8, 9	• uneven economic growth and prosperity	• diversified economy
5	• declining number of farms, farmers, and farmland • decline of tobacco culture and farming • environmental costs of the Eastern Shore poultry industry	• agricultural innovations such as no-till cultivation and a young wine industry • tremendous growth of poultry industry
6	• depleting resources of the Chesapeake Bay • long-standing Maryland-Virginia border dispute	• active marine science and research to save the bay • interstate cooperation to save the bay
7, 4	• many issues of environmental degradation • decline of steel industry • decline of coal mining and associated environmental degradation	• strong governmental and private efforts to address environmental issues • economic transformation to a more white-collar, information economy • long minerals heritage dating to the colonial period
8	• decline of manufacturing employment	• new manufacturing jobs and the rise of the high-tech industry • strong knowledge economy
9, 12	• heavy demands for recreational and open space as population grows	• tremendous inventory of natural resources: state and national parks, greenways, the bay, proximity to Washington, DC
9	• significant population growth; high population density; changing geographic patterns of population, diversity, housing, migration, wealth and poverty, crime, education	• wealthy state overall • well below average poverty • highly skilled population • high education levels to meet needs of the knowledge economy
9, 10	• degradation of infrastructure • decline of the Port of Baltimore • lack of health care for many	• proximity to Washington, DC; many state and federal institutions generate jobs and income • centrality within megalopolis
12	• decaying urban centers • urban sprawl • rising crime, homelessness	• forward looking smart growth policies • public awareness of sprawl issues • time and resources to address problems of sprawl

cline in manufacturing jobs, Maryland is reinventing itself as a producer of high-technology products in knowledge-based industries that produce for the regional, national, and international markets. The decay of older urban centers and older suburbs has been addressed aggressively by experimenting with approaches such as smart growth, which emphasizes higher-density urban development over expanding into formerly rural areas. Marylanders generally recognize that Baltimore City and the Port of Baltimore form a huge economic engine that drives the Maryland economy. The degradation of urban infrastructure, crime rates, homelessness, and crises in education are some of the most difficult challenges facing Maryland today. Even so, at the dawn of the twenty-first century, families in Maryland had the highest median family income in the United States.

There are numerous environmental challenges facing Marylanders, ranging from deforestation to the recovery of former mining areas and the health of the Chesapeake Bay. The Chesapeake Bay and Port of Baltimore are tremendous assets that funnel large volumes of trade into and out of Maryland while supporting thousands of jobs. Depleting the bay's resources—especially its fishery—is of major concern to Marylanders. The Chesapeake is the largest estuary in the United States and a national treasure. Paul Sarbanes, a former US senator from Maryland, once described it as the soul of the state. Any list of unifying iconic features in Maryland is likely to have the Chesapeake Bay at the top.

Perhaps one of the wisest decisions ever made by the Maryland forefathers was to donate a block of land along the Potomac River to the federal government on which to build the nation's capital. The social, economic, and political benefits that come from Maryland's intimate proximity to Washington, DC, are immense. It is has been said about Maryland's economy that it is built upon "feds, meds, and eds"—the federal government, the medical and biotechnology industries, and higher education. This trio certainly contributes in significant ways to modern Maryland.

PART I

Shaping the Landscape

1 Maryland's Past in Today's Landscape

To understand the contemporary physical and cultural landscapes of Maryland, one must place them in a historical-geographical context. The present cultural landscape of Maryland has evolved over five historical phases: (1) pre-European American Indian occupance, (2) colonization in the seventeenth century, (3) agrarian development in the eighteenth century, (4) the early urban industrial period of the nineteenth century, and (5) the mid-industrial period of the late nineteenth to the early twenty-first century. Population, transportation, economy, culture, and environment were features of each of these historical periods, and their evolution through the centuries can help us understand the emergence of mid-twentieth-century Maryland's isolated, rural, conservative Eastern Shore; the rural tobacco land of southern Maryland; the rugged mining area of western Maryland; the prosperous, mixed farming area of the Piedmont; and the metropolitan Washington–Baltimore corridor.

Upon first seeing the countryside around the Chesapeake Bay in 1608, Captain John Smith was recorded to have exclaimed, "heaven and earth seemed never to have agreed better to frame a place for man's habitation" (Vokes 1968). Captain Smith's travels did not take him to the land west of the bay, but had he traveled there, he would have found a region just as suitable for European settlement. After finding the environment of his first royal land grant in Newfoundland too harsh for settlement, Lord Baltimore secured a royal charter in 1632 from King Charles I for a proprietary colony in the northern Chesapeake area. Lack of knowledge of this new land and environmental physical difficulties were profound, yet there were compensating factors that allowed the Maryland colony to take root. Primary among these factors were similarities of climate and vegetation to the colonists' homeland. The late cultural geographer Carl Sauer once stated, "It would be impossible, indeed, to cross an ocean anywhere else and find so little that is unfamiliar on the opposite side." Little did Lord Baltimore realize that he was founding a future state in a new nation—a state that was to become a focus in the new country, a gateway to the West, a border state between north and south, and part of a great megapolitan urban region.

Pre-European American Indian Occupance

Three archaeological periods constitute the long span of human prehistory in eastern North America: (1) the Paleo-Indian Period, 11,000 to 8000 BCE; (2) the Archaic Period, 8000 to 1000 BCE; and (3) the Woodland Period, 1000 BCE to 1600 AD. Much of what we know about these prehistorical periods is based on cultural traits reflected in artifacts. The environment, too, exerted strong influences on the evolution of Native American culture in Maryland; dramatic environmental changes over the past 14,000 years have transformed the climate, vegetation, fauna, topography, and bodies of water of the region. The Indians of the region skillfully adapted to these environmental changes.

The Paleo-Indian Period: 11,000 to 8000 BCE

Approximately 14,000 years ago, the ancestors of American Indians migrated to North America, coming across the Bering Strait from Asia to Alaska. These people spread across the continent, reaching the Middle Atlantic region about 10,000 to 11,000 years ago. Although the fact of this migration was for a long time considered by anthropologists and cultural geographers as conjecture and guesswork, the recent Genographic Project of the National Geographic Society has uncovered definitive genetic proof that the ancestors of North American Indians indeed originated in Asia.

These earliest arrivals were nomadic hunters and are today known as Paleo-Indians, or "ancient Indians." When they arrived in what is now Maryland, the continental climate was much colder than it is today. The colder climate of the most recent Ice Age, also known as the Pleistocene Epoch, was ideal for many large animals, including the mammoth, great bison, ground sloth, caribou, and saber-toothed tiger. Yet evidence indicates that in the Middle Atlantic region the Indians hunted white-tailed deer, caribou, and elk. They also hunted smaller animals; collected edible plants; and ate fish, nuts, and berries.

Great glaciers covered the northern part of the continent, stretching southward into northern Pennsylvania. Although Maryland was not directly covered by ice, it was greatly affected by it. Because so much of the Earth's water was trapped in these continental glaciers, the sea level was much lower than it is today; by geologists' estimates, about 325 feet (99 meters) lower. Indeed, the Chesapeake Bay did not yet exist (Porter 1983). Before there was a Chesapeake Bay, small, seminomadic bands of Paleo-Indians lived along the ancient Susquehanna River and its tributaries. Little evidence is left of these people except for fluted spear points and other tools such as steep-edged scrapers used for scaling fish and cleaning animal hides to make clothing.

The Archaic Period: 8000 to 1000 BCE

As the Pleistocene began to come to an end about 10,000 years ago, the climate gradually became warmer and drier. Glaciers melted and retreated northward, and the meltwaters carved out new drainage patterns on the landscape. The lower ancient Susquehanna River flowing southward through Maryland began to increase in size as it was supplied with vastly greater amounts of freshwater from the upper river as well as salt water from the rising level of the sea. Mixing fresh and salt waters drowned the lower river and formed the Chesapeake Bay. As the climate warmed, big game animals did not adapt well and eventually died out. Trees began to grow in the former grasslands, and before long, forests of oak and hemlock replaced spruce and fir trees throughout Maryland. Smaller animals, such as deer, became important sources of food for the Archaic Indians.

During the Archaic Period, the Indian population of Maryland changed in many ways. Although they still occasionally moved about by season, they developed more permanent villages. There was by now a limited dependence on the fish and shellfish of the Chesapeake Bay. Most settlements were concentrated on inland sites near swamps that attracted game animals and provided edible plants and freshwater. Indians continued to hunt forest animals with their notch-tipped spears, the most important prey being the white-tailed deer. They used stone axes and scraper tools to clear away forests and to build shelters and log canoes. The Archaic Indians had not yet developed domesticated animals, nor did they grow their own food. They traveled by season within the Chesapeake Bay region in search of a livelihood. There is little evidence that they extensively harvested oysters, crabs, or other marine life from the bay and its tributaries during the late Archaic Period. Instead, Indians of this period worked the bay in crafted dugout canoes, which were cut and cured from tree trunks, and then had their insides burned out. Finally, the canoes were shaped with sharpened stone tools and shells.

Several significant environmental changes in the late Archaic Period led to adaptations by the Indians. As the climate became drier, and oak, hickory, chestnut, and pines dominated forests, acorns and other nuts became more important in the Indian diet. Unlike the early part of Archaic Period, when the sea level rose rapidly and the Chesapeake Bay emerged, the late part of this period witnessed a slower rate of sea level rise. The estuary and its tributaries began to stabilize, and populations of various species of fish and shellfish grew. The bay became a source of dependable and abundant food. Major Indian settlement locations began to shift from inland sites to the coasts of the estuary. Archaeological evidence (deposits of shell middens) dates the earliest Indian village sites to about 2000 BCE. These people had more access to a wider variety of food because they were still hunting and gathering plants and nuts.

The Woodland Period: 1000 BCE to 1600 AD

During the late Archaic Period, Maryland Indians were leading an increasingly sedentary way of life that became more pronounced over time. In the early Woodland Period, small groups lived in kin-based hamlets. By the end of the period, larger villages were established. Maryland Indians began to make pottery from local clays, and the bow and arrow were introduced around 800 AD. (Maryland Geological Survey 1976). The various Indian tribes of the Chesapeake Bay region established wide-ranging trading networks as far away as the Ohio River area. These tribes participated in cooperative trade and security arrangements, evidenced by the presence of large, fortified villages that emerged during the late Woodland Period. As the Indian tribes began to develop their agricultural skills, the need for more arable land sometimes led to conflict. The causes for conflict between tribes were most likely related to the greater competition for land and other resources as the Indian population and settlement size reached their apex during this period. Populations began to disperse as small- and medium-sized settlements emerged in the lowland floodplains inland from the coast.

Around 900 BCE, the climate began to warm even more. The Medieval Warm Period brought drier and warmer conditions. During this time, Maryland Indians began to cultivate more plants, especially maize (corn), beans, and squash. The drier climate likely reduced natural food resources, leading to the need for greater cultivation. Settlement patterns had an impact on food production, too; with populations spread over a wider area, density and pressure on resources were reduced, whereas the larger, higher-density villages put greater pressure on smaller areas for resources.

By about 1300 AD, Maryland Indians were again changing their settlement patterns. Larger, fortified villages began to appear along the coastal plain, especially near the fall line on the Western Shore of the Chesapeake Bay. Conflicts arose between coastal Indians and the piedmont Indians to the west. Populations were growing, and groups fought over land and resources. As agriculture became more intensive, it required a more permanent settlement regime. The land had to be cleared, and crops had to be planted, cared for, and protected. In the springtime, some Indians prepared the fields while others traveled to the coast to harvest oysters. A complex division of labor emerged, and tasks were divided by gender and age. The women worked the fields, raising crops. They also gathered wild plants, berries, and nuts and perhaps also harvested shellfish in the bay and its tributaries. Men did the hunting to provide meat; they also fished. Overall, women provided most of the food.

The annual cycle of life for Indians living in eastern Maryland during the late Woodland Period was based on a strategy to secure ample and reliable resources. Indians had a deep understanding of their geography, and knowledge of the seasons and when resources became available in each part of their ecosystem was essential to their survival. From late winter into early spring, food stocks were low. The Indians gathered clams and oysters, and hunted waterfowl on their annual migration to the bay. Various anadromous saltwater fish migrating to freshwater tributaries of the Chesapeake Bay to spawn were important parts of Indians' springtime diet. Another early spring activity was clearing and planting fields. Maize, the most important crop, provided as much as 50 percent of the calories in Indians' diets. Other crops such as beans, sunflowers, squash, and melons were planted in the spring. As crops grew over the summer, Indians gathered seafood from the bay, including crabs, turtles, and fish. By fall, the most abundant season of the year, all the crops were harvested, and celebrations with feasts were held. By late autumn, the men set out inland toward the fall line to hunt deer.

When the first Europeans visited Maryland in the late sixteenth and early seventeenth centuries, they found a growing Indian population with complex social and religious institutions. There were about forty tribes in the region in the early seventeenth century (fig. 1.1), most of which belonged to the Algonquin linguistic group. A federation of tribes on the lower Western Shore included the Choptico, Mattawoman, Patuxent, Piscataway, Yaocomaco, and other tribes. The Piscataways were the dominant tribe in the region west of the Chesapeake Bay, and their leader was the *tayac,* or emperor, of the federation (Chappelle et al. 1986). Dominant east of the bay were the Nanticokes. Numbering about 1,500, the Nanticokes provided the leadership in the Eastern Shore region. At the head of the Chesapeake Bay lived the powerful and feared Susquehannocks of the Iroquois Nation. Estimates of the entire Indian population of the region in the early seventeenth century are highly variable (from anywhere between 8,000 and 10,000 to between 30,000 and 45,000). Transportation was mostly by water in bark or dugout canoes, but there were several major foot trails that connected villages and hunting grounds as well as provided trade routes.

Colonization

In the seventeenth century the Indian tribes encountered by early Marylanders were truly peaceful, loosely governed, and fairly weak in comparison to the powerful Virginia Algonquians under Chief Powhatan (Brugger 1988). The newly landed Maryland settlers befriended local Indian tribes, promising them protection from the nearly 2,000 warlike Susquehannocks who lived in a cluster of villages about 40 miles (64.4 kilometers) up the Susquehanna River. Indian tribes in the region generally welcomed the European settlers and generously

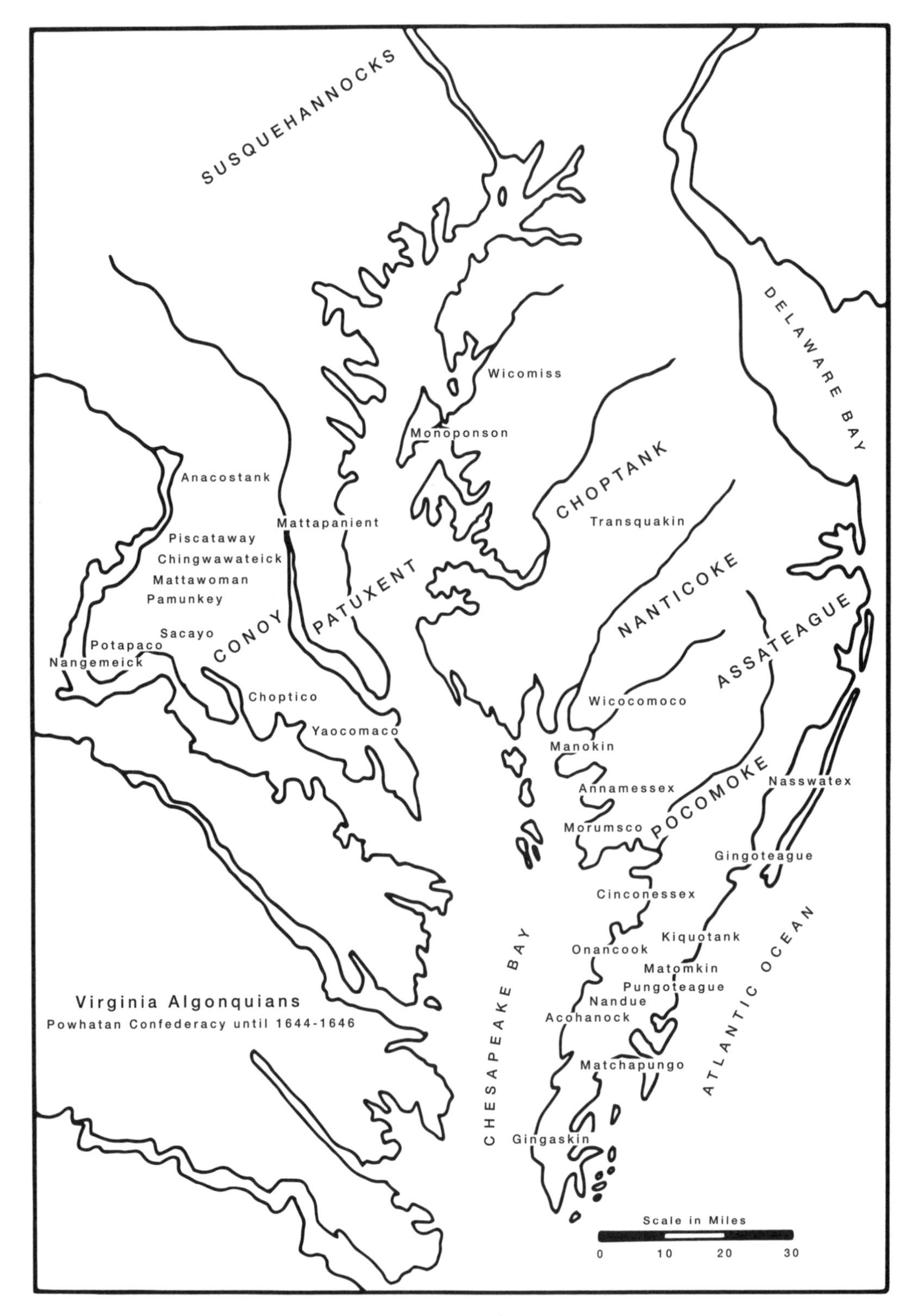

Figure 1.1 Maryland Indian Tribes of the Early Seventeenth Century

NAMES ON THE LAND

Maryland Indians gave names to places on the landscape; many of those place names (called *toponyms* by geographers) are still in use today. A place receives a name because there is a feeling that it possesses an individuality and character that differentiates it from other places, and because it is useful and worth naming. Sometimes a place has a makeshift identification that evolves into a more proper name. In other instances, there is a purposeful act of bestowing a place name. In Maryland, Indians identified and named many topographical features. When the English settlers arrived, they began to give their own place names to villages, towns, and counties.

European explorers and settlers usually employed one of three methods in the naming of a feature or settlement. Places could be named after locations in their homeland, such as Kent or Salisbury. They could be named after a town founder, or another important person, as was Annapolis—literally "Anne's City"—named after Princess Anne, soon to be Queen of England in 1702. Settlers sometimes adopted existing Indian place names, often in corrupted form; for example, Anacostia evolved from Nacotchtanck, and Potomac from Patawmeck.

About 315 major Algonquin place names survive in Maryland (Kenny 1961). The most common Indian place names are those of rivers and streams, reflecting the importance of water bodies for transportation and food supply (see table below). There are twice as many (forty-five) Indian stream names than any other class of Indian names in the state. A great majority of the Indians of Maryland were fishers who lived along the shores of the Chesapeake Bay and its tributaries. When the English settlers arrived in Maryland, they accepted and adopted most of these Indian names, but they were often unintelligible and difficult to pronounce. Through folk etymology—the process of modifying strange or unusual sounds into understandable words—Maryland settlers altered and adopted Indian place names. The name Chesapeake is of special interest given its prominence in this region; an Anglicanization of the Indian phrase *kehtc-as-apyaki* "great shellfish bay," the name was adopted by Maryland colonists as Chesepiooc and later transformed and standardized to its current spelling.

The colonists often adopted Indian place names without regard to their meaning. By simplifying spelling and pronunciation, they altered the original sense of the words. Potapaco, for example, which is thought to mean "a jutting of the water inland," was corrupted to produce Port Tobacco in Charles County. Although a great amount of tobacco was once shipped from this town, the original Indian name had nothing to do with the plant. Settlers often tried to find English words that would phonetically mimic the Indian sounds. Acchawmake, the Indian name for the Eastern Shore, has been retained in the name of Accomack County, Virginia. Today, many Indian names for water features usually precede an English generic term, such as river or bay.

Maryland settlers also brought with them familiar place names from England. The names of cities, towns, and political subdivisions created by the Europeans reflected their own cultural heritage. Some towns received Indian names, however, usually reflecting an association with a natural feature. For example, Lonaconing has its origins in the Indian *w'l-anahkw-aning* "where there is a beautiful summit," which is probably a reference to Dan's Mountain, located near the town. In naming the counties of Maryland, Indian names as well as the names of famous English and American individuals were used. Of Maryland's twenty-three counties,

Selected Indian place names in Maryland

Place name	*Derivation*	*Meaning*
Acchawmake	Indian for Delmarva	land beyond waters
Antietam Creek	*andi-etan*	the swift current
Assateague Bay	*acaw-tuk*	river across
Catoctin Mountain	*ketag-adene*	speckled mountain
Chesapeake Bay	*keht-as-apyaki*	great shellfish bay
Chincoteague Bay	*chingua-tegw*	large stream
Choptank River	*keht-ap-ehtan-ki*	flows back strongly
Conococheague Creek	*guneu-chin(g)we*	at the rapids
Conowingo	*ka-hna-wa-ga*	at the rapids
Honga River	*kahangoc*	goose
Kittamaquindi	*kittamaquindi*	place of old great beaver
Linganore	*link-anwi-wi*	melts in springtime
Lonaconing	*w'l-anahkw-an-ing*	a beautiful summit
Matapeake	*matta-peag*	junction of waters
Monocacy River	*menagassi*	too many large bends
Nanticoke	*unalachtgo*	tidal people
Patapsco River	*pat-apsk-ut*	at the rocky point
Patuxent River	*paw-e-teku-senk*	at the little falls
Piscataway	*pesketekwa*	branch of a stream
Pocomoke River	*pocquemoke*	place of clams
Pomonkey Creek	*pyami-ckw*	river twisting in land
Port Tobacco	*potapaco*	jutting water inland
Potomac River	*pya-taw-ameki*	where goods are brought in
Sinepuxent River	*esseni-pakwesen*	stones lie shallow
Susquehanna River	*susque-hanna*	smooth flowing stream
Vienna	*unnacocassinon*	chief of the Nanticokes
Wicomico River	*wighcocomoco*	pleasant village
Youghiogheny River	*yo-wiyak-agan*	muddy flow stream

Source: Kenny (1961)
Note: All place names are from the Algonquin language group.

twenty are named for people, one a geographical location, and two Indian words or tribes. Of the twenty county names that commemorate people, one is the name of a saint, one a president, two senators, and five women (see table below).

The original meanings of these names, obscured and blurred by the passage of time, are keys to the understanding of important aspects of the lives of Maryland's first inhabitants. Although not fully understood by most Marylanders, these rich place names are nonetheless

Origins of Maryland county names

*Type**	*County*	*Origin*
I	Allegany	Indian for "beautiful streams"
W	Anne Arundel	after Anne Calvert (née Anne Arundell), wife of Cecilius Calvert, 2nd Lord Baltimore
C	Baltimore	after Cecilius Calvert, 2nd Lord Baltimore
C	Calvert	after Cecilius Calvert, 2nd Lord Baltimore
W	Caroline	after Lady Caroline Eden, wife of Maryland's last colonial governor, Robert Eden
PO	Carroll	after Charles Carroll, signer of the Declaration of Independence
C	Cecil	after Cecilius Calvert, 2nd Lord Baltimore
CF	Charles	after Charles Calvert, 3rd Lord Baltimore
CF	Dorchester	after Earl of Dorset, family friend of the Calverts (early seventeenth century)
CF	Frederick	after Frederick Calvert, 6th Lord Baltimore, or possibly Frederick Lewis, Prince of Wales (the records are not clear on which one)
PO	Garrett	after John W. Garrett, president of Baltimore and Ohio Railroad
CF	Harford	after Henry Harford, son of Frederick Calvert, the last Lord Baltimore
PO	Howard	after John Eager Howard, fifth governor of Maryland
P	Kent	after County Kent, England
PO	Montgomery	after General Richard Montgomery, Revolutionary War era
PO	Prince George's	after Prince George of Denmark (in 1696), the husband of Princess Anne and heir to the throne of England
W	Queen Anne's	after Queen Anne of England (1706)
W	Somerset	after Lady Somerset, daughter of Thomas Arundell, 1st Baron of Wardour, sister of Anne Calvert, Baroness of Baltimore (née Anne Arundell), wife of Cecilius Calvert, 2nd Lord Baltimore
St	St. Mary's	after the Virgin Mary; considered the "mother county" of Maryland
W	Talbot	after Lady Grace Talbot (in 1661), sister of Cecilius Calvert, 2nd Lord Baltimore
Pr	Washington	after President George Washington
I	Wicomico	Indian for "a place where houses are built"
CF	Worcester	after the Earl of Worcester

Source: Compiled by the author from various Maryland county sources

*Abbreviations are as follows: C, Cecilius Calvert, 2nd Lord Baltimore; CF, Calvert family and friends; I, Indian name; P, place; PO, prominent others; Pr, president; St, saint; W, women.

a gift to Maryland from its original people. By the nineteenth century, most of Maryland's Indian culture was destroyed and lost. The knowledge of these people—who had witnessed dramatic environmental changes over thousands of years and survived by learning to live in harmony with the natural cycles of the land—was largely discarded. For over 12,000 years, Maryland Indians survived by adapting. Their success was based on knowledge of the environment, mobility, and flexibility. In the twenty-first century, as modern Marylanders struggle with serious environmental issues (see chapter 4), perhaps they could learn something from the prior caretakers of this land.

shared land, food, and labor. Initially, their relationship was good, but in the long run the Indians were seen as an obstacle to the colonists' further settlement and expansion. The territorial expulsion or physical elimination of the native Indians of Maryland seems to have followed the same general pattern as the removal of Indians in the other coastal colonies. Although Maryland's proprietors were historically more enlightened in their treatment of the local Indians, the end result was the same as elsewhere (Brugger 1988). By the end of the seventeenth century, Maryland's colonist population of 33,000 had replaced nearly all of the Indians existing in 1630.

In 1634, Governor Leonard Calvert and other first-wave colonists established the initial Maryland settlement, landing at what they called St. Clement's Island in southern Maryland. Calvert purchased land from the local Yaocomaco Indians and founded Maryland's first capital, St. Mary's City. A Virginian, William Claiborne, established an earlier settlement on Kent Island consisting of a trading post, church, and fort. Claiborne had arrived in Virginia in 1621 and, shortly after his arrival, sailed up the Chesapeake Bay and recognized its potential riches. He selected Kent Island (today the eastern anchor of the two bay bridges) as the site for a fur-trading post. The island was well drained and ideally located to control the fur trade in the middle and upper bay. By 1631, Claiborne and other Virginia settlers had built a stockade, church, and a store and had cleared land and planted crops. When the Maryland settlers arrived in 1634, Claiborne dismissed the inclusion of Kent Island in Lord Baltimore's land grant, as it had been settled prior to the grant, and resisted Maryland jurisdiction. After several clashes with armed Maryland vessels on the nearby bay, Claiborne's fur exports began to decline, and his financial backers in London finally replaced him with another agent. The Virginian Kent Islanders finally agreed to throw in their lot with the Calvert colony, but not until a Maryland expeditionary force landed in 1637 to subdue the rebel leaders.

The elaboration of Maryland's boundaries was slow, owing in part to a lack of precise geographical information about the land and its resources. From St. Mary's City, the population spread north and west along the Potomac River and the Western Shore (fig. 1.2). The fall line severely restricted travel inland by water, so planters established themselves on the many tidewater peninsulas that form the shores of Chesapeake Bay. There, each planter's boat landing provided sufficient commercial access to the outside world. In this natural and socioeconomic environment, commercial towns had no great function, therefore never developing on the same scale as in New England, for example. This settlement pattern is reflected in the absence of large- and medium-size commercial towns in southern Maryland even today. Likewise, Maryland's transportation in the 1600s was locally oriented. The Chesapeake Bay and its many tributaries were its main highways; roads were few, consisting mainly of short, narrow, muddy paths leading from plantations down to the river.

Cecilius Calvert, the second Lord Baltimore, had a vision of a feudal settlement system for his Maryland proprietary colony, a manorial system with a stable social order. But it was not to be. Rapid economic

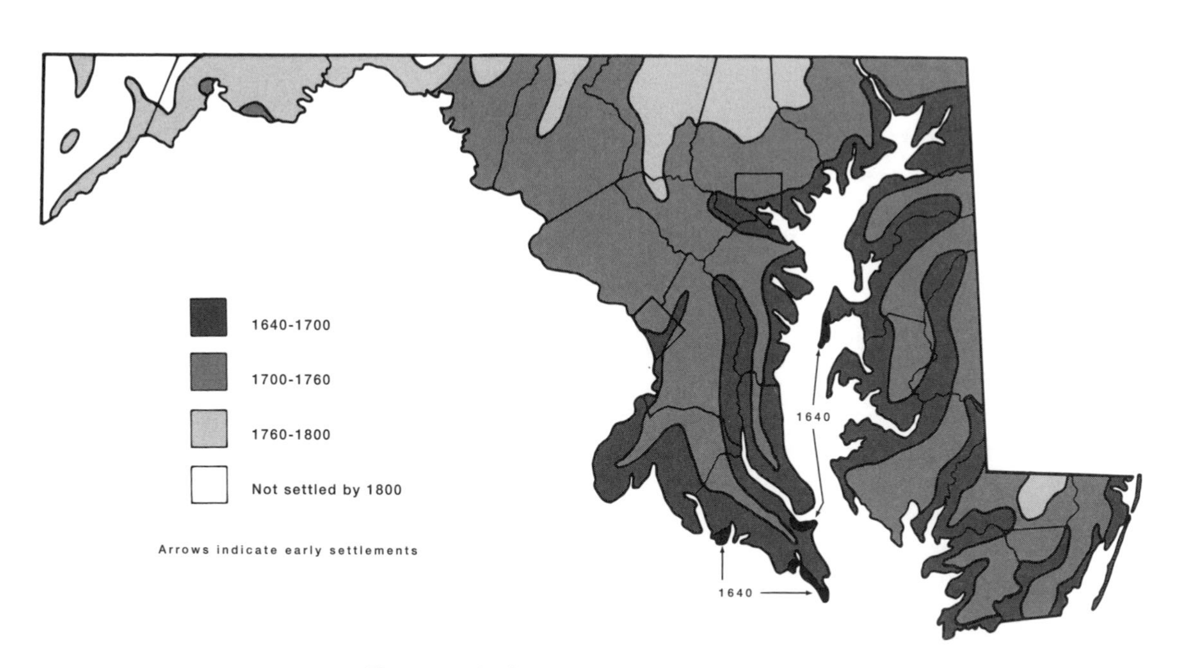

Figure 1.2 Settlement Expansion, 1640–1800

development amid the primitive conditions of a newly settled frontier meant that lower-class men were able to acquire land and wealth, thus disrupting the social order that Calvert envisioned. Likewise, his vision of a dependent, proprietary colony composed of nuclear families living in villages within a manorial system supported by agriculture and some manufacturing proved unsuitable for the Chesapeake Bay region. Instead, what evolved were Roman Catholic planter-gentry who dominated Protestant tenants, all widely dispersed in small settlements along the many inlets of the Bay shoreline.

By 1642, 83 percent of settled land in Maryland was located in sixteen manors concentrated in the hands of the Jesuit order and six Catholic families (Mitchell and Muller 1979). Many immigrants were former indentured servants who had become freehold farmers after approximately five years of labor. The traditional European manorial system, which did not succeed on the primitive frontier, evolved into a New World variant: the plantation. Initially, Maryland plantations ranged from 250 to 500 acres (101.3 to 202.6 hectares); in addition, many planters owned or rented scattered parcels of land. Eventually the term *plantation* came to identify a method of working and controlling land, as well as a way of life.

Isolated farms and plantations with a few small port towns specializing in the export of locally grown crops dotted the seventeenth-century Maryland landscape. Farmers readily adopted the cultivation of tobacco, which the local Indians had long grown. Tobacco was at first a risky crop to grow because the markets for it fluctuated widely. Nevertheless, new land was cleared and many new small, independent farms were created to meet growing demand. Exports of tobacco rose dramatically from 50 tons (45.4 tonnes) in 1640 to 8,000 tons (7,256 tonnes) in 1690.

In the 1630s and 1640s, the population remained clustered in southern Maryland. Thereafter it shifted slowly northward along the Western Shore and also along the Eastern Shore. Between 1650 and 1700, the population of the Eastern Shore grew from only 11 percent of the colony's population to nearly 50 percent. During this period, Eastern Shore farmers began to diversify from tobacco to growing grains and raising livestock (Clemons 1974). To better serve the needs of a more dispersed population, the capital was moved in 1695 from the somewhat isolated St. Mary's City to Annapolis, which was more centrally located.

Another significant demographic change occurred in the late 1600s: tidewater Maryland's economy began to rely on slavery. Most of the early laborers had been indentured servants, often teenage boys from the slums of London. By the late 1600s, as conditions in England improved, this source of labor proved too costly and eventually dried up. Soon, Maryland planters, following the lead of the Spanish and the Dutch in the New World, turned to slavery to supply the intensive la-

MARYLAND'S TERRITORIAL EXTENT

As one of the smaller states of the nation, Maryland ranks forty-second in terms of area. The state covers 12,204 square miles (31,865 square kilometers), which can be separated into 9,775 square miles (25,574 square kilometers) of land; 1,726 square miles (4,470 square kilometers) of Chesapeake Bay; 106 square miles (275 square kilometers) of Chincoteague Bay; and 597 square miles (1,546 square kilometers) of inland and tidal waters. For the purposes of comparison, Maryland is six times larger than Delaware, and a little less than one-third the size of Pennsylvania. Laying a rectangle upon an image of the state, one could measure the east-west extent as stretching 238.14 miles (383.24 kilometers); the north-south maximum distance as 125.99 miles (202.76 kilometers); and the diagonal distance from the northwest corner of Garrett County extending to the southernmost corner of Worcester County, where Maryland meets Virginia at their common barrier island, as 255.78 miles (411.59 kilometers). The north-south extent varies greatly, from 125 miles (201.16 kilometers) in the east to a mere 1.9 miles (3.06 kilometers) at the wasplike waist near Hancock in Washington County.

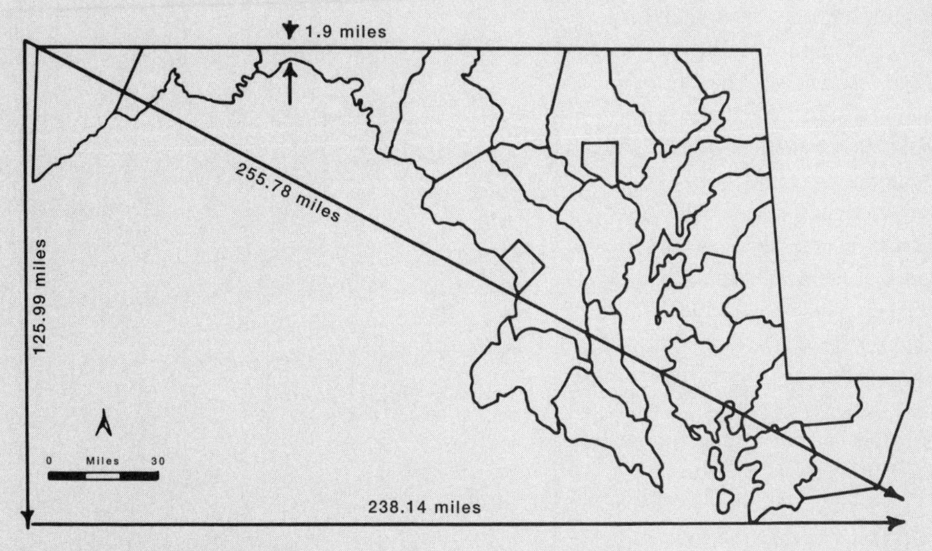

Distances across Maryland

The present shape of Maryland has evolved since the original grant to Lord Baltimore in 1632. Today, the northern, western, and most of the eastern boundaries are straight lines. The irregular southern boundary follows the southern shore of the Potomac River, which lies entirely within Maryland. The Charter of Maryland, issued on June 20, 1632, by King Charles I of England, described the boundaries of the colony as extending from the Potomac River north to the 40th parallel (including the present site of Philadelphia) and west from the Atlantic Ocean to the first "fountain" of the Potomac. Unfortunately, officials in London had little accurate geographic information concerning the interior of the region. In 1680, a part of the same territory was granted to William Penn, and another part of the Maryland grant, which had been settled by Swedes (present-day Delaware), was granted to the Duke of York, who, after becoming King James II, ceded the area to Penn in 1682.

Endless boundary disputes plagued Maryland until the close of the nineteenth century, as Pennsylvania and Virginia were reluctant to recognize the legality of Lord Baltimore's grant. Arguments centered on the unclear terminology of the Maryland charter of 1632 (e.g., the grant of "land hitherto unsettled" from the Potomac to a line, "which lieth under the fortieth degree north latitude from the quinoctal," and extended westward to a line due north from the "first fountain of the Potomac"). In 1760, an English court set the Maryland-Delaware-Pennsylvania corner, and three years later two Englishmen—astronomer Charles Mason and surveyor Jeremiah Dixon—were contracted to demarcate these boundaries. It took them until December 1767 to run the northern boundary, marking each mile with a stone from a quarry on the Isle of Portland, England. The northern side of each stone marker displayed a "P" for Pennsylvania and the southern side an "M" for Maryland.

Virginia disputed the western border of Maryland, and after the Civil War differences arose with the new state of West Virginia. It was not until the US Supreme Court decision on May 7, 1912, that the western boundary of the state was firmly established. The southern boundary with Virginia was disputed as late as 1930. At that time, a commission established by the governors of the two states agreed to the low watermark on the Virginia side of the Potomac River as the Maryland-Virginia boundary. In 2004, the US Supreme Court once again considered a case concerning the boundary dispute, deciding that Virginia had the right to draw water from the Potomac River to supply the growing population in Northern Virginia, although the river was recognized as being entirely within the borders of Maryland.

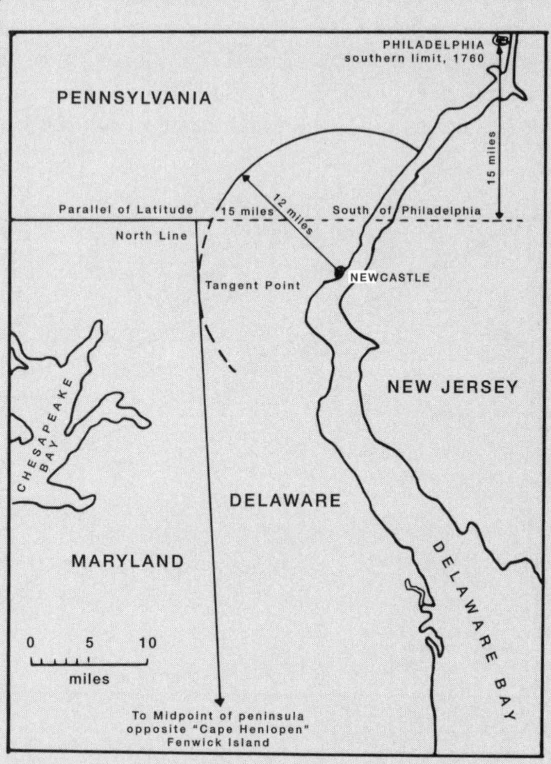

Maryland-Delaware-Pennsylvania Corner as Fixed in 1760

Maryland lost all of its major boundary disputes with neighboring colonies (and later with states). Other colonies closed in on Maryland and settled the periphery; territorial retreat followed each court defeat. The Calvert family, which founded the first colonial settlement in Maryland, did not aggressively pursue its land grant. Much of their failure to protect the integrity of Maryland's boundaries was due to a lack of geographical information, but we can understand it in another context. The Calverts were a prominent and wealthy Roman Catholic family. The early seventeenth century was not a pleasant time for Catholics in England and its colonies, and the Calverts did not desire to stir up controversies with the largely Protestant colonies surrounding Maryland. The "Papists" in Maryland were barely tolerated by their neighbors in these early years.

bor needed to grow tobacco. An act was passed in 1671 encouraging the import of slaves from West Africa. In the 1670s, indentured servants outnumbered slaves four to one, but by 1690 there were four times as many slaves as indentured servants. In 1658, there were only one hundred slaves in southern Maryland, accounting for 3 percent of that region's population; by 1710, there were over 3,500 slaves, comprising almost a fourth of the region's population (Mitchell and Muller 1979). During the 1680s, Maryland's nonslave population began to experience positive rates of natural increase, eventually leading to the creation of an American-born Maryland society.

Agrarian Development

In the preindustrial eighteenth century, land was the principal source of wealth in Maryland. The average size of landholdings ranged from 255 acres (103.3 hectares) in Worcester County and 282 acres (114.2 hectares) in St. Mary's County to 370 acres (149.9 hectares) in Frederick County, 372 acres (150.7 hectares) in Cecil County, and even 473 acres (191.6 hectares) in Anne Arundel County (Giddens 1933). Freeholding farmers were more numerous on the Piedmont and to the west than on the Western Shore. Eighteenth-century Maryland had an economy dominated by tobacco, grains, and livestock, with some lumber and iron production. Unlike Virginia to the south, which remained heavily dependent on tobacco, Maryland diversified earlier, as food processing and small manufacturing were added to the economy.

By 1755, there were four distinct regions on the Maryland landscape: the Baltimore area, the Eastern Shore, southern Maryland, and western Maryland. Historian David Curtis Skaggs noted that "while population and economic power moved northward and westward, the center of social and political influence remained in the tidewater homes of tobacco planters and the drawing rooms of Annapolis" (Skaggs 1973). Of these four regions, the Eastern Shore had the most diversified economy and a growing population size. Population there increased from 16,000 in 1750 to 108,000 in 1790 (50% of the state's population). The counties with the best soils (southern Cecil, Kent, Queen Anne's, and Talbot) produced tobacco, wheat, and corn. Marginal soils in Somerset, Worcester, and parts of Caroline and Dorchester were used for lumbering, especially for export to England. During the American Revolution, the northern counties of the Eastern Shore sent wheat, flour, corn, cattle, and salted fish to Philadelphia; these commodities left through Oxford and Chestertown by water for Elkton at the head of the Chesapeake Bay, where they were loaded onto wagons (Arnett et al. 1999). By the end of the eighteenth century, the export trade of the Eastern Shore was dominated by wheat, shipped mainly to Baltimore.

Tobacco and corn dominated in southern Maryland, where the population reached 89,000 by 1790 (25% of the state's population). Yet Annapolis was unable to hold off commercial competition from Baltimore and became a quiet market town serving a rural hinterland. Near the end of the century, soil erosion became evident in the coastal areas where tobacco and corn were grown, and silting began to affect the navigability of small ports such as Port Tobacco in southern Maryland and Joppa north of Baltimore. Although Port Tobacco was still used during the Civil War, neither it nor Joppa are ports today. In 1608, when Captain John Smith first saw the Patapsco River, whose main estuary forms the inner harbor of Baltimore, the limit of open water was 7 miles (11.3 kilometers) farther inland than it is today.

Throughout the eighteenth century, the rural-agricultural Maryland landscape evolved into a more complex socioeconomic system. By the first US federal census of 1790, the population had increased to 320,000. At this time, Maryland's upper Eastern Shore was the geographic center of population in the newly created United States. In 1790 there were as many people in the United States living north of this point as south of it, and as many living east of it as west of it. The selection of nearby Washington as the site for the new national capital was appropriate for the time. But the westward shift of the geographic mean center of the population would come to reflect the overall change in migration and settlement of the country in successive decades.

During the 1700s, Maryland went from being a royal colony of England to becoming a state in the new nation. As European markets demanded more tobacco and other products, the economy of Maryland grew. Inducements to settle the West were given to settlers in the form of tax reductions, and settlement concurrently expanded westward as farmers needed more land to plant new crops and to raise livestock (Main 1977). During the 1730s, the population increased steadily in the tidewater area and expanded westward into the Monocacy Valley. As the Piedmont was settled, increasing numbers of non-English speaking immigrants arrived. German-speaking immigrants settled along the fertile Monocacy Valley, while Scots-Irish and English farmers pushed farther westward into the agriculturally less desirable but lovely rolling hill country. Some of the new settlers in the Monocacy Valley came south from Pennsylvania.

In 1748, Frederick County was formed. Its population increased so rapidly that by 1755 it was the second most populous county. Frederick Town soon replaced Annapolis as the largest town in Maryland. But by 1756, Indian raids during the French and Indian Wars had driven many settlers from the area west of Frederick until hostilities ceased. The American Revolution once again slowed westward movement and development, but the far western reaches of the state were inevitably settled by the late eighteenth century. The emergence of an ethnically mixed, yeoman-dominated population raising wheat and livestock west of the fall line added a new dimension to the regionalization of Maryland during this time.

As it opened up to settlement throughout the eighteenth century, western Maryland developed at a phenomenal rate. In the 1740s, this region of only 8,000 people had less than 5 percent of the state's population. By 1790, it had 85,000 inhabitants constituting over 25 percent of the state's population. Growth was clustered in several fertile lowlands, especially the Monocacy Valley in Frederick County and the Hagerstown Valley in Washington County. The agricultural emphasis was on livestock and grains, particularly wheat and rye; tobacco did not become an important crop in this region. Despite the major culture hearth in the Chesapeake tidewater area, the settlers of western Maryland were mainly influenced by the cultural characteristics of southeastern Pennsylvania's Midland culture hearth (Mitchell 1978). These were yeomen freeholders practicing mixed agriculture of wheat, rye, flax, and cattle on 200- to 400-acre (81- to 162-hectare) farms. Through Maryland, the Chesapeake and Midland cultural traditions penetrated to influence the Midwest and upper South during the nineteenth century.

Maryland's slave population increased from 4,400 in 1700 to over 111,000 in 1790, with slaves eventually constituting one third of the state's population. By 1790, the black population was unevenly distributed, reflecting the prevailing economic land uses found in the state. Almost 42 percent of all blacks lived in the plantation-dominated parts of southern Maryland, and another 39 percent resided on the Eastern Shore. The ties of slavery to tobacco were weakening somewhat, however, as 13,000 blacks (90% of whom were slaves) lived in western Maryland where little tobacco was grown.

The urban population was also growing. After the 1740s, significant changes in the low-order urban hierarchy began to occur. In 1750, Annapolis had only 800 residents, Baltimore merely 150, and Frederick 1,000. But by 1776, Baltimore was an urban settlement of over 5,000, followed in size by two relatively new western towns, Frederick and Hagerstown, with 1,800 and 1,500 residents, respectively. Baltimore's growth, centered on its port, continued rapidly, and by 1790 it was the fifth-largest city in the United States (table 1.1). Enlightened and visionary Baltimore businessmen united to consolidate the position of their port city as the leading trade center of the Chesapeake region. Just as the rise of towns in western Maryland was tied to the tobacco and wheat trades and their accompanying needs for transportation, storage, and processing (Earle and Hoffman 1976), the evolving economic-geographical pattern was directly related to Baltimore's growth as a major port city.

The rise of the Baltimore area, evident on the eighteenth-century Maryland landscape, was linked to the changing economic landscape of other regions of the state. In 1711, a flour mill was established at Jones Falls, and in 1723 iron furnaces were built at the mouth of Gwynns

Table 1.1. Twenty largest US cities by population, 1790–2012

Rank	*City*	*Rank*	*City*	*Rank*	*City*
	1790		*1830*		*1870*
1	New York, NY	1	New York, NY	1	New York, NY
2	Philadelphia, PA	2	**BALTIMORE, MD**	2	Philadelphia, PA
3	Boston, MA	3	Philadelphia, PA	3	Brooklyn, NY
4	Charleston, SC	4	Boston, MA	4	St. Louis, MO
5	**BALTIMORE, MD**	5	New Orleans, LA	5	Chicago, IL
6	Salem, MA	6	Charleston, SC	6	**BALTIMORE, MD**
7	Newport, RI	7	Cincinnati, OH	7	Boston, MA
8	Providence, RI	8	Albany, NY	8	Cincinnati, OH
9	Gloucester, MA	9	Brooklyn, NY	9	New Orleans, LA
10	Newburyport, MA	10	Washington, DC	10	San Francisco, CA
11	Portsmouth, NH	11	Providence, RI	11	Pittsburgh, PA
12	Brooklyn, NY	12	Richmond, VA	12	Buffalo, NY
13	New Haven, CT	13	Pittsburgh, PA	13	Washington, DC
14	Taunton, MA	14	Salem, MA	14	Newark, NJ
15	Richmond, VA	15	Portland, ME	15	Louisville, KY
16	Albany, NY	16	Troy, NY	16	Cleveland, OH
17	New Bedford, MA	17	Newark, NJ	17	Jersey City, NJ
18	Beverly, MA	18	Louisville, KY	18	Detroit, MI
19	Norfolk, VA	19	New Haven, CT	19	Milwaukee, WI
20	Petersburg, VA	20	Norfolk, VA	20	Albany, NY
	1910		*1950*		*2012*
1	New York, NY	1	New York, NY	1	New York, NY
2	Chicago, IL	2	Chicago, IL	2	Los Angeles, CA
3	Philadelphia, PA	3	Philadelphia, PA	3	Chicago, IL
4	St. Louis, MO	4	Los Angeles, CA	4	Houston, TX
5	Boston, MA	5	Detroit, MI	5	Phoenix, AZ
6	Cleveland, OH	6	**BALTIMORE, MD**	6	Philadelphia, PA
7	**BALTIMORE, MD**	7	Cleveland, OH	7	San Antonio, TX
8	Pittsburgh, PA	8	St. Louis, MO	8	San Diego, CA
9	Detroit, MI	9	Washington, DC	9	Dallas, TX
10	Buffalo, NY	10	Boston, MA	10	San Jose, CA
11	San Francisco, CA	11	San Francisco, CA	11	Detroit, MI
12	Milwaukee, WI	12	Pittsburgh, PA	12	San Francisco, CA
13	Cincinnati, OH	13	Milwaukee, WI	13	Jacksonville, FL
15	New Orleans, LA	15	Buffalo, NY	15	Austin, TX
16	Washington, DC	16	New Orleans, LA	16	Columbus, OH

Table 1.1 (*continued*)

Rank	City	Rank	City	Rank	City
	1910		*1950*		*2012*
17	Los Angeles, CA	17	Minneapolis, MN	17	Fort Worth, TX
18	Minneapolis, MN	18	Cincinnati, OH	18	Charlotte, NC
19	Kansas City, MO	19	Seattle, WA	19	Memphis, TN
20	Seattle, WA	20	Kansas City, MO	20	Boston, MA
				21	**BALTIMORE, MD**

Source: James Vance, "Cities in the Shaping of the American Nation," *Journal of Geography* 75, no. 1 (January 1976): 41–52. With permission of the National Council for Geographic Education. Updated from US Census of Population Estimate, 2012

Falls. As roads were opened into western Maryland and a public tobacco inspection warehouse was sited in the city, Baltimore's centrality strengthened. After reaching a population of 13,500 in 1790, a period of rapid growth followed. By 1800, the population had more than doubled to 31,500 (Bernard 1974).

The Early Urban Industrial Period

During the nineteenth century, a number of basic structural changes, part of a national process, occurred on the Maryland landscape. The last frontier in Garrett County was finally settled. Maryland merchants began to encounter stronger domestic and foreign commercial competition. Farmers faced new and increasing competition from the newly opened western lands. In response, some farmers took advantage of their better relative geographic location and diversified to meet the needs of the growing urban areas. Some people from rural areas moved to the cities to work in new and growing industries and were joined by an influx of foreign-born immigrants. Baltimore was the center of a vortex of forces, and it grew into the stereotypic mid-nineteenth-century American commercial city—large, bustling, and heterogeneous. By 1830, Baltimore had become the second-largest population center of the country (table 1.1). As the population grew, so did Baltimore's domination of Maryland's manufacturing activities, especially shipbuilding, food processing, weaving, ironworks, and printing. The chief exports from Baltimore remained tobacco and grains for the European market.

The cultural, economic, and social regional contrasts of Maryland came to be more sharply defined during the nineteenth century. The initial thirty years of the century witnessed stagnation in Maryland's still dominantly agricultural economy. This period witnessed the Embargo of 1807, War of 1812, and the depression of 1819–22. Outdated farming methods were accompanied by soil erosion and exhaustion, resulting in poor crop yields (Van Ness 1974). The population increased by only 127,000 between 1790 and 1830 as people left Maryland for destinations west. Baltimore City accounted for 53 percent of this increase, while the upper Eastern Shore lost population and southern Maryland remained stable.

After the 1830s, a number of changes came to Maryland: agricultural reform, transportation improvements, coal mining, immigration, and industrial development. A sharp rural-urban dualism became clear by 1850. Differences in the character of farming and fishing on the rural Eastern Shore, tobacco plantations in southern Maryland, burgeoning Baltimore, and the western coalfields became ever more clear.

From 1830 to 1860, Maryland's population increased by more than a third, reaching 687,000, a gain of 240,000. Baltimore City accounted for over 50 percent of this growth. Population data show a considerable decline in the population of Baltimore County between 1850 and 1860, but this was largely due to the carving out and separation of Baltimore City from Baltimore County in 1851. During the 1870s, every county except Allegany increased in population; this, too, was due to the creation of a new political entity when Garrett County was carved out of Allegany County. Baltimore City and Washington, DC, were both growing rapidly during the late nineteenth century.

In 1860, the number of free blacks nearly equaled the number of slaves, and together all blacks now accounted for 25 percent of the population of Maryland. Because of the influx of immigrants, mostly to Baltimore and areas west, more and more Marylanders were unfamiliar with slavery and black people. In southern Maryland, however, slaves still accounted for 45 percent of the population at the onset of the Civil War in 1861. After the Civil War, Baltimore attracted large numbers of free blacks from Southern states, and between 1870 and 1900 Baltimore had one of the largest clusters of urban blacks in the United States. Between 1830 and 1860, rapid population growth occurred in Baltimore City, Cecil County, the Piedmont, and western Maryland. In terms of population growth among immigrants and blacks, western and north-central Maryland were gaining on the Eastern Shore and southern Maryland. Baltimore, with its influx of Irish Catholics and Germans, was becoming exotic to the rest of Maryland.

A major scene linked to nineteenth-century industrialization was developing in Allegany County. The coming of the railroad and canal to western Maryland opened up coal production, and more miners came to Maryland, mainly from the British Isles and Germany. By 1860, nearly 20 percent of Allegany County's population was foreign born. The coal company towns, mines, and their foreign-born workers contrasted with the dispersed rural mountain folk culture of western Maryland.

Baltimore was also rapidly industrializing. By 1860, the city had over 1,100 industrial establishments employing nearly 20,000 workers. New industries developed, such as ready-made clothing, ironworks, and railroad shops. Travelers were impressed with Baltimore's urbanity, modernity, density, diversity, and dispersal (Alexander 1975). It was during this time that Baltimore's red brick row houses began to appear—today a defining characteristic of the city's personality.

With the population growing and its demographics changing, as well as with rapid industrialization, Maryland had an ever-increasing need for improvements to its transportation system. But it was not until the mid-1800s, following the adaptation of steam power to water and rail, that important changes to transportation came to Maryland. In the early 1800s, turnpikes were built from Baltimore to Frederick and north into Pennsylvania, but by the 1820s canals and railroads were diverting attention from roads. By 1830, turnpikes were no longer the predominant lines of long-distance transportation, but rather had become feeders to the railroads and canals. Responding to the beginning of coal production in western Maryland and the desire of mercantilists to break the isolating trade barrier of the mountains, the Maryland legislature passed a bill to canalize the Potomac River to Cumberland. Fearing that the proposed Chesapeake and Ohio (C&O) Canal would enable Georgetown to rival their city, Baltimore merchants chartered the Baltimore and Ohio (B&O) Railroad to be built westward from Baltimore. Construction on both the C&O and the B&O began in 1828. The B&O Railroad reached Cumberland in 1842. By the 1850s, Baltimore had rail connections to the Ohio and Mississippi valleys. The 13-mile-long (20.9-kilometer-long) Chesapeake and Delaware Canal opened in 1829, connecting the Chesapeake Bay with the Delaware River, contributing to the growing centrality of Baltimore (Hungerford 1928). When completed, the C&O Canal moved flour, agricultural produce, and coal to Georgetown, but Baltimore was secure as the leading commercial center of the region, thanks to the efforts of its businessmen and its geographical position. In the 1870s and 1880s, use of the C&O Canal declined after being damaged from a number of floods.

As transportation improved and the costs of movement decreased, more farmers had access to Maryland's urban markets. Agricultural improvements were employed in Maryland, such as the use of lime and guano fertilizers and crop diversification. Dairy farms and orchards were developed on the Piedmont and in western Maryland. Fruit and vegetable production on the Eastern Shore led to the establishment of canneries, while truck farming of perishable fruits and vegetables for the urban markets developed around Baltimore and Washington, DC. Amid all this change, tobacco farming and slavery persisted in southern Maryland. Although there were many miles of roads in Maryland, it was still rough going for travelers. By 1900, about 90 percent of Maryland's 14,483 miles (23,303 kilometers) of roads were still unpaved; most of the rest had been improved with stone, gravel, and shells (Thompson 1977).

The National Road

When the American Revolution officially ended in September 1783, the newly created United States was vast, extending from the East Coast to the Mississippi River Valley. The Northwest Territory, estimated by US Geographer Thomas Hutchins to have covered some 220 million acres (88 million hectares) of government land, much of it fertile (Raitz 1994), lay along the right flank of the Ohio River. Realizing its potential, eastern mercantilists—including prominent Marylanders—and politicians were anxious to open up this region to connect it to the population centers of the east, and to sell off public lands to pay down the federal debt. Still remote and isolated, the Northwest Territory's only transportation connections were over primitive trails across Pennsylvania and Maryland. The great Ohio River flowed west and then south into the Mississippi River to New Orleans. Baltimore businessmen were among those displeased to see commodities of the Northwest Territory flowing southward to their competitors in New Orleans. Arguing that it would be more efficient and less costly to ship over land to the closer ports of the East Coast, the mercantilists secured the support of Treasury Secretary Albert Gallatin, Henry Clay, and Thomas Jefferson to fund a transportation construction program of canals and roads of which the proposed National Road would be part.

Authorized by Congress in 1806, the National Road was the first federally planned and funded roadway in the United States. It provided access to the Northwest Territory and thus enabled more local regional subsistence economies, such as in western Maryland, to link into an expanding national economy. Spatial interaction extended well beyond the confines of Maryland. The road began in Cumberland, Maryland, and extended westward through Uniontown and Washington in Pennsylvania, Wheeling in West Virginia (then still part of Virginia), and across Ohio and Indiana, ultimately reaching Vandalia, Illinois, in 1852 (fig. 1.3). Its route was selected to allow for geographic expediency. It would link from Cumberland back to Baltimore, which was closer to the Ohio River in direct overland distance than Philadelphia or Richmond (Raitz 1994). Trade grew as farm produce and herds of cattle and swine made their way along the National Road to Baltimore. Wagons returned westward with manufactured goods and newly arrived immigrants. Migrants included Scots-Irish from the north of Ireland, Germans, Quakers, and others from the British Isles. Baltimore had become a major gateway to the settlement of the Midwest.

By the 1830s, canals across New York and Ohio offered alternative means of travel. Soon railroad crews were laying track alongside the National Road westward out of Baltimore, and long-distance overland trade over the road declined. Years later, the Ford Model T would re-

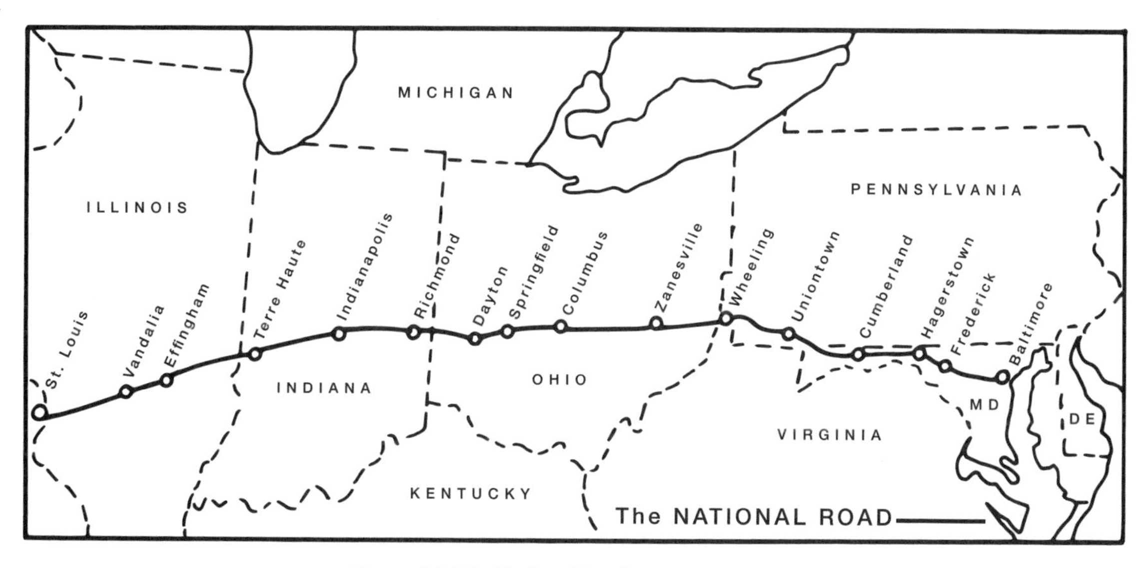

Figure 1.3 The National Road, 1818–38

vive the National Road, but by then it was in poor shape owing to years of neglect. In 1906, the federal government passed the Federal Aid Road Act to build a hard-surfaced cross-country highway network, and the National Road became a key element of that road system. In 1926, the United States adopted a system of uniform route numbers for the nation's highways, and the National Road became US Route 40. Funding from the Federal Aid Road Act allowed for many improvements; the new concrete road had curves straightened and grades flattened. New trussed iron and steel bridges bypassed the older humpbacked, smaller stone bridges. Once again, long-distance traffic along the National Road increased. New businesses such as restaurants, service stations, hotels, and tourist camps were built along the road close to the towns to accommodate travelers.

The landscape along Route 40 in Maryland and beyond was changing as the financial success of these ribbon developments was realized. Automobile travelers often liked to avoid the congestion of central towns and cities, and during the 1930s and 1940s, long sections of Route 40 were widened to four lanes and bypasses were built around small towns. With these changes, the decline of businesses along Main Street began. In the 1960s, Interstate 70 bypassed Route 40, and many roadside businesses failed. But as modern-day travelers seek attractions off the beaten path (and clogged interstates), many of the nearly abandoned towns along Route 40 have been rediscovered in recent decades. Grand old buildings that once housed banks, barbershops, and drugstores now showcase boutique businesses. East of Frederick, in the town of New Market, a cluster of antique shops exemplifies this transformation.

The National Road is a trail of history and geography. Its story of construction, decline, re-creation, and conversion goes on. More than a link between otherwise disparate and distant places, the road is a route of cultural diffusion and a symbol of both Maryland and national identity.

The Mid-Industrial Period

The Civil War had a disrupting effect on the economic and social structures of this border state. During the war, the federal government had found it necessary to keep Maryland loyal to the Union. In order to keep Washington, DC, from becoming an isolated Union enclave in the South, Union troops occupied much of Maryland, especially the key railroad links in Baltimore connecting Washington to the North. The war-depressed farm economy finally picked up as the federal government's need for food, supplies, and transportation grew.

After the Civil War, Maryland agriculture changed. Tobacco production was more costly without slaves, and wheat farmers were faced with competition from the western plains. As industrialization continued, farmers turned more to perishable fruits, vegetables, and dairy products to supply the Baltimore and Washington urban markets along the expanding railroads. Tobacco farming retreated to the better soils of southern Maryland. Fruits and vegetables became popular on the Eastern Shore and in the counties of Prince George's and Anne Arundel. Piedmont farmers turned to forage crops and dairying; orchards were planted in the western mountain valleys and on the slopes. With new crops, few staple products, smaller farms, and more machinery and out-migration, the rural landscape of Maryland experienced significant changes (Mitchell and Muller 1979).

Even though agriculture was still the leading economic activity in Maryland until 1900, other sectors of the economy developed strong basic infrastructures for future growth. After the disruption of the Civil War, agriculture slowly declined in importance, and by 1900 more Marylanders were employed in industries than in agriculture. The shift in employment emphasis from primary activities, especially agriculture, to secondary manufacturing activities is a part of the development process that is well documented by regional economists. Today, this continually unfolding development process has given Maryland an employment structure with a greater emphasis on tertiary retail and service as well as government.

By 1910, Maryland had nearly doubled its 1860 population, and the majority of Marylanders lived in towns and cities of 2,500 or more. The state's cities and towns settled largely into the mold of a central place system. According to the 1910 US Census, there were fourteen settlements with a population over 2,500, and half of them had fewer than 5,000 people. The Chesapeake Bay ports of Cambridge and Crisfield developed oyster packing and fish processing, while Salisbury developed flour milling, vegetable canning, and timber products. West of the Catoctin Mountains, only Hagerstown and Cumberland experienced industrial growth, chiefly iron foundries, railroad repair yards, and machine shops. Coal transportation was the basis of Cumberland's manufacturing and commercial activities.

Most of the new immigrants stayed in urban areas (the rural population was only 4% foreign born). The number of blacks in rural areas declined between 1860 and 1910, many moving to Baltimore and Washington, DC. Still, southern Maryland remained 40 to 50 percent black, the Eastern Shore 25 to 33 percent, while the Piedmont and western Maryland had only small black populations. By 1910, nearly 43 percent of all Marylanders resided in the increasingly culturally pluralistic city of Baltimore. The ethnic neighborhoods of Baltimore continued to grow and develop, embracing nearly 75 percent of Maryland's foreign-born population. The Germans formed the largest group, with Eastern European Jews, Irish, Italians, and Poles also growing significantly in size.

By 1900, Baltimore City had become the center of the state in many ways. Some canning of vegetables, timber processing, and seafood processing remained in rural parts of Maryland, but by 1900 most of the industry was in the Baltimore area. By 1920, Baltimore had clearly become the central focus of manufacturing. It had over 40 percent of the state's population, 33 percent of the black population, 75 percent of the foreign-born population, 66 percent of industrial workers, and 60 percent of the total value of production. Manufacturing became the leading activity in Baltimore by 1910, exceeding the value of foreign commerce. New large industries such as steel and petroleum refining were established. Smaller industries requiring semiskilled hand labor, such as manufacturing men's clothing, remained in the city, while new heavy industry located in surrounding suburbs. Baltimore was largely a branch-plant town rather than an industrial headquarters. Although the fire of 1904 destroyed most of its downtown, Baltimore rebuilt, and by 1910 this sprawling metropolis had over 500,000 people, making it the seventh-largest city in the nation (see table 1.1). Industrialization brought with it congested housing, poor sanitation, diseases, and poor working conditions, which were accompanied by political corruption. Baltimore may have entered the twentieth century with serious challenges to its economic and social structure, but the quality of its neighborhood communities and varied delights of its heterogeneity still made it attractive to businesses, workers, and residents.

Although the story of the Maryland landscape points to the growing differences among the regions of the state, at a macrolevel the regions became structurally interlinked and focused on Baltimore. The legacy of Maryland's past geography is still present in the distinct and diverse—yet connected—regions of the state. Each region with its own characteristics and personality contributes to the physical and cultural landscape called "Maryland."

2 The Mosaic of Maryland

Maryland's people and their many socioeconomic activities are not confined within the state's political boundaries. As is common in the information age of the twenty-first century, Marylanders are connected to and interact with people and places in numerous ways. The high degree of spatial interactions, both within and beyond its borders, is indicative of the hyperactivity of the Maryland landscape and its people.

The Regional Setting

To understand how Maryland fits into this increasingly interconnected global community, it is useful to look at how geographers define and describe the various wider regions that include Maryland and various aspects of its location. One way of presenting Maryland's contemporary location is by describing its *absolute location* (i.e., its position on the Earth's latitude/longitude grid). Maryland lies between parallels 37°43'26" and 39°53' north latitude, and between meridians 75°4' and 79°29'15" west longitude. But this information alone is not that helpful for most people. Geographers have become much more interested in what they call *relative location* (i.e., the location of a place as it spatially relates and interacts with other places). Maryland's general relative location within the eastern US market is central and therefore within easy reach of many other large cities.

The geographical diversity of Maryland is reflected in the numerous regions included by geographers attempting to describe its relative location: the American Manufacturing Belt, Appalachia, megalopolis, and several physiographic regions including the Appalachian Highlands, Coastal Plain, and Piedmont. Regions are areas of the Earth defined in terms of specified characteristics, and they extend as far as those characteristics are distinctive. By looking at the relative location of Maryland and briefly looking at the characteristics of these various regions, one can see the multifaceted character of Maryland itself.

Megalopolis

Maryland lies at the southern part of a noticeable region of dense urban population running in a 500-mile (800-kilometer) arc northeast-southwest along the East Coast from southern Maine to southeastern Virginia. This urban corridor was termed *megalopolis* by Jean Gottmann in his classic 1961 study, an outgrowth of twenty years of research on the area. Gottmann included central and eastern Maryland as well as metropolitan Washington, DC, in his definition of megalopolis, which originally extended from southern New Hampshire to Northern Virginia. Over the past fifty years, however, this megalopolis has grown to extend from Portland, Maine, to the Portsmouth-Norfolk area of Virginia. Gottmann recognized the national dominance of this region when he wrote, "Megalopolis provides the whole of America with so many essential services of the sort that a community used to obtain in its downtown section that it may well deserve the nickname 'Main Street' of the nation" (Gottmann 1961).

The dominant feature of megalopolis, its urbanism, is reflected in its intense spatial organization. Although the region is small in area by world standards—only about 50,000 square miles (130,000 square kilometers)—it contains over fifty million people, or 17 percent of the US population. This is an impressive figure when coupled with the fact that megalopolis covers only about 1.5 percent of the land area of the United States. According to geographer Tom McKnight, "It is the premier region of economic and social superlatives to be found in North America. It represents the greatest accumulation of wealth and the greatest concentration of poverty; it has the greatest variety of urban amenities and the greatest number of urban problems; it has the highest population densities and most varied population mix; and it is clearly the leading business and government center of the nation" (McKnight 1997). But this region of cities and supercities is not entirely covered by urban centers. Only about 20 percent of it is in urban land use; the remainder contains extensive open space and farmland, forested, and wetland/coastal areas.

There are nine metropolitan areas within megalopolis that exceed one million people, including the core of megalopolis, the New York–northern New Jersey–Long Island metropolitan area, which is home to nearly twenty million people (table 2.1). When Gottmann wrote his classic study in 1961, the Baltimore and Washington, DC, metropolitan areas were separate and distinct. Over the past fifty years, however, these two metropolitan areas have grown together and are now considered by the US Census Bureau to be a single, coalesced metropolitan region. The Baltimore and Washington metropolitan areas together have over eight million people, making it the fourth-largest greater metropolitan region in the United States (after New York, Los Angeles, and Chicago, in that order).

The Washington–Baltimore metropolitan area, which extends westward to include Frederick and Hagerstown, combine two very different core cities. Baltimore was founded in 1729 at a waterpower site along the fall line. It developed as an important port with excellent rail connections to the interior, diverse local industries, and a productive

Table 2.1. Major metropolitan areas in megalopolis, 2010

Metropolitan area	*US rank*	*Population (thousands)*
New York (NY-NJ-PA)	1	19,070
Philadelphia-Camden-Wilmington (PA-NJ-DE)	5	5,968
Washington (DC-MD-VA-WVA)	8	5,476
Boston-Cambridge-Quincy (MA)	10	4,589
Baltimore-Towson (MD)	**20**	**2,691**
Norfolk-Virginia Beach-Newport News (VA-NC)	36	1,674
Providence-Fall River-Warwick (RI-MA)	37	1,601
Richmond-Petersburg (VA)	43	1,238
Hartford (CT)	45	1,196
Albany-Schenectady-Troy (NY)	58	858
Allentown-Bethlehem-Easton (PA)	62	816
Worcester (MA)	64	804
Springfield (MA)	74	699
Harrisburg-Lebanon-Carlisle (PA)	96	537
Portland (ME)	99	517
Lancaster (PA)	101	508
York (PA)	113	429
Reading (PA)	123	407
Atlantic City (NJ)	166	272
New London-Norwich (CT)	169	267

Source: Statistical Abstract of the United States, 2011

surrounding hinterland (region). Baltimore is located farther inland than any other major seaport along the East Coast and is the center of the rich agricultural district surrounding the Chesapeake Bay. Baltimore has been a major food-processing center for many years; it also exports grain, coal, and manufactured items, and it is a major national importer of copper, sugar, and petroleum. These imports are the basis of many processing industries, including sugar refining, copper smelting, commercial fertilizers, and formerly steel making. (Built in 1887, the Sparrows Point tidewater steel plant was closer to the Midwest than New York or Philadelphia, meaning that rail freight rates through Baltimore were consistently lower than through other East Coast megalopolitan ports. But by 2013 the steel mill was closed after failed attempts by a succession of owners to keep it operating.) Baltimore's locational advantage changed with the deregulation of the railroads in the 1980s. Since then other ports, such as Newport News in Virginia, have been able to offer lower rates, thus cutting into the business of the Port of Baltimore.

After World War II, Baltimore experienced a long period of decay. Since the early 1970s the city has undergone revitalization, reflected in several major developments: the Charles Center, Convention Center, Harbor Place, National Aquarium, Oriole Park at Camden Yards, and a biotechnology agglomeration centered on the University of Maryland at Baltimore. Even with these developments, however, Baltimore is still a troubled city with stark contrasts. It shares with many other American cities the serious problems of crime, a school system in crisis, and racial tension. Contrasting the sparkling showpiece revitalization projects are nearby poverty-stricken neighborhoods. Baltimore remains a highly segregated city, with about two-thirds of its census tracts being either more than 90 percent white or more than 90 percent nonwhite. Culturally, Baltimore sits along an imprecise cultural-geographical dividing line. It has been called "the most southerly Northern city and the most northerly Southern city."

The numerous internal and external linkages of megalopolis are complex and essential to the functions of this region, which has a high degree of interchange with other regions of the United States and with other parts of the world. Megalopolis is the western terminus of one of the world's busiest ocean routes across the North Atlantic from Western Europe. The convoluted coast of megalopolis contains many fine natural harbors, including Baltimore. In addition, many of the large cities of megalopolis have good access to the interior. This accessibility has been important to the development of Baltimore and Maryland. Goods, people, and ideas have entered megalopolis from overseas and have been absorbed, changed, and sent into the interior. There are also flows from the interior of the country to megalopolis and then overseas. For this reason, Gottmann referred to megalopolis as the "hinge" of the continent. In many ways, this is where the action is, where many national and international economic and political decisions are made. Megalopolis is the nerve center of the country. It also contains the continent's largest city (New York) and the nation's seat of government in Washington, DC.

The American Manufacturing Belt

Although manufacturing activity spread to many other areas over the past one hundred years, the core manufacturing region of the United States remains in the northeastern quarter of the country. The American Manufacturing Belt was defined over eighty years ago as a quadrangle, with corners in the Northeast in Portland, Maine; in the Northwest in Milwaukee, Wisconsin; in the Southwest in St. Louis, Missouri; and in the Southeast in Baltimore, Maryland. Since then, the definition has undergone many refinements, but all remain centered on the old quadrangle.

There are about 11.7 million people employed in manufacturing within the United States today, and their geographic distribution is shown in table 2.2. Manufacturing employment is still heavier in the eastern part of the United States (56.4%) than in the West (43.6%), where it is spread out more widely. The western region, however, has been growing steadily. In 1900 it accounted for only 11.5 percent of manufacturing employment, but it has steadily grown to its current 43.6 percent. Of the 56.4 percent of employment found in the eastern region, about 37 percent of the jobs are found in states east of the Mississippi River and north of the Ohio River (New England, Middle Atlantic, and east-north central); this area roughly coincides with the American Manufacturing Belt quadrangle. This Northeast region also accounts for 43 percent of all value added in manufacturing in the United States.

Within the Northeast quadrant there are several manufacturing subregions: the Great Lakes, the Middle Atlantic (including Maryland), New England, the New York metropolitan area, and the Ohio Valley. Each subregion possesses a unique personality, reflecting different source areas for raw materials; different historical development patterns; different labor, market, energy, and transport factors; and different relative locations.

Maryland is located in the Middle Atlantic subregion along with Delaware, eastern Pennsylvania, and southern New Jersey. Diversity

Table 2.2. Percentage of US labor in manufacturing

Region	*Selected years*					
	1899	*1929*	*1954*	*1970*	*2000*	*2012*
	East of Mississippi					
New England	18.1	12.4	9	7.7	5.6	5.2
Middle Atlantic	34.1	29	26.6	21.9	11.5	12.1
East-north central	22.8	28.8	28.6	26	23.2	19.4
South Atlantic	9.7	10.3	11	13.5	15.9	13.4
East-south central	3.8	4.3	4.5	6.1	7.9	6.3
Total	88.5	84.8	79.7	75.2	64.1	56.4
	West of Mississippi					
West-north central	5.6	5.4	6	6.3	8.2	13.3
West-south central	2.4	3.4	4.5	6.2	9.3	10.6
Mountain	0.9	1.2	1.1	1.9	4	4.8
Pacific	2.6	5.2	8.7	10.4	14.4	14.9
Total	11.5	15.2	20.3	24.8	35.9	43.6

Source: US Bureau of Census, *Census of Population, Census of Manufacturing* (Washington, DC: US Government Printing Office, various years)

is the hallmark of manufacturing in Maryland as it is within the rest of the region—nearly every type of industry in the United States is represented in this region in at least small amounts. A variety of manufacturing takes place in Maryland, ranging from heavy industries such as shipyards and petroleum refineries to lighter industries such as electronics, biotechnology, and food processing. Although the states east of the Mississippi River still contain over two-thirds of the manufacturing labor in the country, the trend in table 2.2 is one of slow decline. The Middle Atlantic region has declined from 34.1 percent in 1899 to 12.1 percent of manufacturing employment in 2012. To compensate for the westward shift in population and markets as well as the higher cost of labor in the Middle Atlantic, Maryland has changed its focus to manufacturing high-value added products that can bear the cost of long-distance transportation. The relative location of Maryland gives it a number of critical advantages. The heavily populated megalopolitan region is a ready market for its products. The Atlantic coastal frontage and fine harbors, in addition to providing easy access to foreign imports, facilitates exporting. The manufacturing centers of Maryland are well connected by transportation lines to the interior and to the entire East Coast. These accessibility factors enable Maryland to interact with other national and foreign places with comparative ease and low cost.

A number of advantages found in urban areas help to explain the interlocking growth patterns of cities and industries. Large urban areas generally provide more accessible markets than do rural areas. Urban areas provide laborers and consumers, transportation and communication networks, utilities, public services such as fire and police protection, schools and recreation for workers and their families, and a host of other amenities. One of these characteristics is called *economies of scale* (i.e., large-scale production that allows a manufacturer to spread the fixed costs of buildings, machinery, land, etc., over more units of production). Improving economies of scale results in lower costs per unit produced, lower prices for consumers, and higher profits for large-scale producers. Urban areas may also benefit from what are known as *agglomeration economies* (i.e., a clustering of related industries in which component parts produced by various firms are brought together for assembly by one firm). Related industries that are clustered can also share mobile skilled labor that can move from business to business. Yet, in some industries, such as computer hardware manufacturing, agglomeration economies have been weakened as a location factor. The global economy, with its greater efficiency in communications and lower transportation costs, has allowed many manufacturers to assemble components manufactured in many parts of the world. Many Maryland industries import component parts and export them nationally and worldwide.

Appalachia

Another facet of Maryland's regional kaleidoscope is found in the Appalachian region of western Maryland, in Allegany, Garrett, and Washington Counties. This western part of Maryland has a physical and cultural character distinct from the rest of the state. Even the most advanced countries often have one or more regions that lag behind in the arenas of economic and social development. These regions fail to keep pace with the rising standard of living and structural economic changes that occur in growth areas of the country. In the United States, Appalachia has been one such region—an extensive area of related rural poverty that stretches southwest along the highland region from central New York State south of the Mohawk Valley to central Alabama.

The definition of Appalachia as a region is complex. First and foremost, it is a physical region possessing the unity of geologic similarity. Although the distinction between "mountains" and "hills" is far from exact, the uplands of much of this region can be considered mountains. Even so, they are by no means the highest mountains of the United States (the Sierra Nevada and the Rocky Mountains have peaks that rise over a mile higher than the highest Appalachian peaks). Appalachia includes three distinct physiographic regions running parallel to each other. When traveling westward from the coast, the first Appalachian region encountered is the Blue Ridge; next comes the Ridge and Valley region, and then the Allegheny Plateau. All of these subregions of the Appalachian system are represented within the borders of Maryland.

Physical characteristics alone do not fully explain the way in which the name *Appalachia* is used; it also carries historical, cultural, and political meanings. Historically, the Appalachian system has been a barrier to east-west movement. Settlers did not cross the Blue Ridge into the highlands until late in the colonial period. In 1732, Cecilius Calvert, second Lord of Baltimore, officially proclaimed western Maryland open for settlement, but it was not until the late 1700s and early 1800s that the area was settled. In the vanguard of the pioneers were many Welsh Quakers and Scots-Irish Presbyterians from Ulster in the north of Ireland, who were followed by many Germans and Highland Scots. As time passed, Appalachia became isolated from other areas. Transportation in Appalachia was difficult. Even the famous Wilderness Road that cut through the Cumberland Gap to the Bluegrass Basin of Kentucky was a difficult passage, and most of the rivers ran too swiftly to be used for transportation. The region was originally heavily forested, and the rugged terrain made road and railroad construction difficult and expensive. As a result, Appalachia developed as a sort of landlocked place largely isolated from the rest of the nation.

Culturally, Appalachia does not fit the description of a unified region with common characteristics throughout. Although it has been vaguely defined as a region on the basis of a common syndrome of poverty, limited opportunities, and stagnated economic development, the cultural variations within Appalachia are great. Appalachians in Maryland may resemble Appalachians in Alabama in education, lifestyle, and income, but they are not linked to each other by any significant commercial trade or migration flows. Economist Edgar Hoover once stated that "Appalachia should be regarded as a succession of hinterlands to various centers located mainly outside the region as officially defined; a row of back yards as it were" (Hoover 1975). Appalachia is often described in ways that emphasize its negatives: farmlands, coal production, and employment declined for many years, and schooling was backward. The better educated, young, and energetic often left, creating a "brain drain" that had a serious negative impact on this region, which already suffered from a scarcity of these human resources. Corporations from outside the region often controlled local businesses such as coal mining and lumbering, meaning that most of the profits left Appalachia.

As a result of regional immobility and isolation, a cultural distinctness has persisted in Appalachia. In his penetrating analysis of the region, *Night Comes to the Cumberlands,* Henry Caudill stated, "In spite of the desirability of migration from certain depressed areas, studies have shown a pronounced reluctance on the part of the people to move. The social attachments to the area in terms of association with churches and clubs, with strong family ties, the distinctive local ways of life, and the difficulties of adjustment to a metropolitan environment all tend to reduce potential out migration" (Caudill 1962). Appalachian people have often been pejoratively called "hillbillies." Geographer Tom McKnight has said that it is tempting to paint this entire region as a hill-people haven, replete with colorful speech, a charming folk culture, and hidden moonshine stills. Although some hill people and mountain folk still reside in the area today, this exaggerated image belongs to another era. With a few local exceptions, the isolation and distinctiveness of Appalachians are things of the past (McKnight 1997). Appalachian people are mostly Anglo-Saxon and Protestant. They represent the largest group of predominantly white low-income people in the United States (about one-third of Appalachian families live below the federally defined poverty level).

Appalachia developed as a region with heavy employment in occupations for which demand has either grown slowly or declined (e.g., coal mining and general farming). As a result, many Appalachians in the labor force today lack experience in more dynamic occupations and are therefore at a disadvantage when seeking work elsewhere. The high rural population densities are not supported by a commercial agricultural system; small farms and coal mines are where most of the labor has traditionally been employed.

When the time came in the sequence of occupance for the transition from the stage of subsistence agriculture to the stage of commercial, market-oriented agriculture, farmers in Appalachia were largely unsuccessful. The average Appalachian farm is small and remote, and the rugged topography, poor soil, and short growing season all work against commercial farming. The efficient use of heavy machinery is often not possible here because of the lack of large, level farm acreage. Farming in Appalachia is generally a mixture of many products; no single crop or livestock type dominates, as corn does in parts of the Midwest or poultry on Maryland's Eastern Shore. In western Maryland, most farming is done on the limited valley floors and immediate slopes, with steeper slopes often being forested. The mixture of farm activities in Appalachian Maryland includes producing syrup from maple trees; tending fruit orchards; maintaining pastureland; dairying; and raising hay, wheat, and poultry.

One feature that has brought some unity to the Appalachian region is coal mining. Its coal became important after the Civil War with the development of the railroads and the steel industry, helping to drive the US industrial revolution. But coal mining had negative impacts, too. It adversely affected the land and workers' quality of life, even while providing employment for many. Then, during the Great Depression and after World War II, the coal industry declined; the resultant economic distress hit Appalachia hard. Unemployment in coal counties soared as mechanization eliminated many jobs. Some counties lost over a quarter of their population to outmigration between 1950 and 1960. Many migrants from Appalachian Maryland, Pennsylvania, and West Virginia were attracted to urban jobs in Baltimore and Washington. They settled in close-knit neighborhoods and retained much of their culture. Strong links with relatives and friends back in Appalachia were usually kept.

John F. Kennedy became aware of the deteriorating economic and social conditions in Appalachia while campaigning for the presidency in 1960. His promise to remedy these great regional disparities was addressed after his death when in 1965 the Appalachian Redevelopment Act was passed as an extension of the Area Redevelopment Act of 1961. The definition of Appalachia adopted by the act was broadly based on vague criteria: high unemployment; low income; substandard levels of housing, health, and educational facilities; dominance of the regional economy by one or two industries in long-term decline; substantial out-migration of labor and capital; and slow growth rate of aggregate output. In the final analysis, it was largely left to the states to decide what Appalachia was. The emergent political administrative development region stretched from New York State to Mississippi, including 373 counties in eleven states.

The Appalachian Redevelopment Act made available $1.1 billion to the Appalachian Regional Commission to be spent on projects agreed upon by federal and state officials. Since 1965, lawmakers have added to the original allocation, and the commission has spent several billion dollars for health centers, vocational training, erosion control, schools, libraries, sewage facilities, recreation facilities, and roads. By far the largest share of the funds (80% of the original allocation) was used to build roads. It was hoped that new roads would decrease the region's isolation and attract industries. Much criticism has been leveled at the commission for committing so much money for roads, as their goal of attracting development corridors along the new roads has not been fully realized. In many instances, the improved roads have allowed motorists to travel through the region faster and have facilitated additional out-migration.

Appalachia was long regarded as an economic anomaly, a large enclave of poverty within one of the richest nations on Earth. This region has experienced three major economic development waves, each financed largely by outsiders, and most of the profits have left the region. The first two waves involved lumbering and mining. Both provided many jobs, most of which were low paying. Lumbering and mining also degraded the natural environment, creating many problems that are still being addressed today. The more recent third wave of development involves recreation and tourism. Once again, many outside investors have purchased scenic land and natural amenities to develop massive housing estates and recreational facilities. This kind of activity can be seen in Garrett County around Deep Creek Lake, the nearby ski resorts, and whitewater rafting facilities along the Youghiogheny and other rivers. A renewed interest in coal after the energy crises of the early 1970s also helped Appalachia to rebound. Future demand for the region's coal by the nearby large markets of megalopolis as well as overseas markets is uncertain. By the last two decades of the twentieth century, Appalachia's population was once again growing, reversing the long trend of net out-migration. Today new factories and jobs in recreation and tourism are attracting people to the area.

Many older Marylanders seek the quiet rural hills of western Maryland as places of retirement. Modern communications and improved transportation have redefined geographic space, and the relative location of Appalachian Maryland has found itself changed in recent decades. With satellites, the Internet, mobile phones, and other devices, this region is no longer isolated. Many farms are even becoming more profitable as they specialize in beef cattle and poultry. But despite all of these positive developments, parts of Appalachia will for a long time to come still have large areas of poverty, many poor farms on eroded soil, marginal coal mines, and low-paying jobs.

Physiographic Provinces

Maryland's landscape displays a cross section of five major physiographic provinces and a number of subdivisions by cutting across their continental northeast–southwest grain (fig. 2.1). From east to west, the provinces are the Coastal Plain, Piedmont (eastern upland and west-

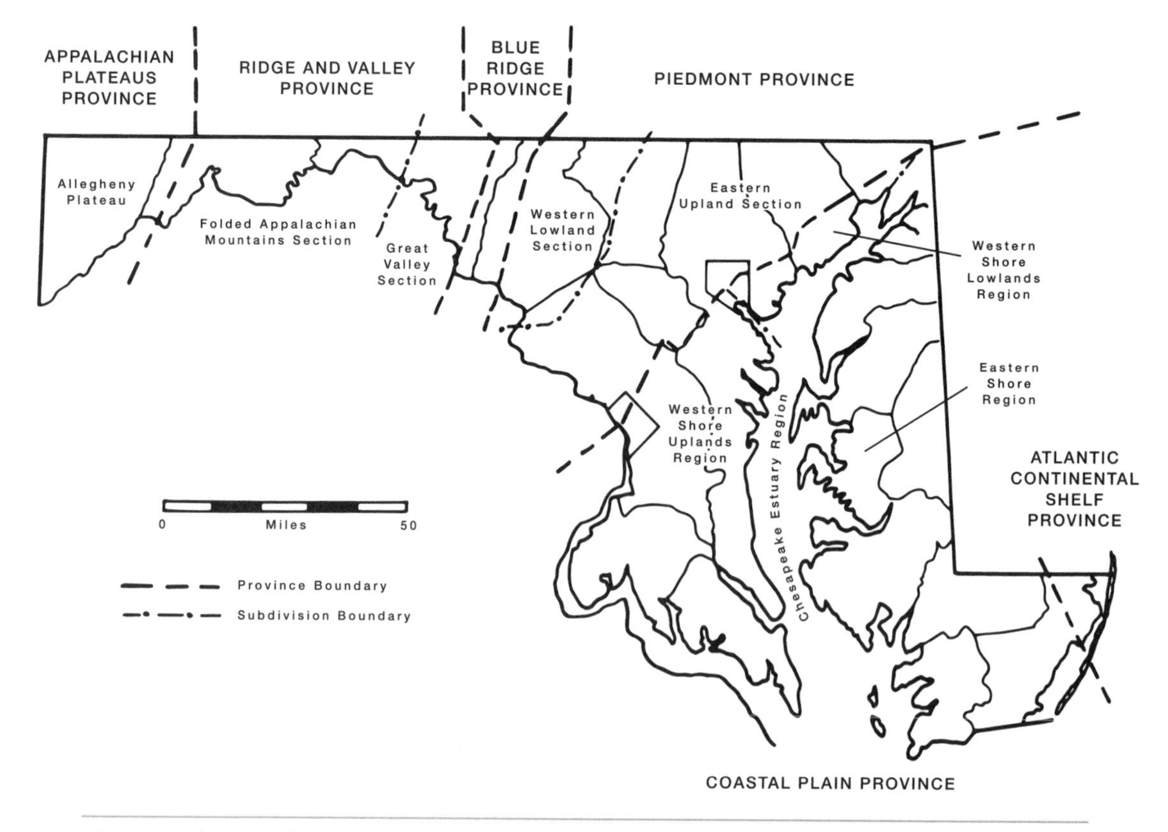

Figure 2.1 Physiographic Provinces and Their Subdivisions

ern lowland divisions), Blue Ridge, Ridge and Valley (Great Valley and Folded Appalachian Mountains divisions), and Appalachian Plateaus.

Coastal Plain

This province covers approximately 5,000 square miles (12,950 square kilometers), which is about half the area of Maryland. The Coastal Plain is an extensive, partially submerged, and undulating to flat surface consisting mainly of recent marine sediment deposits characterized by a number of seaward dipping beds of unconsolidated clay, sand, and gravel. It is also a depository for materials washed by rain and rivers from the Appalachian Highlands as a consequence of millions of years of erosion. The Coastal Plain extends from below sea level on the continental shelf westward to the fall line zone, a general boundary separating the higher-elevation crystalline rocks of the Piedmont from the unconsolidated sediments of the Coastal Plain. Streams frequently have falls or rapids where they flow from the older, more resistant metamorphic uplands of the Piedmont onto the Coastal Plain; a line on a map connecting these falls would approximate the fall line. It is most likely that the fall line was first mapped by boat, because the rivers are generally wider and deeper on the seaward side of the fall line, with the falls and rapids marking the upper limit of navigation from the sea. During the early European settlement period of Maryland, these cargo break-bulk places became the sites of a number of settlements, including Baltimore. Other attractive factors along the fall line included a ready source of waterpower for milling and manufacturing, and proximity to natural harbors.

The Coastal Plain is a complex network of stream drainage and broad, branching bays and estuaries at low elevation. The Chesapeake Bay divides the Coastal Plain into the Eastern and Western Shores. "Eastern Shore" was the name used by Captain John Smith in 1608 to describe all of what is now the Delmarva Peninsula; the Eastern Shore of Maryland is found east of the Chesapeake Bay and south of the Elk River. Elevations on the Eastern Shore range from sea level along the coast and riverine locations to approximately 60 feet (18.3 meters) in the north. As a result, the streams of this area have a gentle gradient with wide and shallow valleys. Low-lying wetlands account for nearly 20 percent of the land of the Eastern Shore. The embayed Coastal Plain becomes somewhat rolling in the north as it approaches the Piedmont and takes on some of the characteristics of that province. Overall, the Coastal Plain nearly lacks relief, which is the range from lowest elevation to highest elevation.

The Western Shore rises to an elevation of 270 feet (82.3 meters). Some major rivers coming off the Piedmont dissect this area; for example, the Gunpowder, Patapsco, and Patuxent. The rivers flowing across the Western Shore in most cases carry much greater volumes of water than the smaller, low-velocity rivers of the Eastern Shore. Many low knobs and ridges characterize the Western Shore region. That part of the region, which is known as "southern Maryland," is a low-relief upland area dissected by the Patuxent and Potomac Rivers, with elevations ranging from 0 to 270 feet (82.30 meters). The Maryland county of greatest relief is Allegany in the west, which has a low elevation of approximately 420 feet (128.02 meters) and a high point of 2,898 feet (883.31 meters), a relief of 2,478 feet (755.29 meters). By comparison, Coastal Plain counties have relatively little relief. For instance, Anne Arundel County has a relief of 300 feet (91.44 meters), Calvert County 168 feet (51.21 meters), Charles County 235 feet (71.63 meters), Dorchester 57 feet (17.37 meters), and Wicomico 73 feet (22.25 meters). In Calvert County, dissection has produced flat-topped ridges and the steep Calvert Cliffs along the Chesapeake Bay.

Geologically, the Coastal Plain has a rich past, which is reflected in the many fossils and sediments in the area. There are many types of fossils found throughout Maryland. Trilobites, brachiopods, mollusks, plants, shark teeth, and dinosaurs are just some of the archeological treasures common in the state. *Astrodon johnstoni* Maryland's state dinosaur, reached a height greater than 30 feet (9.14 meters). The first fossil described from North America, *Ecphora quadricostata*, was found in deposits of the St. Mary's Formation (Maryland Geological Survey 1976) and was illustrated in a work on Mollusca published in England in 1685. Deposits are exposed in cliffs up to 100 feet (30.48 meters) high between Chesapeake Beach and Drum Point and constitute the most complete section of Miocene deposits in the eastern United States. The skeletal remains of such land fauna as tapirs, mastodons, rhinoceros, horses, and dogs have been found here. The Miocene was a period of uplift in the middle part of North America and in the Antillean region, which was accompanied by folding of the Earth's crust and volcanism. Erosion of the recently uplifted areas produced extensive deposits of clay, sand, and marl. Some of these sediments were consolidated, forming shale and sandstone. During this time, North and South America were united, and the island of Florida joined the Georgia mainland. The North American continent assumed approximately its present outlines. When the great Pleistocene glaciers melted about 10,000 years ago, the level of the sea rose and drowned the stream valleys of the western shore to the fall line. On the Eastern Shore, the rivers were drowned 10 to 40 miles (16.1 to 64.4 kilometers) from the present shoreline. The highly irregular coastline of Chesapeake Bay is an indication of the submerged nature of this embayed area of tidal estuaries.

A series of beaches extending south from Cape Henlopen, Delaware, along Maryland to Virginia are economically important and ecologically significant to the Atlantic coast. In Delaware, the barrier beaches are attached to the mainland, but in Maryland the barrier islands are separated from the mainland by tidal lagoons, locally called "bays." The beaches are composed of material eroded at Cape Henlopen and carried south by strong longshore currents that cause littoral drift. The great, long barrier islands of Fenwick and Assateague are found along the Maryland coast. Fenwick Island is separated from the mainland by Assawoman Bay, Assateague Island by Sinepuxent Bay in the north, and Chincoteague Bay in the south. Fenwick and Assateague Islands were at one point one island until a hurricane breached it in 1933, producing what is today called the Ocean City Inlet. The inlet has now been regularized and the sides stabilized with rocks to prevent it from closing. In Maryland and Virginia, the 37-mile (59.5-kilometer) stretch of Assateague Island has been designated the Assateague Island National Seashore.

Piedmont

Between the Coastal Plain and the Blue Ridge lies the Piedmont, or "foot of the mountain." The eastern boundary of the Piedmont is the fall-line zone; the western boundary is Catoctin Mountain. The Piedmont is a province of varied topography with low, wooded hills separated by well-drained fertile valleys.

The entire Piedmont covers approximately 2,500 square miles (6,475 square kilometers), one-fourth of Maryland's area. It has two geologically and topographically distinct subdivisions: the eastern division, called the Piedmont upland, and the western division, called the Piedmont lowland. The Piedmont upland is underlain by metaphoric rock; its diversified, broadly undulating topography is low to moderate in relief. Parrs Ridge, 880 feet (268.2 meters), and Dug Hill Ridge, 1,200 feet (366 meters), are the highest areas of the Piedmont upland. They form a drainage divide that separates the flow east to Chesapeake Bay from the flow west to the Potomac River (and eventually to the Chesapeake Bay as well). Gold is found in rocks of the Piedmont Plateau, a belt of metamorphic rocks extending from New York to South Carolina. As the *Maryland Journal* reported in May 1901, "Many persons will be surprised to know that within easy walking distance of the National Capital there are no less than a half-dozen gold mines in actual operation. Prospecting is now a rather extensive industry along the banks of the Potomac, from a point near Georgetown up the river, past Great Falls, a distance of perhaps ten miles" (Kuff 1987). Major streams deeply dissect the Piedmont upland, and a number of dams have been built across narrow stream valleys to meet the need for water, electricity, and recreation of the large, nearby population concentrations of the urban corridor of northern Delaware, Maryland, southeast Pennsylvania, and Washington, DC. Reservoirs have been created along several rivers, including Triadelphia and Rocky Gorge Reservoirs on the Patuxent River, Liberty Reservoir on the Patapsco River, and Prettyboy and Loch Raven reservoirs on Gunpowder Falls River (table 2.3).

The Piedmont lowland centers on the Frederick Valley, extending north from the Potomac River to the vicinity of New Midway, Frederick County. Because it is underlain by limestone, the area was worn down more rapidly than the more-resistant metamorphic rock to the east. The Monocacy River drains the fertile Frederick Valley, a prosperous dairying area with a low relief averaging 300 feet (91.4 meters). To the north, west, and south is that part of the Piedmont called Triassic lowland, an area of sedimentary rock of Triassic age with a relief averaging 500 feet (152.4 meters). A prominent feature along the boundary of the Piedmont upland and lowland in southeastern Frederick County is Sugarloaf Mountain, an erosional remnant called a *monadnock*, with an elevation of 1,280 feet (390 meters).

Blue Ridge

West of the lower Piedmont is the Blue Ridge, named for the blue haze that often hangs over the mountains. This mountain belt consists of three separate ridges: Catoctin Mountain, South Mountain, and Elk Ridge. Catoctin Mountain lies a bit east of the Blue Ridge and is considered a prong, or outlier. It is separated from South Mountain (con-

Table 2.3. Maryland's largest reservoirs

Reservoir	*County*	*Affected river*	*Primary purpose*	*Surface area (acres)*
Conowingo	Cecil and Harford	Susquehanna	hydroelectric	8,563
Deep Creek Lake	Garrett	Deep Creek	hydroelectric and recreation	3,900
Liberty	Baltimore and Carroll	Patapsco	Baltimore City water	3,106
Youghiogheny River	Garrett	Youghiogheny	flood control and hydroelectric	2,800
Loch Raven	Baltimore	Gunpowder	Baltimore City water	2,400
Prettyboy	Baltimore	Gunpowder	Baltimore City water	1,500
Jennings Randolph Lake	Garrett	Potomac	flood control	965
Triadelphia	Montgomery	Patuxent	hydroelectric	800
Rocky Gorge	Montgomery	Patuxent	Washington, DC, water	773

Source: Maryland Geological Survey, "Maryland's Lakes and Reservoirs." Baltimore: Maryland Geological Survey, 2004

sidered the beginning of the Blue Ridge proper) by the Middletown Valley. Located in this area is the 10,120-acre (4,099-hectare) Catoctin Mountain National Park, which includes the historical presidential retreat named Shangri La by President Franklin Delano Roosevelt. Later, President Dwight Eisenhower would rename the retreat Camp David after his grandson. Camp David lies west of Thurmont, Frederick County, and is 60 miles (96.5 kilometers) from Washington, DC. The National Park Service has jurisdiction over this area.

The crests of the Blue Ridge were long ago removed by weathering and erosion, leaving only erosional remnants. The Potomac River has cut a spectacular water gap in the ridges of the Blue Ridge, which lies perpendicular to the river's course. Steep, sharp slopes characterize the Potomac's valley as it cuts through the metamorphic quartzite of the area, and the channel of the Potomac is strewn with rock outcrops forming rapids in the southeastern part of the physiographic region from Harpers Ferry, West Virginia, to Point of Rocks, south of Frederick in Maryland.

Ridge and Valley

Extending approximately 65 miles (105 kilometers) west from South Mountain and Elk Ridge to Dan's Mountain (the local name for the Allegheny Front) is the Ridge and Valley division. The two distinct subdivisions of this region are the Valley-Ridge section in the west and the Great Appalachian Valley in the east, which is locally called the Hagerstown Valley in Maryland, Cumberland Valley in Pennsylvania, and

Shenandoah Valley in Virginia and West Virginia. The Hagerstown Valley lies between South Mountain and Elk Ridge on the east and Powell and Fairview Mountains on the west. This broad, rich valley contains many farms with fruit orchards being prevalent. The lowland has an average elevation of 500 to 600 feet (152 to 183 meters), increasing from 400 feet (122 meters) at the Potomac River to 800 feet (244 meters) at the Maryland–Pennsylvania border. The two main rivers draining the Hagerstown Valley are Antietam Creek in the east and Conococheague Creek in the west; both have their source in Pennsylvania and flow south into the Potomac River. Conococheague Creek is the most prominent feature in the valley, as it is incised approximately 100 feet (30.5 meters) into the surface of a belt of soft shale that is 22 miles (35.4 kilometers) long and 3.5 miles (5.6 kilometers) wide, and it has many pronounced meanders along its course.

From Powell and Fairview Mountains west to Dan's Mountain is a series of northeast–southwest trending ridges separated by narrow valleys, resulting from a series of alternately weak and resistant sedimentary rock. These ridges run parallel to each other and are nearly even in elevation. One of these narrow valleys is the Cumberland Narrows looking west from the city of Cumberland, Maryland. Sideling Hill, at an elevation of approximately 1,615 feet (492.25 meters), is 6 miles (9.66 kilometers) west of Hancock on Interstate 68. The 380-foot (116-meter) near-vertical cut removed 4.5 million cubic yards (3,440,497 cubic meters) of rock to expose 350-million-year-old marine sediments overlain by younger, thick-bedded sandstones, conglomerates, and dark coaly shale. The rock is tightly folded in a syncline nearly 810 feet (246.89 meters) in height.

Appalachian Plateaus

Dan's Mountain, immediately west of Cumberland in Allegany County, marks the eastern edge of the Allegheny Plateau, the regional name for this portion of the Appalachian Plateau. This physiographic division includes western Allegany County and all of Garrett County. Although it is called a plateau, the area is far from being an elevated tableland. To the west are a series of gentle, open folds of compressed sedimentary rock that have been eroded to produce a coarse, grained surface with ridges standing 500 to 800 feet (152 to 244 meters) above the land surface.

From east to west, the Allegheny Plateau can be subdivided into Georges Creek Valley (dissected along its axis by Georges Creek), Deer Park Valley (dominated by the Savage River system), Casselman Basin, Accident Dome, and the Youghiogheny River Basin. Along the Maryland–Pennsylvania border, drainage flows north by the Casselman River system; tributaries of the Youghiogheny drain the southern area. Sandwiched between the Casselman and Youghiogheny Basins is Accident Dome. Big Savage Mountain is in the eastern part; its southern extension, Backbone Mountain, north of Kempton, is the highest point in Maryland at 3,360 feet (1,024 meters). Maryland's highest mountains are located in this westernmost section of the state.

The narrow water gap of the Youghiogheny is ideal for the development of reservoirs for water, hydroelectricity, and recreation. Today a dam impounds the Savage River at the narrow gap of Savage and Backbone Mountains. A small dam near the confluence of Deep Creek and the Youghiogheny River has produced Deep Creek Lake, the largest lake in Maryland, covering 4,000 acres (1,620 hectares). Like Deep Creek, all of Maryland's lakes are human-made. A remarkable feature of this area is the Youghiogheny River Valley, with its deep, steep walls and many falls. This physiographic region is home to many of Maryland's highest waterfalls (table 2.4).

All of Maryland's diverse physiographic regions have in common a constant assault by the elements. Its physical topography is largely a reflection of the interplay of erosional and depositional forces affected by dominant climate and weather patterns. Climate change plays a major role in influencing long-term weather patterns affecting the regional mosaic in Maryland, and how these landforms evolve over time. It is within this complex morphological background, where the seashore blends with the mountains, that Marylanders have sculpted a cultural landscape.

Table 2.4. Maryland's highest free-falling and cascading waterfalls

Waterfall	*Approximate height*	*Type*	*Location*
Muddy Creek Falls	54 feet (16.46 meters)	free falling	western Garrett County (Allegheny Plateau) in Swallow Falls State Park
Kilgores Rocks	17 feet (5.18 meters)	free falling	north-central Harford County (Piedmont Plateau) on Falling Creek
Cunningham Falls	78 feet (23.77 meters)	cascading	north-central Frederick County (Blue Ridge) in Cunningham State Park
Unnamed	17 feet (5.18 meters)	cascading	east-central Baltimore County (Piedmont) near head of Dulaney Valley Branch
Swallow Falls	16 feet (4.88 meters)	cascading	western Garrett County (Allegheny Plateau) in Swallow Falls State Park

3 The Physical Environment

To understand the full scope of the geography of Maryland, one must first understand the land itself, for it is upon the land that people have left their mark over the centuries. This imprint is known as the cultural landscape and includes features such as settlements, transportation routes, agriculture, manufacturing, and recreational facilities.

Maryland's physical environment is delightfully varied. Its low-lying areas of eastern Maryland bordering the Atlantic Ocean along the 32-mile (51.2-kilometer) coastline of Worcester County, and Chesapeake Bay on the west, have climates with a strong maritime influence. West of the Bay, beyond the fall line, are the low, rolling, and pitching hills of the Piedmont. These hills give way to the relatively undeveloped ridges, valleys, and upland plateau of the Appalachians of western Maryland, which escape the influence of the ocean to offer a more extreme "continental" environment. These regions are connected by a system of rivers, most of which flow through every part of the state all the way to the Chesapeake Bay. In fact, only the rivers in the westernmost extremity, the northeastern corner (along the Delaware border), and the region adjacent to the Atlantic Ocean do not drain across the land and into the Chesapeake Bay. This diverse physical environment—climate, soils, vegetation, and water—interacts with other parts of a much larger environmental system—including landforms, bodies of water, headwaters, and watersheds—to influence Maryland's climate.

Climate

The first volume of the Maryland Weather Service's annual report was produced in 1899 by Johns Hopkins University Press under the joint auspices of Johns Hopkins University, the Maryland Agricultural College, and the US Weather Bureau. The volume noted that the first instrumental weather observations within Maryland and Delaware were made by Richard Brooke, starting on September 1, 1753 (Brooke and Baker 1759). But John Smith's rich descriptions, made in the winter of 1606–7, are the first widely known accounts of the Maryland bay region's climate:

> The sommer is hot as in Spaine; the winter colde as in Fraunce or England. The heat of sommer is in June, July and August, but commonly the coole Breezes asswage the vehemencie of the heat. The chiefe colde is extreme sharpe, but here the Proverbe is true, that no extreme long continueth.
>
> The winds here are varieable, but the like thunder and lightning to purifie the aire, I have seldome either seene or heard in Europe. From the Southwest come the greatest gusts with thunder and heat. The Northwest winde is commonly coole and bringeth faire weather with it. From the North is the greatest cold, and from the East and Southeast as from the Barmudas, fogs and raines.

Smith also made observations about the "natives," noting that they "Divided the yeare into 5 Seasons. Their winter some call Popanow, the Spring Cattapeuk, the sommer Cohattayough, the earing of their corne Nepinnough, the harvest and fall of the leafe Taquitock. From September untill the midst of November are the chiefe Feasts and sacrifice. Then have they pleanty of fruits as well as planted as naturall, as corne greene and ripe, fish, fowle, and wilde beastes exceeding fat." Smith's account underscores the importance of long-term understandings, rather than year-by-year descriptions, for comprehending climate patterns. Climate is the long-term pattern of weather conditions typical of a place. The two obvious climatic elements are temperature and moisture, which directly influence agriculture, water supply, human comfort, industry, recreation, and much more. Climate is an important aspect of the natural resource base of Maryland, which has a moderate midlatitude climate that can be described as *mesothermal*. Climates to the north and west are more harsh, while those to the south are milder.

Maryland's climate is defined both by spatial variations (place-to-place) and temporal variations (season-to-season), which can be measured in great part thanks to data collection stations across the state, which have been observing weather and climate patterns for many decades. Climatic characteristics sometimes vary significantly at different geographical scales. At the continental scale, the general climate within Maryland does not display significant variations. But when viewed at a more local scale, one can see distinct climatic variations across the state. Slight variations in temperature and precipitation throughout the state are significant to the people living on this land and the livelihoods they pursue. These variations affect farming, forestry, and recreation, among other activities. For instance, the climatic variation in Maryland includes ski resorts in Garrett County and cypress swamps in Calvert County.

Elevation, the extent and juxtaposition of land and water surfaces, maritimity, continentality, human settlements, and latitudinal location are all factors in the state's climatic differences. The Appalachian Highlands in western Maryland have a climate that is significantly different from the Coastal Plain area in the east, in part because the

Coastal Plain is adjacent to two major bodies of water, the Chesapeake Bay and the Atlantic Ocean, which have a moderating (maritime) effect. There are also significant local differences in the Baltimore and Washington, DC, metropolitan areas owing to their different sizes and human activities.

Temperature

Mean winter temperatures in Maryland range from 47°F (8.3°C) in the high elevations of Garrett County to 59°F (15°C) on the Coastal Plain. Mean January temperatures (fig. 3.1) are about twenty degrees lower than the annual mean temperature, and mean summer temperatures (fig. 3.2) about twenty degrees higher. Urban areas have noticeable microclimates that are warmer than their surrounding hinterlands and as such are referred to as *heat islands*. This phenomenon results from the intense discharge of heat from industry and impervious surfaces and pollutants from various urban activities. One of the root causes of the heat island effect is impervious surface covers that do not allow water to penetrate. Such surfaces are known for their capacity to cause flash flooding; to increase the speed and volume of runoff into streams, causing erosion; and to heat water as it runs off of surfaces, contributing to thermal pollution of water and in-stream habitats. Impervious surfaces tend to trap heat more efficiently than natural surfaces, and then release that heat more slowly than surrounding natural surface areas. While temperatures may drop rapidly at night in surrounding rural areas, urban areas cool more slowly. In fact, the release of warm air into cool air, and the atmospheric instability that accompanies it, has been blamed for a high incidence of evening thunderstorms in some large southeastern US cities.

The heat islands of Baltimore and Washington, DC, show up clearly on mean annual temperature maps. Yet the patterns are not as clear as they were just a few decades ago, when Baltimore was home to a larger volume of heavy industry. Heat islands are especially noticeable during the summer, when winds are weaker and heat-pollution domes build up over cities. The heat island effect is not as pronounced in the winter. The Baltimore heat island is more intense than Washington's, a difference explained by its industrial activities; Baltimore has a strong mixture of heavy and light industries that release much more heat into the atmosphere. Washington, a service-oriented city, has little industry producing waste heat.

Table 3.1 displays the mean annual temperature at six climatological stations at various elevations in Maryland. Baltimore City has a mean annual temperature of 58.2°F (13.1°C), while Baltimore-Washington International Thurgood Marshall Airport, located south of Baltimore just beyond the heat island, has a mean annual temperature of 55.1°F (11.6°C). The difference between the mean annual temperatures of Snow Hill and Baltimore, both below 100 feet (33.5 meters) in eleva-

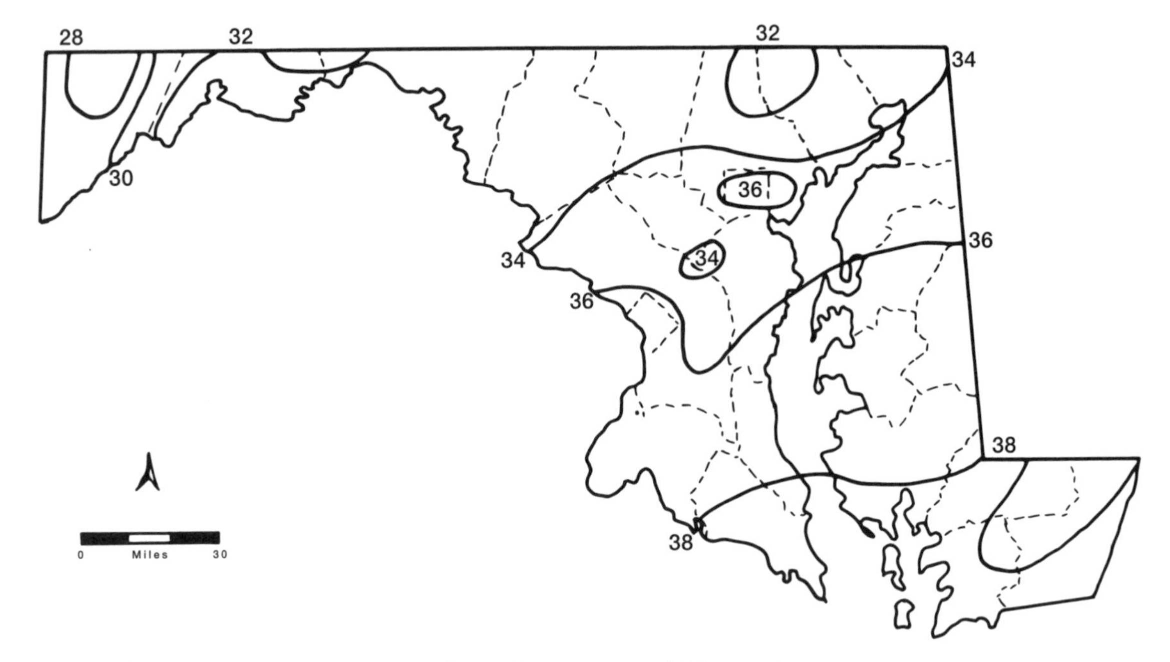

Figure 3.1 Mean Winter Temperature (December, January, and February, in Degrees Fahrenheit)

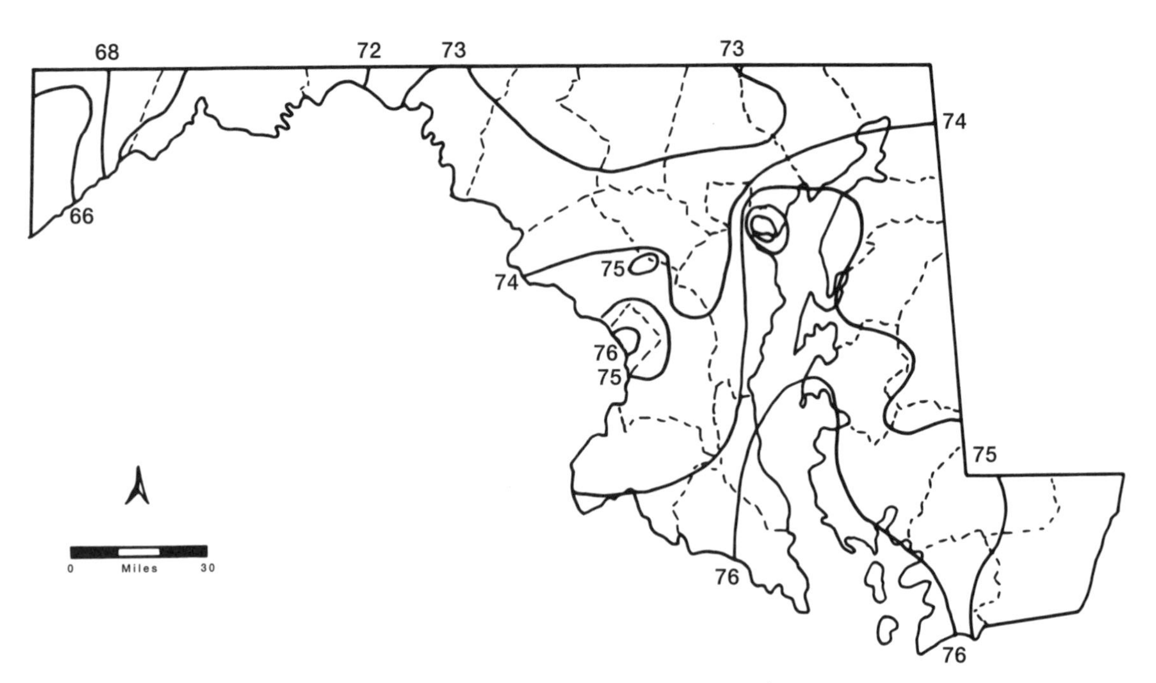

Figure 3.2 Mean Summer Temperature (June, July, and August, in Degrees Fahrenheit)

Table 3.1. Selected climatological stations in Maryland

Station	*Latitude north*	*Longitude west*	*Elevation (feet/meters)*	*Mean annual temperature*	
				°F	°C
Baltimore City	39° 17'	76° 37'	91/27.74	58.2	13.1
BWI Airport	39° 11'	76° 40'	196/59.74	55.1	11.6
Cumberland	39° 39'	78° 45'	730/222.50	52.5	11.4
Oakland	39° 24'	79° 24'	2,420/737.62	47.3	7.7
Snow Hill	38° 11'	75° 24'	28/8.53	57	12.5
Westminster	39° 35'	77° 00'	765/233.17	53.5	10.8

Source: NOAA, *Climatography of the U.S.*, No. 81, *Maryland & D.C.* Washington, DC: NOAA, 2002

tion, is partly explained by the fact that Snow Hill is not a major city with a heat island. But the difference in elevation of Westminster, Carroll County, at 765 feet (234.7 meters), and Oakland at 2,420 feet (737.6 meters), in the extreme western part of Maryland, explains to a great extent the 6.2°F difference in the mean annual temperatures of the two towns.

These temperature variations from place to place reflect four major factors: latitude, elevation, urbanization, and proximity to large bodies of water. In figures 3.1 and 3.2, the isotherms (solid lines connecting points of equal mean temperature) trend north–south over the mountains of western Maryland and less so over the Chesapeake Bay area. In central Maryland, the influences of the mountains and the bay are not as pronounced; the isotherms trend more east–west along lines of latitude. The overall pattern of the isotherms shows more variation from west to east than from north to south. Although latitudinal differences do play a role in the climatic variations across Maryland, the range of latitude is small and this effect is minimal. The strong effects of elevation, bodies of water such as the Chesapeake Bay and Atlantic Ocean, and urban heat islands are clearly reflected in these maps.

Daily temperatures in Maryland are also affected by the degree of cloudiness. The greatest daily ranges (diurnal ranges) occur on cloudless days in the spring and fall. The absence of clouds during the day allows a great amount of solar insolation to reach the surface and warm the atmosphere. Conversely, during the night, the absence of clouds allows heat to rapidly escape the atmosphere. Cloudiness has the opposite effect and reduces the daily temperature range, with snowy and rainy days often having only a slight temperature variation. The degree of cloudiness is an important factor for temperature and thus for agriculture.

The highest temperature ever recorded in Maryland was 109°F (42.7°C) at Keedysville, Washington County, on August 6, 1918, and at

Cumberland, Allegany County, and Frederick, Frederick County, on July 10, 1936. The coldest recorded temperature in the state was –40°F (–40°C), at Oakland, Garrett County, on January 13, 1912. All of these locations are in western Maryland, away from the Atlantic Ocean and Chesapeake Bay's moderating influence on temperature. These large bodies of water store heat during the summer and slowly release it during the winter to moderate nighttime cooling. Places close to these bodies of water have fewer days with a minimum temperature below 32°F (0°C). In the summer, the water absorbs heat, and the result is fewer days with a maximum temperature exceeding 90°F (32.2°C) over the adjacent landmass. Thus the Eastern Shore and the Western Shore near the Chesapeake experience milder winters and slightly cooler summers with higher humidity. Farther inland to the west, the marine influence gives way to a continental influence, which produces more extreme seasonal temperatures.

Of great importance to Maryland agriculture is the length of the growing season, or the period of time between the last frost in the spring and the first frost in the fall. The number of days in the growing season is based on the period of time between the mean spring and fall frost dates (fig. 3.3). It is difficult to define "killing frost" because plants vary in their temperature tolerance, but 32°F (0°C) is often used as an indicator of frost occurrence. Severe freezes at temperatures at or below 24°F (–4.44°C) cause heavy damage to most plants as the ground freezes solid. Between 32°F (0°C) and 24°F (–4.44°C), damage occurs to Maryland's fruit blossoms and tender and semihardy plants. The longest growing seasons in Maryland are found near the Chesapeake Bay and along the ocean, owing to the moderating influence of the water, and in the Baltimore and Washington, DC, areas, which as urban heat islands tend to have slightly warmer climates.

Frosts may occur earlier or later in any given year; in Baltimore, frosts have occurred as late as the beginning of May and as early as the beginning of October. They have also occurred as late as the third week of June and as early as the third week in August in Oakland. Local topography can also affect the occurrence of frost, so low-lying valleys and hollows may experience a frost later in spring and earlier in fall than hilltops because they are shielded from the sun and also because warm air tends to rise along the walls of valleys. For this reason, valley hillsides are often used for growing certain types of fruit trees.

Moisture

In Maryland, temperature is much more predictable than precipitation. Precipitation amounts do not vary widely throughout Maryland, but they do follow a general pattern. Annual precipitation totals vary from 40 inches (101.6 centimeters) on the Eastern Shore to over 48 inches (122 centimeters) in southern Maryland; the mean for the state is 43.98 inches (111.71 centimeters). At a microscale, there are lower

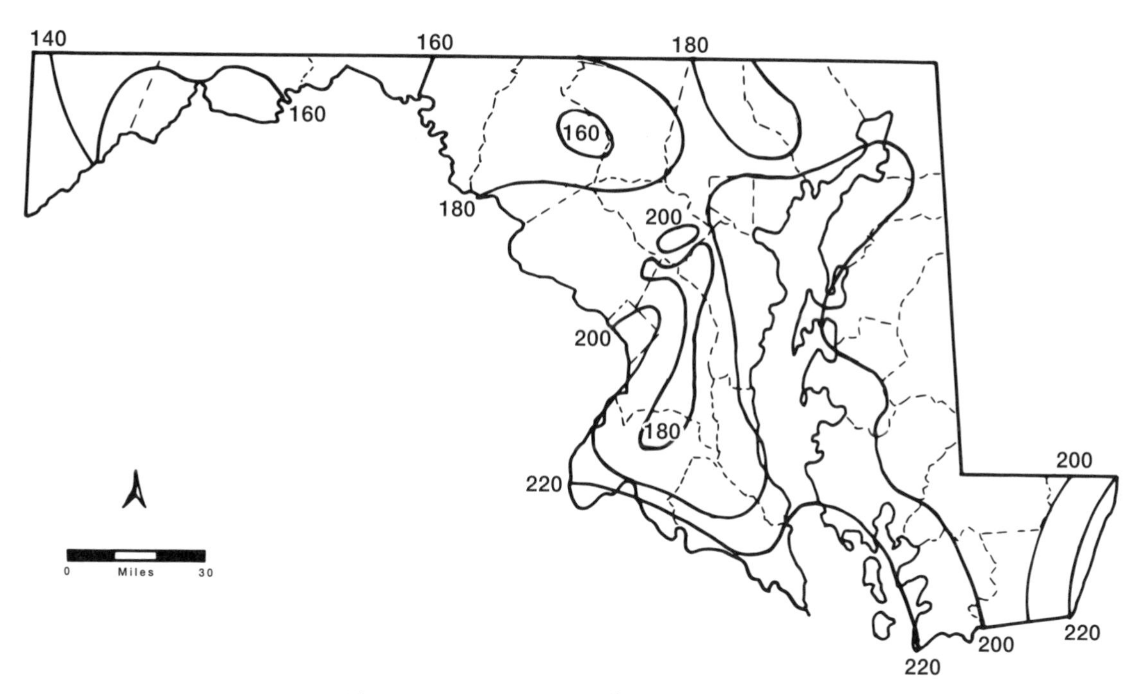

Figure 3.3 Growing Season (Mean Number of Days)

precipitation averages in the area around and east of Cumberland in Allegany County, known as the Ridge and Valley section of the Appalachians, which lies at a lower elevation than the Allegheny Plateau immediately west of Cumberland. Because the winds generally come from the west and northwest, Cumberland and the area to the east experience a noticeable "rain shadow" effect. A rain shadow is an area that experiences a relatively lower amount of rainfall because of its position on the leeward (eastern) side of a hill or mountain range. Uplift on the windward (western) side of the hills cools the air and decreases its ability to hold water content, often leading to precipitation. As the air continues traveling east, it descends on the leeward (eastern) slope. Descending air warms, increasing its capacity to hold water; this drying effect produces the rain shadow. Oakland, Garrett County—which lies at 2,420 feet (737.6 meters) elevation on the Allegheny Plateau, just 40 miles (64.4 kilometers) west of Cumberland—receives more precipitation than most of the rest of Maryland. Maryland's mean annual snowfall varies greatly, from over 100 inches (254 centimeters) on the Allegheny Plateau in Garrett County to less than 10 inches (25.4 centimeters) in the extreme southeastern part of the state. Throughout most of Maryland, snow usually remains on the ground for a short period of time.

Precipitation is greatest in early summer, mostly from convectional thunderstorms. Evapotranspiration is near zero all winter, but it increases with warming temperatures. During June, July, and August, evapotranspiration usually exceeds precipitation in all of Maryland except Garrett County, where precipitation exceeds evapotranspiration all year. From April through September, much of Maryland's rain comes from showers and thunderstorms; from October through March, much of the precipitation comes from frontal storms. Overall, Maryland has a moist climate with no significantly dry months. In this water-surplus climate, annual precipitation normally is greater than evapotranspiration (evaporation of surface water or soil moisture plus plant transpiration), and extended droughts are infrequent. Short drought periods that occur more often can be serious if they occur during the growing season. Recent short-term droughts occurred in the summers of 1995, 1999, and 2002 (Koontz et al. 2000).

Winds and Storms

Although the winds do vary somewhat by season, as global wind patterns change, Maryland's latitude places it in the prevailing westerly wind belt. From October to April, winds come primarily from the northwest; from May to September, from the south and southwest. In the mountains of western Maryland, west-northwest winds are common all year. The mean wind velocity is greatest in March and lowest in August. Highly destructive winds from tornadoes and hurricanes are infrequent in Maryland, but they have occurred. From time to time microbursts—strong downdrafts caused by powerful thunderstorms—can occur anywhere in Maryland. As a microburst fans out at the surface it can result in wind shear, which is particularly dangerous for aircraft taking off or landing.

Hurricanes (tropical cyclones), their remnants, and their associated storm surges, while infrequent, do affect the state, but not all Maryland's counties face the same risk. Storm surges are normally the most damaging and costly aspects of these weather events for coastal areas. Koontz et al. (2000) define a storm surge as: "an abnormal local rise in sea level that accompanies a tropical cyclone. The actual height of the surge is the deviation from mean sea level and could reach over 25 feet (7.62 m). A storm surge is caused by the difference in wind and pressure between a tropical system and the environment outside the system. The end result is that water is literally pushed onto a coastline. The most devastating storm surges occur just to the right of the eye of a land-falling hurricane." From 1886 to 1999, seventy-two hurricanes or their remnants traveled within 60 miles (96.6 kilometers) of the Maryland coast. Only one of these storms made landfall while still being powerful enough to be classified as a hurricane. The fact that North Carolina protrudes into the Atlantic Ocean, where it has stood in the path and absorbed many cyclones, has reduced most hurricanes to tropical storm or depression status by the time they reach Maryland. Although no longer classified as hurricanes upon reaching

Maryland, tropical storms Agnes (1972), Fran (1996), and Isabel (2003) caused flooding of historic dimensions as far inland as central and western Maryland. The great hurricane of 1933 offers a prime example of the threat that storm surges pose in Maryland. The storm generated a 7-foot (2.13-meter) surge and waves over 20 feet (6.10 meters) high, cutting the inlet to the Atlantic Ocean across Fenwick Island at Ocean City.

In contrast to storm surges, which approach Maryland from the eastern coastline, most normal weather patterns approach Maryland from the west. Cold air masses that bring cold winter days and cool summer days originate over central Canada and travel over the Midwest to Maryland. Although warm, dry air masses originating over the southwestern United States seldom reach Maryland, warm, humid air masses from the Gulf of Mexico and Atlantic Ocean are common occurrences.

Much of the human activity in the eastern part of Maryland occurs in and around the Chesapeake Bay, which generally experiences fewer dangerous high winds than any other large body of water in the United States. The inlet at the southern end of the bay is at a right angle to the main trend of the bay, so the Eastern Shore, despite its low profile, acts as a windbreak from the main force of Atlantic Ocean storms. The bay's dimensions average about 15 miles (24 kilometers) east–west and 180 miles (290 kilometers) north–south. Since the prevailing winds are from the west, the narrow east–west dimension is not great enough to favor the frequent development of high, dangerous, wind-driven waves. In the summer, the Chesapeake Bay produces good air circulation over eastern Maryland, and southeast land breezes often refresh the Western Shore, including Baltimore, on summer afternoons.

Climatic Zones

The climatic factors in Maryland combine to produce five general climatic zones: highland, valley, central hills, metropolitan, and oceanic/maritime. The characteristics of each of these zones are summarized in table 3.2. Although the climatic differences throughout Maryland are not dramatic at a macroscale, microanalysis does reveal some differences, which have particularly important ramifications on agriculture in the state.

Soils and Geology

What goes on beneath the surface of the Earth is of vital importance to the activities of every state. Soil and geological characteristics, for example, directly affect the ability to access groundwater. Understanding how different soils drain under various circumstances allows for the proper modeling of septic tank sites, determining ahead of time whether they will function properly. There are numerous reasons why understanding soils and geology is so critical. But because soils and ge-

Table 3.2. Climatic zones of Maryland

Zone	*Moisture*	*Climate*	*Locations*
Highland	very wet	cool summer, cold winter	Garrett County eastward to central Frederick County, except for major valleys
Valley	dry to moderate	warm summer, cold winter	valleys of Allegany and Washington Counties
Central hills	moist	warm summer, cool winter	Piedmont region
Metropolitan	moist	hot summer, mild winter	centered on Baltimore and Washington, DC
Oceanic	moist (dry on seashore)	cool summer, mild winter	southern Maryland and the Eastern Shore south of Kent County

ology are difficult to survey, much of the data are generalized and not frequently updated.

Soil scientists examine each horizon, or layer, within a given soil profile (all the horizons from the top layer down to unaltered rock) in order to identify properties such as depth, color, texture, structure, density, consistency, porosity, and other characteristics. Soil sampling and examination lead to the classification of a particular soil and an associated understanding of its potential and constraints for human activity. The letters O, A, B, C, and R are used to denote the major horizons from the surface through the bottom layer; Arabic numerals are added to the first three letters when subdivisions are present. These layers can be removed through human activity such as bulldozing or erosion caused by the creation of impervious surfaces upstream. Although their removal can occur in a short time, it usually takes thousands of years for these soil horizons to develop.

Soil is composed of mineral and organic matter, air, and water. How these elements vary in proportion and distribution throughout a given soil profile (as well as in their chemical and mineral composition) contributes to differences in soil types. The mineral component is made up of various sizes of particles such as sand, silt, and clay that have been born of physical and chemical reactions on the "parent material." The proportions in which these particle sizes occur, as well as the kind of particle, greatly affects the properties of the soil. The organic component of soil consists of the decayed remains of plants and animals. The soil also contains a pore space that occurs between solid particles and aggregates of particles; this space is occupied by water and air.

Prime soil conditions for crop growth normally result when approximately 25 to 35 percent of the soil is occupied with water, 25 to 15 percent with air, 5 percent with organic matter, and the rest with mineral material. How the layers are arranged, how they erode, and the topography they overlay are also important factors. The mineral fraction of the upper horizon is composed of the more resistant minerals because this layer has endured intense weathering. The least-resistant minerals either have leached downward through the profile as a solution or have been transformed into more resistant forms of soil clays. These resistant secondary materials are found in most of Maryland's subsoils. All soils, regardless of their exact composition, have "structure" and "texture" affecting their durability, drainage, and other characteristics.

Maryland Soils

Excluding the Eastern Shore, local relief varies greatly around the state, with slopes of up to 20 percent. When used for urban development or cropland, slopes of 3 to 10 percent need special treatment to mitigate erosion; slopes greater than 10 percent are not normally used for these purposes. Activities such as forestry and recreation (when conducted in an environmentally sensitive manner) may be appropriate on slopes of over 20 percent, where the soils are usually extremely thin.

The largest area of steep slopes and thin soils in Maryland occurs west of Frederick in the Blue Ridge, Ridge and Valley, and Allegheny Plateau region. There are also some scattered steep slopes on the Piedmont of central Maryland and the Coastal Plain of southern Maryland. Large areas of gentle slopes and thicker soils are found on the Eastern Shore, the western periphery of the Coastal Plain, and in the valleys around Frederick and Hagerstown. The sediment soils of the Coastal Plain vary from excellent to very poor. The soils of the area known as the Sassafras Association are excellent and are well drained. Othello soils are "good," while tidal soils are unproductive and sandy (Miller 1967). Soil category, quality, and suitability for agriculture are all vital to the success of farming.

The soils in the various regions of Maryland differ greatly by type. The geographic distribution of soils is highly generalized (fig. 3.4), as there are 225 soil series types in Maryland, each with distinct characteristics. In terms of land use, the agricultural capability of soils is more important than the classification by soil types. Soils best suited for agriculture are deep, fertile, and well drained, with a high moisture-holding capacity. Soils least suited for cropland are thin, shallow, stony, and are either poorly or excessively drained. Prime agricultural land accounts for less than one-third of the total land area of Maryland; less than half of this land is currently used as cropland.

Underlying rock types and sediments in Maryland are shown in figure 3.5. The Coastal Plain consists of relatively unconsolidated layers of sand, silt, clay, and some gravel over underlying crystalline rocks. Sediment depth generally increases from west to east. Along the Atlan-

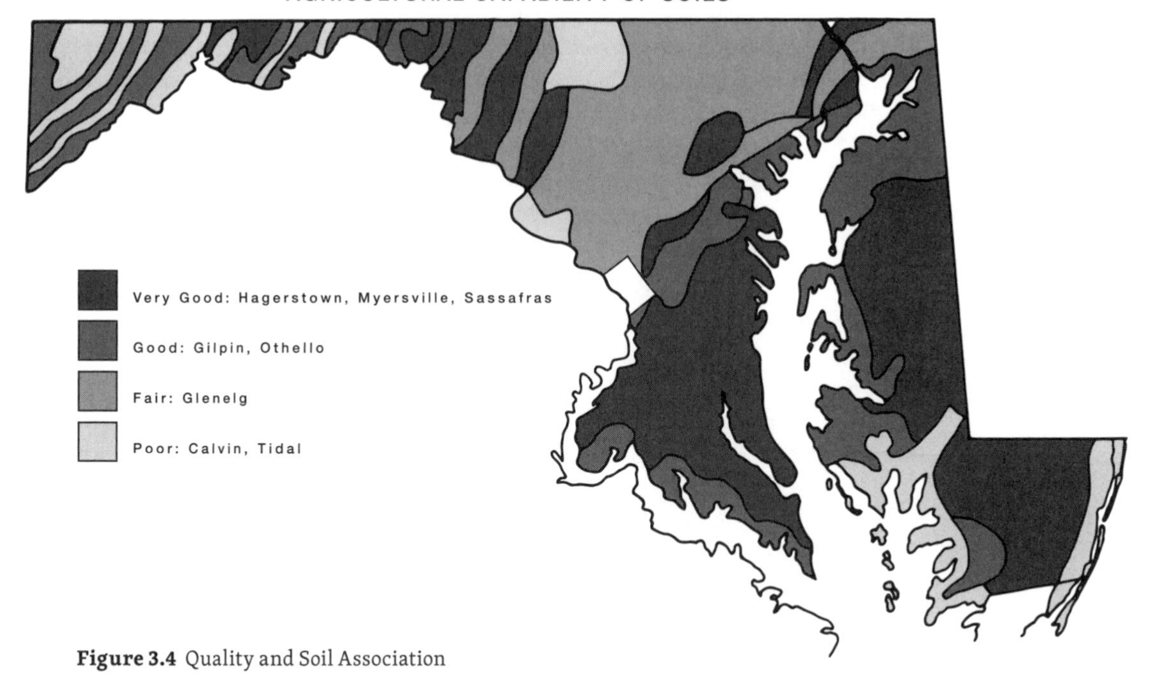

Figure 3.4 Quality and Soil Association

tic coast in Ocean City, wells often pass depths of 8,000 feet (2,438.40 meters) before reaching bedrock. A comparison of figures 3.4 and 3.5 reveals that there are productive and unproductive soils in each of the major physiographic regions of the state.

The sedimentary rock area of western Maryland has areas of excellent soil developed on limestone; this deep, well-drained, fertile loam soil is known as the Hagerstown Soil Association. The area of "good" soil in western Maryland is called the Gilpin Soil Association; it is formed on shale and sandstone, and it is deep and well drained. The Calvin Soil Association is thin and stony; it is found on steeper slopes and is formed mainly of sandstone. On the Piedmont of central Maryland, which is characterized by its rolling hills and relatively deep, narrow valleys cut by streams, the best soils are the Myersville Soil Association, and the "fair" soils are known as the Glenelg Soil Association. To be useful for agriculture, both types require heavy applications of lime and fertilizer. Geologically, the bedrock underlying the eastern division has been metamorphosed, or structurally and mineralogically changed by tremendous heat and pressure on sedimentary and volcanic-igneous rocks deep within the Earth's crust. Exposed over a long period of time, these include gneisses, slates, phyllites, schists, marble, serpentine, and granitic and gabbroic rocks. Differential erosion, or uneven effects on different types of rocks, results in varied topography.

On the Coastal Plain, streams have eroded the sediments at a greater rate than on the more resistant Piedmont rock. Where the more resistant crystalline rock of the Piedmont meets the softer Coastal Plain, there are numerous waterfalls and rapids. Farther west, limestone that was folded but not strongly metamorphosed underlies the Frederick Valley. The relatively more rapid weathering of this limestone has produced a wide valley with soil good that is for agriculture. The Piedmont Upland, underlain by more resistant crystalline rock, is on the eastern end of this region. The Piedmont dips to the west, where the Piedmont Lowland is underlain by softer, less resistant rock.

Just west of the Frederick Valley is the Appalachian region. Catoctin and South mountains and Elk Ridge constitute the Blue Ridge area, one of three distinct Appalachian areas. The ridges are composed of large sandstone, shale, slate, quartzite, and other metamorphic rocks. Between these ridges, rhyolites and basalts underlie the bottom of the Middletown Valley. Between the Blue Ridge and Allegheny Front (Dan's Mountain) is the Greater Appalachian Valley, locally called the Hagerstown Valley; it is underlain by limestone and shale. The Allegheny Ridge and Valley region west of the Hagerstown Valley is underlain by huge sandstone and quartzite strata, with the valleys between created by differential erosion of less resistant shale and limestone beds. In the westernmost area of Maryland, near the West Virginia border, the Allegheny Plateau covers all of Garrett and western Allegany Counties.

Figure 3.5 Rocks and Sediments

Maryland's highest elevations are found here, where bedrock consists of a series of sedimentary deposits such as gray and red sandstone, acid and calcareous (lime-containing) shale, chert, limestone, and conglomerate. Coal deposits formed in layers between these sediments.

Exposure of the various rock types to nature's elements over many years has resulted in distinct soil types. Maryland's humid and warm climate is a productive catalyst for soil creation. (Soils are thin and sparse in cold, dry climates such as deserts and polar regions.) A combination of five major soil-creating factors determines the characteristics of a soil profile:

1. parent bedrock material (chemical and mineralogical composition, and texture and structure);
2. climate (especially temperature and precipitation);
3. living organisms (especially native vegetation);
4. topography; and
5. time.

Vegetation

Maryland's patterns of natural vegetation vary with latitude, elevation, slope, climate, and soils. Before Europeans colonized it, Maryland was covered with natural vegetation that was quite different from today. The Eastern Shore was an area of oak, pine, cypress, and gum forests. Oak, hickory, and pine forests covered the Western Shore. The Piedmont area was forested by chestnut, walnut, hickory, oak, and pine. In western Maryland, pine and chestnut grew on the hilltops; oak, poplar, maple, and walnut in the valleys. Today, only a fraction of this original pattern remains. In creating the cultural landscape, people have cut, cleared, plowed, urbanized, and otherwise changed the landscape in a number of ways. Even forest growth that exists today is often not original forest.

Forests of the Past

Most of the forests of Maryland have been cut and regrown three to five times since European settlement began in the early 1600s. Prior to the arrival of the Europeans, the Native Americans had cleared small areas around their villages, mostly around the Chesapeake Bay and along its tributaries. Although most of the same species of trees are found in Maryland today as in the seventeenth century, the composition has changed. In earlier times, hardwoods dominated the forests. Today, softwood pines are prevalent, primarily owing to reforestation efforts and the reversion of farmland to forest.

Just as in earlier times in Europe, the settlers coming to Maryland viewed forests as an obstacle to agriculture and travel as well as homes to dangerous wild animals. Because it was plentiful, timber had little

LAND USE

Land-use data are normally derived from either aerial photos, satellite imagery, or other observations of land cover (e.g., forest, urban, agriculture, water, wetlands). Maryland has a dominating urban concentration extending from along the upper half of the Western Shore of the Chesapeake Bay through Baltimore City to the area surrounding Washington, DC. Most of Maryland's wetlands and a good deal of its agriculture reside on the Eastern Shore. Agriculture is also a dominant theme across the Piedmont in the central and western regions of the state, though it is certainly not limited to these areas. Forest cover accounts for an exceptionally large percentage of the far-western reaches of Maryland. An easy way to access land-use maps of Maryland is on the websites of various Maryland state agencies, such as the Maryland Department of Planning.

A word of caution when using generalized land-use maps: a grossly generalized map at a small geographical scale gives only a basic indication of dominant land use or land cover across the state. Regional patterns certainly emerge, however, which is the prime value of these maps. The figure below presents a general picture of major land use for the state of Maryland.

Factors including soil and geology; topography; roads; relative distance to other features such as markets, climate, and demographic characteristics; and economic influences all underlie land use. There is a positive correlation between urban growth and the location of major transportation routes, such as Interstate 95, which connects large cities along the Eastern Seaboard of the United States. Planners attempt to take such factors into account when creating master development plans, but such characteristics are in the end quite evolutionary, and balance in land-use mix is difficult to ensure. Change and development are difficult to predict and control, especially given that each jurisdiction has significant latitude in its governance.

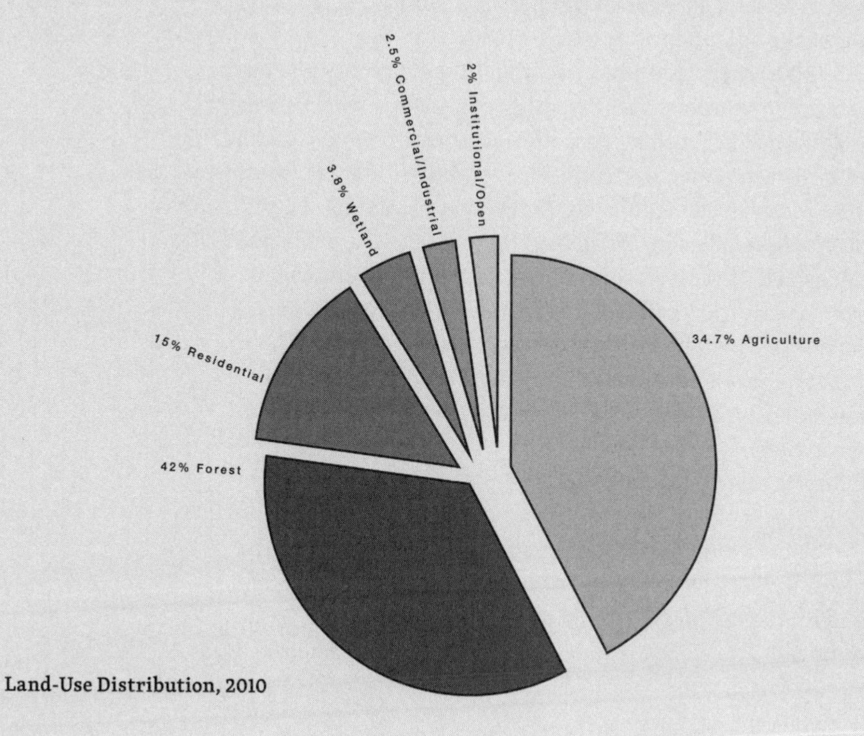

Land-Use Distribution, 2010

or no commercial value. Forests were burned and cleared to grow tobacco and other crops, leading to massive erosion. As a result, many deep harbors around the Chesapeake Bay silted up. After 1732, Lord Baltimore opened the lands to the west of the tidewater for settlement. Since wood was needed for buildings and fuel, more forests were cleared. Iron was in short supply in the Maryland colony, and so early iron furnaces were built. Because the coal of western Maryland had not yet been exploited, charcoal—which is made from hardwoods, especially oak—was used as the fuel to smelt iron ore. It took large amounts of charcoal to operate these early furnaces. In 1732 the Maryland Assembly granted 100 acres (40 hectares) of land to anyone building a furnace. The first furnace in Maryland, the Principio furnace, was built in Cecil County in 1719. It operated for about one hundred years, during which time it consumed about 10,000 acres (4,047 hectares) of hardwood forest.

In the early 1800s, the invention of the steam engine facilitated the clearance of forests at an even faster pace. By midcentury, steam-powered sawmills could produce lumber more rapidly. The railroads were built westward and logging expanded, eventually to the far reaches of Maryland. The B&O Railroad and the C&O Canal transported lumber to Baltimore and other markets. Some lumber was marketed in other states or exported. Although the major product was lumber, other forest products included pulpwood (for papermaking) and tanbark from hemlock and chestnut oak (for tanning leather). Following the Civil War, as many people left Maryland's farmlands for jobs in its cities, forests began to reclaim abandoned farmlands. A similar reclamation took place after the Great Depression in the 1930s.

In the late 1800s, Maryland was caught up in the fervor of the national conservation movement led by such prominent Americans as John Muir, Gifford Pinchot, and Theodore Roosevelt. The Maryland State Board of Forestry was created in 1906, mainly to control forest fires. The first Maryland state forester, Fred Besley, inventoried every 5-acre (2-hectare) woodlot in Maryland. His survey was published in 1916 as Maryland's first forest inventory. Soon after, the Maryland State Forestry Agency (now called the Department of Natural Resources Forest Service) was established to improve the management of both public and private forest resources. Today, the Maryland Forest Division operates extensive programs to supply seedlings for reforestation and to assist owners of forested areas.

By 2010, forests covered about 41 percent of Maryland's 2.7 million acres (1,092,675 hectares), a percentage that varies from 24 to 35 percent in the urban and suburban counties of central Maryland and the upper Eastern Shore. The lower Eastern Shore is about 54 to 61 percent forested, and southern Maryland is 37 to 51 percent forested. The far western counties of Maryland are today about 73 percent forested.

Forests of Today

A traverse along the nearly 260 miles (433 kilometers) from the very southeast to the northwest corner of Maryland cuts through three distinct forest zones that roughly correspond to the physiographic regions of Maryland. The nearly flat Coastal Plain region generally has a sandy soil with a mild climate heavily influenced by the surrounding bodies of water. This forest zone is the northernmost limit of trees found to the south, particularly bald cypress and loblolly pine. Oaks (red, white, and chestnut), yellow poplar, ash, and some pine dominate the forests of the Piedmont of central Maryland. Westward from the Blue Ridge, the mountains and valleys are forested by oaks and hardwoods (beech, birch, and maple). The elevation and colder climate of this region fosters the growth of these northern hardwoods. There are over 150 tree species in Maryland. By far, the dominant forest type in the state is oak-hickory, covering 60 percent of its forests. Other forest types include: loblolly-shortleaf pine (12 percent), oak pine (12 percent), northern hardwood (6 percent), and various other types (10 percent).

Forest Ownership

About 90 percent of Maryland's commercial value forest is privately owned (2.4 million acres or 0.96 million hectares; see fig. 3.6). Only 16 percent of the forests of Maryland are owned publicly. Private buyers purchase forestland for numerous reasons: investment, timber production, recreation, and aesthetics. As large properties continue to be subdivided in the state, the number of private owners grows. Most of these individuals are nonindustrial owners, and 55 percent of them own less than 10 acres (4 hectares). It is usually the case that fragmentation of forested areas corresponds with a decrease in active forest management and declining health of forests. Because of the importance of forests in efforts to improve air quality, reduce pollution and erosion, and protect wildlife habitats, it is important that the state provide assistance to forest owners for proper management.

Forest Loss

Forests have great value in terms of jobs, lumber products, increased land values, improved public health through carbon absorption, and mitigation of erosion by wind and water. They also play a primary role in protecting watersheds. High forest volume and quality are often correlated positively with high water quality. For a small state, Maryland has a remarkable diverse forest cover. The variety of climates, topography, and soils are all reflected in this diversity of native species of trees ranging from the bald cypress in the lower Eastern Shore to the white pines and maples of the mountainous west.

From 1985 to 1990, developed areas in Maryland grew by 3.9 percent annually, exceeding most projections. During that five-year pe-

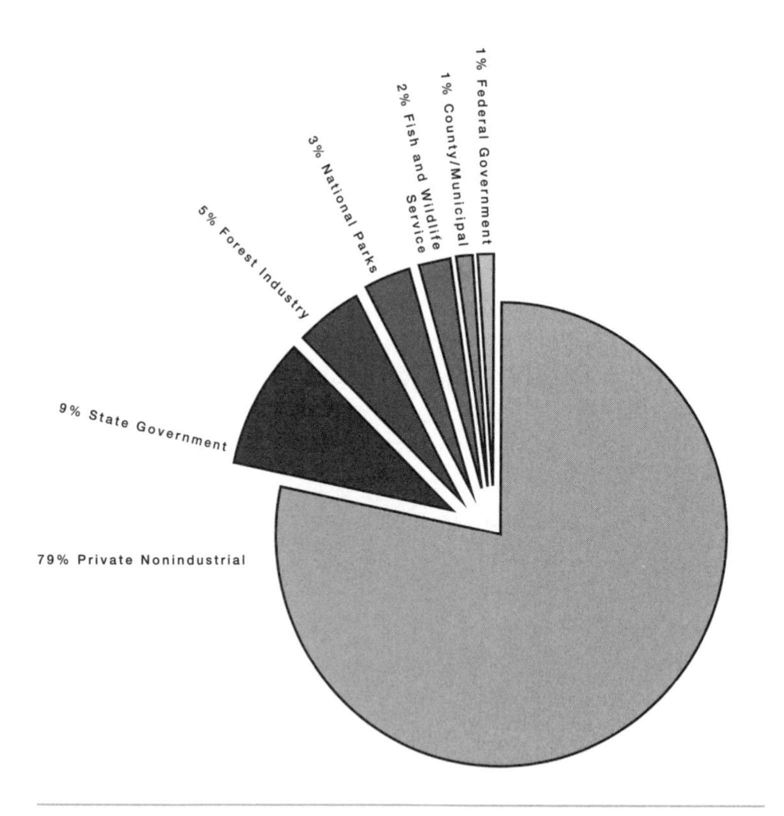

Figure 3.6 Maryland Forest Ownership

riod, 144,500 acres (57,800 hectares)—nearly equivalent to the area of Howard County or three times the size of Baltimore City—were developed in the state. About 50 percent of this newly developed land had been forested. The heaviest loss of forestland has been in the counties around the Chesapeake Bay, especially Anne Arundel, Calvert, Charles, Howard, Montgomery, and Prince George's. Degradation of the bay is directly linked to forest depletion. The watershed of the bay is already an overstressed area and yet its population is growing rapidly. The Maryland Department of Planning has projected that by 2015 the area dominated by urban development (mainly on the western shore and urban corridor) will increase to a total of 1.5 million acres (0.6 million hectares), a 48 percent increase in developed area since 1990 accounting for about 25 percent of all developed areas of Maryland.

Maryland's forestland base is decreasing by about 10,000 acres (4,000 hectares) per year, mainly because of development. Because much of the loss is occurring in urbanized areas where forests are needed most, careful forest management must be considered as part of the overall growth process. One acre of healthy young trees can absorb 2.5 tons of carbon dioxide gas and give off 2 tons (1.8 tonnes) of oxygen in a year. As the trees age and become unhealthy, they begin to rot, and the process is reversed (i.e., oxygen is consumed and carbon dioxide given off). This cycle is why forestry management is critical—many trees must be harvested when they are mature, but before they become unhealthy. Reforestation is therefore a critical element in this process.

Water Resources

Maryland's plentiful water resources serve a multitude of concurrent uses. For example, the flow of western Garrett County's upper Youghiogheny River is controlled by an upstream hydroelectric plant. At the same time, it is a scenic trout stream, and its Class 5 rapids make it a favorite for kayakers on the hunt for white water in Maryland. Equally scenic views can be found along waters at lower elevation, such as at the mouth of the Susquehanna where it pours into the Chesapeake. Although many maps show aquifers and surface water contained within the state, most of these features extend well beyond Maryland's borders. For this reason, Maryland must cooperate with the federal government and neighboring states to monitor and manage its water resources.

Watersheds

Defined as an area on the landscape extending from ridge to ridge, a watershed catches precipitation that subsequently drains into groundwater, marshes, lakes, and rivers. It is believed that the term comes from the German *wasserscheide* or *wasser* (water) and *scheide* (divide, parting).

Topography primarily determines where and how water flows from one watershed to the next. Each large watershed can be broken into smaller drainage basins with their own topographic and hydrologic characteristics; these are called subwatersheds, or subsheds for short. These smaller watersheds are nested within larger ones. Large features such as continental divides significantly influence the flow of water (and whatever it carries with it). Watersheds come in all shapes and sizes. Because they are natural features, watersheds often cross county and state boundaries. This chapter considers only the major watersheds of Maryland (fig. 3.7), but keep in mind that each of these can be subdivided into numerous smaller watersheds. Human activity in watersheds dramatically affects people and the environment "downstream," regardless of administrative borders, often on a large scale with long-lasting implications.

Nearly all of Maryland is drained by rivers that flow into the Chesapeake Bay; the state is unique in that more than 90 percent of its rivers and streams drain into this single body of water. The Chesapeake Bay watershed extends far beyond the borders of the state, underscoring why it is important for Maryland to cooperate with other states in its efforts to manage the resources of the bay. A small area in the extreme western part of the state is drained by the Youghiogheny River, which

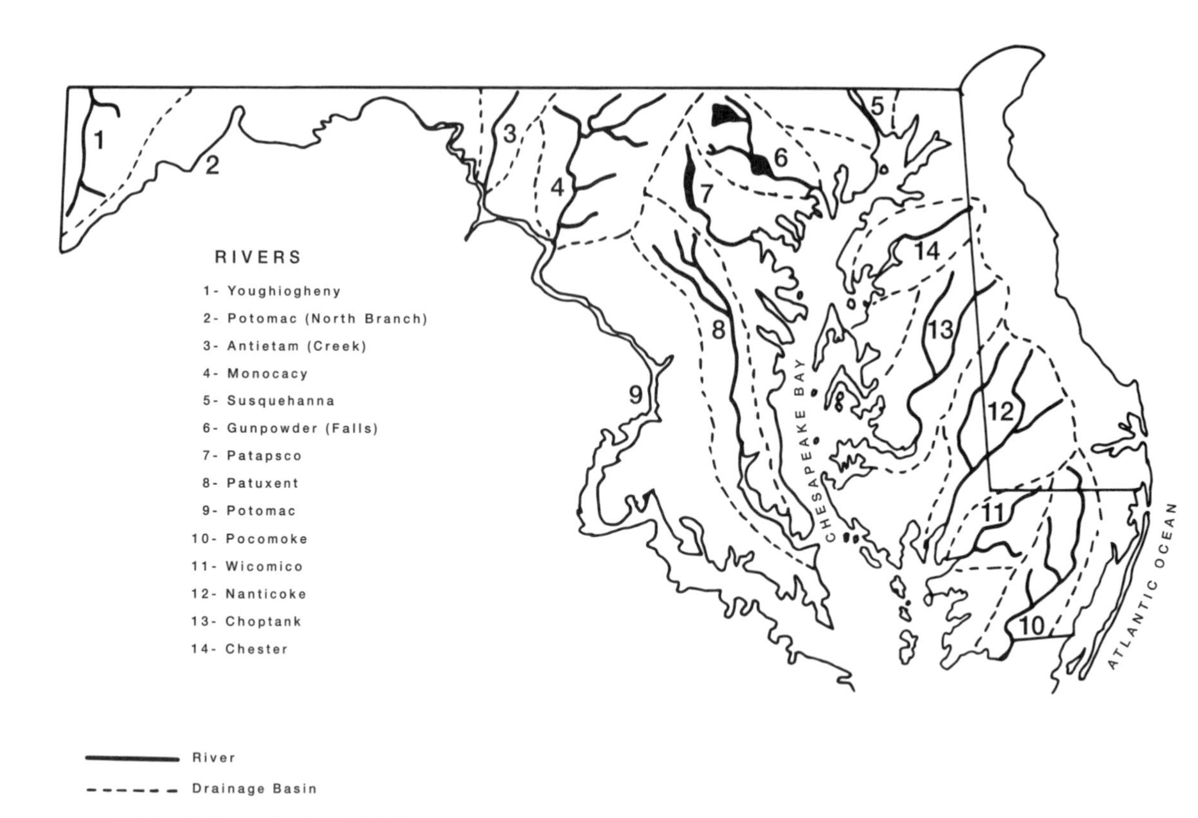

Figure 3.7 Major Rivers of Maryland

flows into the Ohio River, a tributary of the Mississippi River. There is also a small area of the state in Worcester County that drains to the Atlantic Ocean. When driving east on Route 50 toward Ocean City, a sign marks the point where the Chesapeake watershed ends. It stands along the "divide," or ridge, separating watersheds.

The Maryland coastal bays are shallow lagoons located inland from Fenwick and Assateague Islands. The Coastal Bays watershed covers only 175 square miles (453 square kilometers) and is characterized by low elevation (nearly at sea level), high water table, poor natural surface drainage, sandy soils, and abundant wetlands. Owing to limited freshwater runoff, the main source of freshwater is deep groundwater.

Rivers

The close tie between rivers, topography, underlying bedrock, climate, soils, vegetation, and other environmental factors reminds us of the need to relate the pieces to the whole in our attempt to create a regional picture.

One of Maryland's major rivers, the Potomac, is formed by the junction of the Shenandoah and North Branch in the Allegheny Mountains at Harpers Ferry, West Virginia. Along with one of its headstreams, the North Branch, the Potomac forms the entire southern border of Maryland. Because the state border follows the south bank of the river, the Potomac lies entirely within the territory of Maryland. Where the river crosses the fall line about 15 miles (24 kilometers) northwest of Washington, DC, the dramatic Great Falls of the Potomac are found (fig. 3.8).

The Patapsco River, whose estuary forms Baltimore Harbor, is another major and important river in Maryland. It rises in the Piedmont and flows eastward to enter the Chesapeake Bay. Where it flows over the crystalline bedrock of the Piedmont and drops to the Coastal Plain, there are rapids and falls. The lower Patapsco River Valley was an early site for industrialization in Maryland owing to its excellent waterpower, and many mills were located along its banks.

The lower Susquehanna River flows through Maryland and enters the head of the Chesapeake Bay at Havre de Grace. It is the largest freshwater contributor to the bay, supplying about 80 percent of incoming freshwater north of the Potomac and about 50 percent of all the freshwater entering the bay. The Susquehanna is fed by tributary rivers as far away as upstate New York.

The Chester, Choptank, Elk, Nanticoke, Pocomoke, and other smaller rivers enter the Chesapeake Bay from the Eastern Shore, traversing the Coastal Plain. Because the Coastal Plain nearly lacks relief, the

Figure 3.8 Great Falls of the Potomac

streams and rivers of this region have a gentle gradient with wide and shallow valleys. Extensive, low-lying wetlands account for about 20 percent of the land area of the Eastern Shore. But the rivers west of the Chesapeake supply the greatest amount of freshwater to the bay. Those on the Eastern Shore are significant, but they do not match flow from the west.

Within the entire state of Maryland there are no natural lakes. Outside the Chesapeake Bay, the largest body of water is Deep Creek Lake in Garrett County. It has a surface area of only 7 square miles (18 square kilometers). It lies on the Allegheny Plateau behind a dam built on a tributary of the Youghiogheny River.

The Chesapeake Bay

The Chesapeake Bay is the largest estuary in the United States and one of the largest estuarine systems in the world. Although defined a number of different ways, an estuary is most commonly understood as a semienclosed coastal body of water that has a free connection with the ocean and within which seawater is measurably diluted with freshwater from land drainage. Most of the world's great estuaries are about the same age, having been formed during the most recent rise in sea level between 10,000 and 12,000 years ago. Because it receives a small inflow of sediment relative to its size and volume, the Chesapeake Bay is longer and larger than most others.

When Spanish explorers first saw the Chesapeake Bay, they described it as the best and largest port in the world and named it Bay of the Mother of God. But the origin of the name Chesapeake is not clear. One legend is that a group of English explorers visiting the area called it "Chesepiuc" after a local tribe of Native Americans. In 1608, Captain John Smith explored and mapped the bay, and on his map he used "Chesapeack," reportedly a Native American word meaning "great shellfish bay." Elsewhere it has been reported that the name was "Chesapiooc" from "Kchesepiock," a Native American word for a "country on a great river." There seems to be little doubt that the name is of Native American origin and refers to the body of water or adjacent land in some manner.

Chesapeake Bay runs approximately 185 miles (298 kilometers) north–south and varies in its east–west width from 3 miles (4.8 kilometers) to 22 miles (35.4 kilometers). The narrowest part of the bay is near Annapolis, where the 4.4-mile (7.1-kilometer), twin-span Chesapeake Bay Bridge extends to Kent Island. The deepest part of the bay, 174 feet (53 meters), is in the ancient Susquehanna River Valley off Bloody Point on the southern end of Kent Island; the mean depth is 21 feet (6.4 meters). The entire Chesapeake Bay covers an area of 3,237 square miles (8,384 square kilometers), of which 1,726 square miles (4,470 square kilometers) are in Maryland—nearly the total area of the state of Delaware. Its extensive, irregular coastline extends nearly 8,000 miles (12,872 kilometers); 4,100 miles (6,597 kilometers) are in Maryland.

About fifty major rivers and over one hundred small tributaries draining land in a 64,000-square-mile (165,760-square-kilometer) watershed feed into the Chesapeake Bay. These water sources come from parts of Delaware, Maryland, Pennsylvania, Virginia, West Virginia, and as far away as south central New York. The major Maryland rivers flowing into the bay are the Back, Bohemia, Bush, Choptank, Elk, Gunpowder Falls, Honga, Middle, Nanticoke, Northeast, Patapsco, Patuxent, Pocomoke, Potomac, Sassafras, Susquehanna, and Wicomico Rivers. Of all the bay's major tributaries, three account for over three-fourths of the inflow: James, 10 percent; Potomac, 19 percent; and Susquehanna, 50 percent. These rivers vary in chemical composition, reflecting the types of land use and other activities of the nearly eighteen million people in the entire watershed.

The symbolic significance of the Chesapeake Bay is not only regional but also national. It was one of the first regions of the country to be explored. The bay and its surrounding region were the scenes of crucial battles in the American Revolution, War of 1812, and Civil War. Clipper ships carrying goods from all over the world traversed the Chesapeake Bay in earlier times. Today, modern cargo vessels and passenger liners are common sights.

Although for many centuries the human population was too sparse to seriously disrupt the bay, the situation gradually changed as Indians built settlements and turned to agriculture, as Europeans arrived in the early seventeenth century with new agricultural methods, and as more concentrated settlement patterns emerged. Tobacco farming led to greater sediment flow into the bay. Improved technology in fishing diminished the bay's fish population. Today, this great ecosystem faces serious environmental challenges. The Chesapeake Bay's geography and vast watershed require that these issues of environmental degradation be effectively confronted by Maryland in cooperation with other states.

4 Environmental Challenges

Humans have the capacity to alter, manipulate, and cause damage to the environment on a scale that no other living beings can match. The use of the natural resource base by people in pursuit of livelihoods often leads to environmental degradation. The concept of "nature's services" connotes the false assumption that the environment exists for humans' exclusive use, and that its resources will always be available.

The Chesapeake Bay

The Chesapeake Bay roughly divides Maryland into two parts, with its water covering nearly one-sixth of the state. But it was not always this way. A coastal plain covered this area until the bay was created at the end of the last Ice Age (the Pleistocene), some 10,000 to 18,000 years ago. As the Ice Age ended, the climate warmed and the Pleistocene glaciers melted. The sea level worldwide rose, the land slowly subsided, and the ocean invaded and drowned the lower stream valleys up to the fall-line zone. The bay's highly irregular coastline is a classic example of its submerged nature.

About 15,000 years ago, sea level was about 325 feet (99 meters) lower than it is today. The Atlantic Ocean shoreline was over 60 miles (97 kilometers) east of the present shoreline, at the edge of the continental shelf. At these lower water levels, the ancient Susquehanna River cut a deep gorge through the Coastal Plain in what is today the central deepest part of the bay. By about 7,000 years ago, the ocean had risen and filled these deep channels to within 40 feet (12 meters) of the present sea level. Over the next 7,000 years, the sea level continued to slowly rise to its present level. Humans have also altered the shape of the Chesapeake Bay and its surrounding areas over time. Excessive forest clearing and poor land management have increased upland erosion, sending tons of sediment downstream. As a result, communities that were once important ports are now landlocked. The lower portions of the river systems flowing into the bay are not true rivers but instead tidal estuaries.

One of the most important characteristics of the Chesapeake Bay is its salinity, ranging from nearly 0 parts per thousand (ppt) in the north to 35 ppt at the mouth of the bay. (Freshwater ranges from 0 to

3.5 ppt; saline water is over 3.5 ppt.) In the bay's general circulation, warmer and less saline water at the surface flows south, and colder, more saline deeper water flows north. A constant flushing and cleansing action occurs between the estuary and the ocean. Salinity in the bay varies through several different dimensions. Salinity is lower in the north and west. Above the Potomac River, the Susquehanna River provides over half of the freshwater flow into the bay. During the spring high-flow period, a large freshwater pool occurs in a zone from Spesutie Island to Turkey Point. The west side of Chesapeake Bay is less saline than the east. A number of major rivers enter on the west and dilute the water, whereas the rivers coming off the Eastern Shore are smaller in volume and have less diluting effect on the eastern side of the bay. Another factor affecting salinity is the Coriolis force, which because of the rotation of the Earth on its axis deflects currents to the right in the Northern Hemisphere. The denser saline waters flowing north are deflected somewhat to the eastern side of the Bay. During winter and early spring, the salinity of the bay decreases and its waters are rich in dissolved oxygen. During the summer and early fall, salinity increases as the water flow into the bay decreases.

The Chesapeake Bay is biologically a special place where both freshwater and saltwater organisms live, and its ecological balance can be upset by changes in salinity. Dry weather in 1914 caused salinity in the bay to increase, and saline water nearly reached the freshwater intake on the Susquehanna at Havre de Grace, while at Turkey Point the salinity increased from 0.1 ppt to 6 ppt. This increase greatly hurt shad and rockfish spawning, which usually occurs in nearly fresh water. In 1972, rains from the severe storm Agnes increased the flow of freshwater into the bay to 15.5 times the normal amount. The decrease in salinity killed 2 million bushels (70,478,144 liters) of oysters, which require saline conditions to survive.

The construction of the 13-mile (20.8-kilometer) Chesapeake and Delaware Canal connecting Chesapeake Bay to the Delaware River and the Atlantic Ocean also increased salinity in the upper bay. The canal first opened in 1829 and was enlarged in 1938. Freshwater flows out of the bay in the upper waters of the canal, as brackish waters enter the bay at lower depths.

The Chesapeake Bay is a rich body of water where many plants thrive because sunlight penetrates its shallow water. The plants in turn provide food to a number of other organisms; these plant "food factories" are eaten by tiny animals that in turn are eaten by larger species, thus forming a food chain. Maryland has 25,000 acres (14,175 hectares) of rich tidal-marsh areas, called wetlands, where much of the plant food grows. The stability of these areas is highly dependent upon the level of point and nonpoint pollution, including sediment from erosion.

Nearly one million ducks and geese winter along the wetlands of Chesapeake Bay, one of the most important areas of the Atlantic Flyway. The bay shoreline is approximately 88 percent privately owned. Local governments, each in need of expanding its tax base, manage most of the wetlands. The destruction of wetlands by diking and filling for new industries or housing is all too common. As public awareness of wetland destruction increased, Maryland wetland laws enacted in 1970 were intended to prevent dredging and filling of wetlands without a permit. Yet the struggle to preserve wetlands goes on as developers continue to seek access to these areas, and successive state administrations have varying commitments to their protection.

The configuration of the Chesapeake Bay is constantly changing as current and tides eat away at the coastline while in other areas deposition creates new coastline. Smith Island, offshore from Crisfield, is one of the last of the bay's inhabited islands. This area is a stronghold of Chesapeake "watermen" and their old way of life, where a dialect echoing earlier times is still spoken. The US Army Corps of Engineers is studying erosion there to determine how to stabilize the shore, but the outlook is not good. It loses up to 50 feet (15.2 meters) of coastline to erosion in some years. In thirty years' time, there may be little left of Smith Island. Like the fictional Devon Island and Rosalind's Revenge of James Michener's book *Chesapeake*, Smith Island and other bay islands will erode and vanish, while new ones appear.

The elevated and more resistant Western Shore of Maryland suffers less erosion. The lower Western Shore has actually been gaining area from deposition. The story is different on the Eastern Shore. It is estimated that the lower Eastern Shore has been losing about 8 feet (2.5 meters) annually since the 1800s. Since 1945, Dorchester County has lost about 8,000 acres (3,200 hectares). The specific configuration of this coastline is indeed ephemeral. It is constantly under attack by wind, water-driven tides, and currents. The rate of erosion at a specific place depends on the configuration of the shoreline, direction and velocity of the prevailing winds, reach of open water over which winds blow, human activity in the area, and composition of the materials that make up the shoreline. Shoreline losses can be great in some areas and negligible in others. Between 1845 and 1942, about 6,000 acres (2,430 hectares) of land were lost to erosion along the Maryland shoreline of the bay; the average loss during this ninety-seven-year span was 2 acres (10.6 hectares) per mile of shoreline.

Poplar Island

In colonial times, Poplar Island, off the coast of Talbot County, was about 1,000 acres (400 hectares) in size. By 2000, it was only about 125 acres (50 hectares) in seven separate remnant islands (fig. 4.1).

William Claiborne, a Virginian, claimed Poplar Island in 1631 after earlier establishing a trading post on Kent Island. Claiborne then gave Poplar Island to his cousin Richard Thompson, who along with his family was killed by Nanticoke Indians in 1637. Over the next few centuries

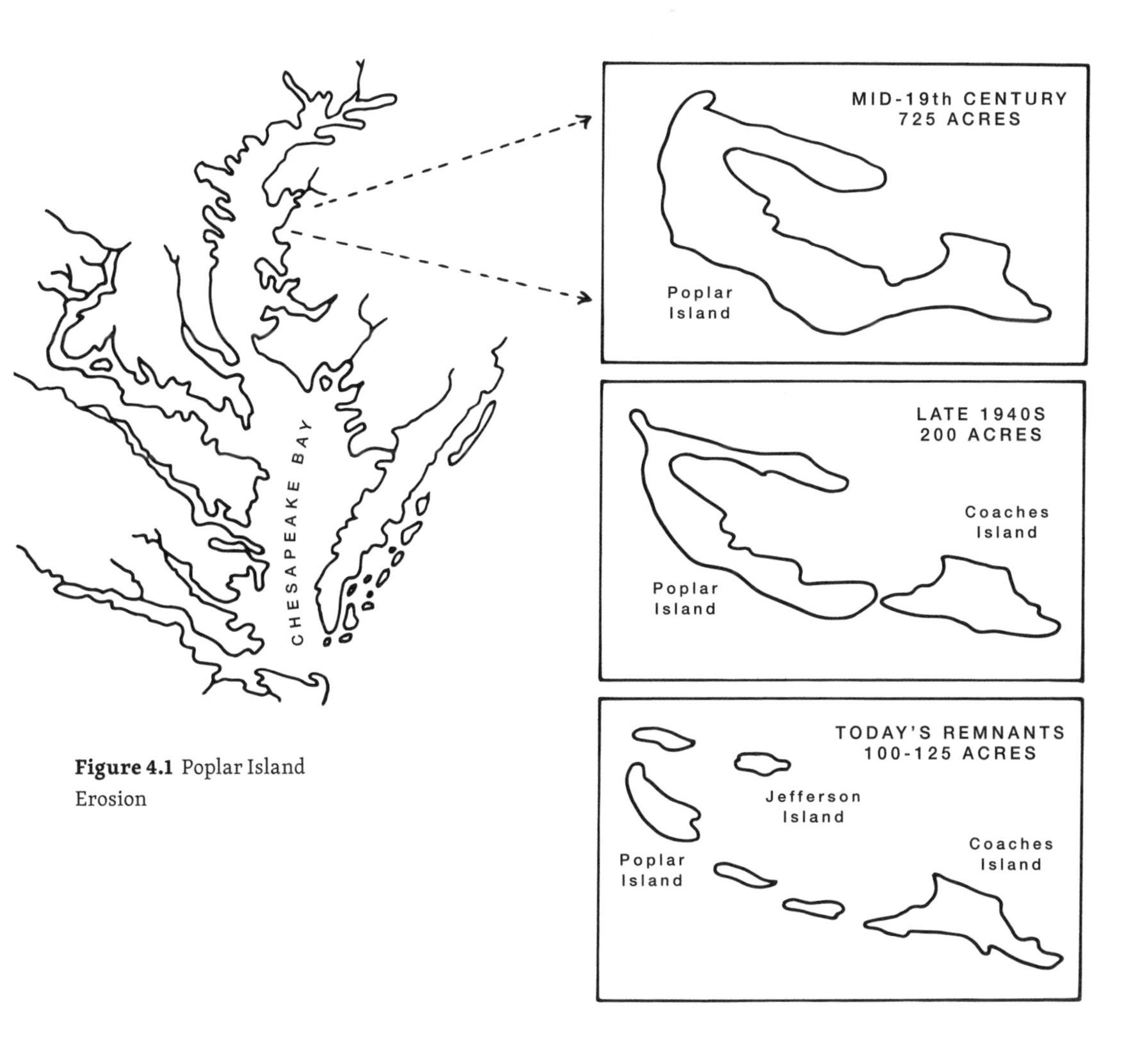

Figure 4.1 Poplar Island Erosion

there were various owners of Poplar Island, including Charles Carroll, whose signature appears on the Declaration of Independence. During the War of 1812, the British plundered the island. By 1900, there were fifteen families on Poplar Island—as well as a general store, post office, church, lumber mill, and several farms. In 1918, the school closed and families began to leave. In 1931, congressional Democrats bought Poplar Island and built the Jefferson Island Club (Presidents Franklin Delano Roosevelt and Harry S. Truman spent time there). The Democrat club burned in 1946, and the islands were abandoned (Meyer 2003). A map from the 1940s shows Poplar in two pieces. In 1965, the owner donated the islands to the Smithsonian Institute. Judging them too costly to save, and questioning the wisdom of trying to halt a natural process, the Smithsonian sold the islands in the 1980s. But that is not the end of the story. In 1995, the US Congress appropriated $15 million to recreate Poplar Island with material dredged from the Baltimore shipping channel. The US Army Corps of Engineers has started the project, now funded with $427 million.

The Poplar Island Reclamation Project brings together a number of organizations that do not often share environmental views. The Maryland Port Authority has a place to deposit (clean) dredge material from the commercial shipping channel. The filled land is protected as a nature preserve; birds have already begun to nest there. The US Fish and Wildlife Service monitors the entire process. Lands to the east and southeast do not erode as rapidly because Poplar Island acts as a barrier to westerly winds, currents, and turbidity from passing ships. Dredge material must be barged 34 miles (54 kilometers) from Baltimore, which is expensive. Nobody knows what will happen to Poplar Island in the future, but the reclamation of one of the Chesapeake Bay's disappearing islands and its preservation as a bird sanctuary have won widespread support, even though the project comes with a high cost. When the restoration project is completed by 2020, another site for dredge material will be needed.

Pollution: Linking Land and Water

Water Resources

In the early seventeenth century, before the early European settlers arrived, Maryland looked quite different than it does today. Since then, changing anthropogenic factors (including demographic, social, economic, and cultural factors) have induced landscape and environmental changes such as deforestation, development, intensive agriculture, and air and water pollution, all of which have contributed to the degradation of Maryland's water resources. The many water resources in the state exemplify how environmental systems intermesh. They truly highlight the value of a "systems-level" approach so often favored in geography. Because water resources shape "upstream–downstream relationships," they highlight the interdependence people have in environmental management.

As industry grew rapidly in the early twentieth century, pollution from industrial and municipal waste became a serious environmental challenge. Raw sewage from Washington, DC, was once dumped directly into surface waters; only recently have efforts to clean up the seriously polluted Anacostia River begun to show some results. In 1938, the Blue Plains Wastewater Treatment Plant, today one of the largest in the nation, was built to serve residents of the Washington area, including many Marylanders. Although several wastewater treatment plants were built in Maryland from the 1930s to 1950s, they were generally inadequate in dealing with water pollution. Wastewater treatment did not remove many of the pollutants, and a dramatic increase in water pollution occurred in the 1950s and 1960s. The federal Clean Water Act of 1972 helped promote significant increases in municipal wastewater treatment. Although the emphasis on water pollution control has been on point sources, efforts to control nonpoint sources of pollution have gained increased attention.

Nonpoint source pollution is one of the most serious threats to Maryland's water resources today; it accounts for a majority of the pollutants that enter the state's waters. Nonpoint source pollution results from runoff caused by stormwater (rainfall or snowmelt) or irrigation water that moves over or through the ground. Runoff picks up and carries pollutants, including nutrients, sediments, pathogens, and toxics. Eventually, these pollutants are deposited into rivers, lakes, wetlands, groundwaters, and the Chesapeake Bay. Sources of nonpoint pollution include agricultural runoff, mining, forestry, urban stormwater runoff, construction, septic systems, and recreational boating. Because it does not come from a single outlet, or point source, it is called nonpoint pollution.

Maryland faces challenges to both its surface and subsurface water. The Potomac River receives nonpoint pollution, such as runoff polluted with cow manure from dairy farms in rural areas, including the many dairy farms of the Monocacy River Valley in Frederick County. There are also point sources such as municipal outfalls. One way that the state has tried to limit pollution runoff into streams is by the use of stream buffers near the banks of water resources. But it is not just surface water pollution that is of concern; groundwater pollution is a serious issue, too. Many pollutants are tracked by various state and federal agencies, but the two most tracked nonpoint sources are the nutrients nitrogen and phosphorus. Maryland's agriculture generates about 38 percent of the nitrogen load and 55 percent of the phosphorus load in the state's waters. Point sources generate about 33 percent of the nitrogen and 29 percent of the phosphorus load.

The Chesapeake Bay

The magnificent Chesapeake Bay is not without major ecological challenges. In the past it was generally believed that the bay had an infinite capacity to break down and scatter waste. "Dilution is the solution to pollution" might have been a convenient logic, but this mentality has led to some severe problems. Scientists, government officials, and environmental groups have serious concerns about how to assure the bay's natural integrity and tremendous ability to produce fish, shellfish, and waterfowl and to break down human sewage.

Every body of water has a limited capability to assimilate nutrients. One sign that the Chesapeake Bay is struggling with an out-of-balance nutrient load is the algae blooms observed in the upper bay, indicating areas where algae are thriving on increased nutrients. Decaying algae consume oxygen in the water, producing a eutrophic environment that has disastrous effects on other forms of life. Shellfish and finfish can die from starvation or suffocation in a water environment that is murky and low in oxygen. Species such as rockfish spawn near the freshwater-saline interface in the upper bay. Menhaden, sea trout, spot, croaker (hardhead), harvest fish, winter flounder, and drum spawn near the mouth of the bay. As the deep water drifts up the bay, it carries their eggs and larvae to the richer, upper bay feeding areas. A condition of nutrient overload and resulting eutrophication in these upper bay feeding areas can lead to high mortality rates for finfish eggs and larvae.

In recent years, upstream land-use patterns have been blamed for the increased numbers of the naturally occurring pathogen known as *pfiesteria* near bodies of water influenced by large poultry farms (which are perhaps more accurately described as poultry factories). Localized cases of fish infected by *pfiesteria* continue to be observed in areas off the lower Eastern Shore and Coastal Plain. *Pfiesteria* has been blamed for large fish kills, and it has also caused memory loss among humans who have handled infected fish. But because the poultry business is a critical part of the economy of Maryland's lower Eastern Shore, some contend that the relationship between poultry waste and *pfiesteria* is not so clear.

Pesticides and herbicides also pose a critical threat to the health of the bay's ecosystem. Estimates are that quantities of agricultural wastes are at least equal to municipal sewage, but treating waters from farmlands at selected points is not feasible because of the contaminants' dispersed, or nonpoint, distribution. One insecticide, Kepone, has levied a heavy toll on the Chesapeake Bay. In 1975, it was found that workers were being poisoned at a plant producing Kepone in Hopewell, Virginia, on the James River. Kepone entered the bay from the river and settled in the bottom sediment of the bay. Its location in bottom sediment means that it is retained in the bay's ecosystem rather than flushed out by water circulation.

The human population in the Chesapeake Bay watershed is estimated to climb to over eighteen million by the year 2030 (fig. 4.2). This population is supported by rich farmland, vast woodlands, and intensely developed industrial areas. An increase in population means that more freshwater—used for, among other things, irrigation and industrial processes as well as the generation of electricity—will be drawn off before entering the bay, thus increasing salinity. During the period from 1955 to 1980, nearly 1.4 million new residents moved into Maryland, while the state lost 1.6 million acres (0.65 million hectares) of open space. This is conversion at the incredible rate of 1 acre (0.41 ha) lost per person.

As the population grows, so will competition for recreational, residential, and commercial space. The challenge is how to reconcile the need for disposal of solid and liquid waste with requirements of other bay ecological "stakeholders." Along with a growing population come a number of major waste disposal problems. Domestic wastes in the form of treated municipal sewage enter Chesapeake Bay at a rate over

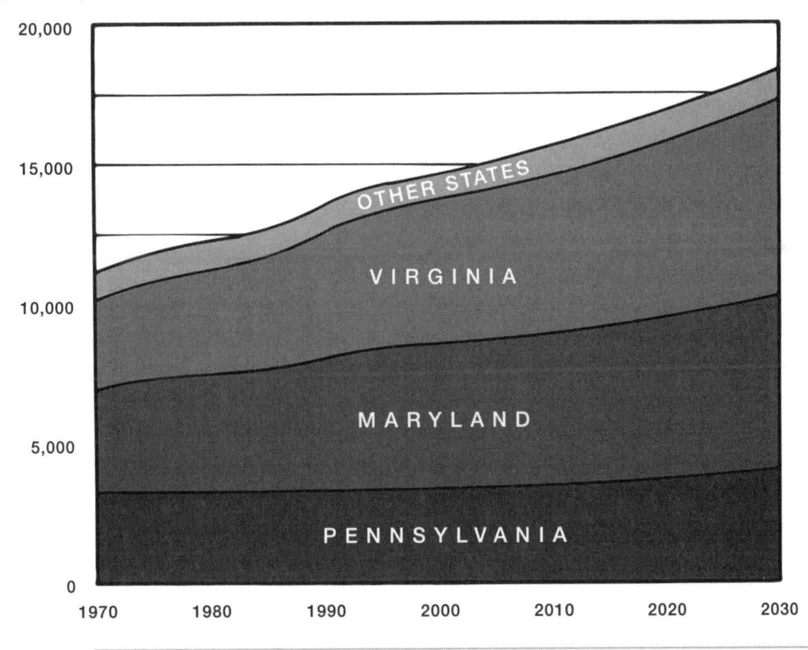

Figure 4.2 Projected Population Growth in the Chesapeake Bay Watershed, 1970–2030

270 million gallons (1.021 billion liters) per day from Maryland counties and over 300 million gallons (1.135 billion liters) more per day from the Washington, DC, Blue Plains Wastewater Treatment Plant on the Potomac River.

Industrial wastes vary from nontoxic rinses or cooling water to harmful chemicals, heavy metals, and oil. Wastes from agriculture include animal excrement, pesticides, herbicides, and fertilizers. Not all of these sources of pollution are visible to the human eye. Both surface runoff and groundwater carry these pollutants into the Chesapeake Bay. Animal wastes and fertilizer steal large quantities of oxygen from the water, carry disease-causing organisms, and accelerate the growth of aquatic pests. To combat such pollution, Maryland and other signatories to the 1987 Chesapeake Bay Agreement created a basin-based strategy to achieve at least a 40 percent reduction of nitrogen and phosphorus entering the main stem of the bay. Governments and the general public formed ten "tributary strategy teams" in the Chesapeake Bay watershed to foster watershed-based plans to reduce pollution. Anyone is welcome to join these teams that bring together experts and citizens at large. Active citizens' groups have been created, including the Alliance for the Chesapeake Bay and the Chesapeake Bay Foundation.

The Chesapeake Bay accommodates vessels ranging from large ships to small pleasure craft. This maritime traffic creates several problems. The amount of raw sewage discharged into the bay by a large vessel in transit is greater than from a community of 20,000 people. Marina and anchorage areas lying near bathing beaches or shellfish harvest areas are particularly susceptible to such waste. Because of its potential impact, an even more serious hazard is oil spills. As the region's population grows, so does its need for oil. Solid waste disposal is a growing challenge for the Chesapeake Bay as well. Because the greatest need for sanitary landfills exists in heavily populated areas where land is valuable, wetlands are often considered as possible new sites. The Chesapeake Bay has become increasingly attractive to those looking for an economically expedient way to dispose of solid waste.

It is predicted that over 80 percent of Maryland's population will continue to be concentrated around the bay. But the bay faces future threats from all states within the Chesapeake Bay watershed. Population growth within the entire watershed translates into an increased need for electricity; most of the thermal waste discharge from additional power plants will impinge upon the bay and its tributaries.

The rivers flowing into the bay, especially those on the Western Shore, carry an enormous amount of sediment. Between 1845 and 1924, over 15 feet (4.6 meters) of sediment accumulated in the Patapsco River near Baltimore alone. Frequent dredging is necessary in many areas. The sediment covers the bottom of the bay and often smothers bottom-dwelling marine life. From 80 to 90 percent of the annual sediment load is carried into the bay during February and March, when spring precipitation and meltwater flow is high. At that time, soils are thawing, rocks and soil have been loosened by frost, and there is less vegetation to hold the soil in place. Human activity such as construction and disruption of vegetative cover strongly influence sedimentation intensity.

Environmental disturbances of many types can destroy fish and shellfish habitats and disrupt life cycles. Natural hazards include extreme variations in salinity caused by drought, extreme sedimentation from severe storms, and pathogenic organisms. Human impacts include destruction of habitats by filling in wetlands, dredging and dumping contaminated materials, diverting freshwater flow, accelerating sedimentation from agricultural and urban development, and generating domestic and industrial pollution (including acid rain deposition from the Midwest).

The air over the Chesapeake comes from a much wider area than just the watershed. This "airshed" varies on a daily basis according to weather conditions. But because the westerly wind belt dominates this region of the world, much of the air over the Chesapeake comes from the Midwest, a region that emits large volumes of air pollution from its industries, power plants, and vehicles. The Chesapeake region emits

its own pollution from the same types of sources, and the air from the Midwest just makes it worse. This pollution—especially oxides of nitrogen—washes out of the air in rainfall, causing additional hazards for the bay and its tributaries.

The Chesapeake Bay is a coupled system. It is linked to the uplands of its entire drainage basin, and through deposition from air pollution it is linked to the Midwest. Symptoms of serious degradation can be seen within the bay, but many environmental scientists believe that we must look to the land for solutions. Most of the surrounding land is a free-market commodity whose use is determined by short-term economics and the path of least regulatory resistance. Although the bay is a "commons," by tradition the land is largely privately owned. Around the Chesapeake Bay, farmland vanishes at a rapid rate as it is rezoned for development. Deforestation and the laying of impervious surfaces pose serious environmental threats to the bay. A balance between growth in the region and the ecological integrity of the bay is necessary. But for most of recent history there has been a strong bias for managing the bay through engineering techniques and by the use of economic growth models, rather than ecological multistakeholder models. If trends continue, portions of the bay will become eutrophic (with little oxygen), especially in the summer. Some species of fish and shellfish have already felt the impact of these conditions. Though certain areas have seen improvement in recent years, much of the shoreline has already been denuded of life-giving grasses. These grasses are of enormous ecological value because they are the breeding ground for many organisms and a vital link in the food chain. Unless the bay is managed more effectively, it is highly likely that oil spills will occur; sediment from development will cloud the water and fill the channels; and the flow of freshwater into the bay will decrease.

Crabs are declining in numbers and as a result are increasingly expensive. Many crab shacks on the Eastern Shore now sell crabs imported from Texas and North Carolina. Some Maryland restaurants are now serving crab dishes made from Asian crabmeat, as well as oysters from Louisiana and Texas. Crab houses in Crisfield, called the "seafood capital of the world," are being razed to make room for condominiums. To a significant extent, imported oysters and crabs are now sustaining the Chesapeake culture. This bleak picture could worsen if current patterns of use and abuse are not reversed, and even then success will be difficult to measure.

A Bolide Hits the Chesapeake Bay Region

One of the most dramatic geological events on the Atlantic coastal fringe of North America took place about thirty-five million years ago near the end of the Eocene Epoch. Worldwide, the sea level was unusually high. The climate was more humid and warmer than today; tropical rainforest covered the land west to the slopes of the Appalachians.

In a brief moment, this ancient environment was suddenly transformed when a giant bolide from far out in the solar system entered the Earth's atmosphere. With a brilliant flash of light, the bolide hit the Earth on the broad lime-covered continental shelf in what is today the lower Chesapeake Bay and Eastern Shore in Virginia (the bay had not yet been formed). The carnage was incredible, as terrestrial plant and animal life were obliterated. Although the bolide did not make direct impact in Maryland, it has ramifications for modern inhabitants of the Chesapeake region. This event from thirty-five million years ago is still of interest today, as its significant environmental impacts have been felt over the long term.

As an extraterrestrial body about 0.62 to 6.2 miles (1 to 10 kilometers) in size that hits the Earth at a speed of 20–70 kilometers per second (Mach 75), a bolide is literally faster than a speeding bullet. It explodes upon impact, creating an enormous crater (fig. 4.3). Bolide collisions have the potential to cause global environmental disasters. The Chesapeake bolide caused the mass extinction of plants and animals in the region. There was a superhot blast wave, a surge of hot debris thrown into the lower atmosphere over hundreds of miles, giant tsunamis, and perhaps even an earthquake.

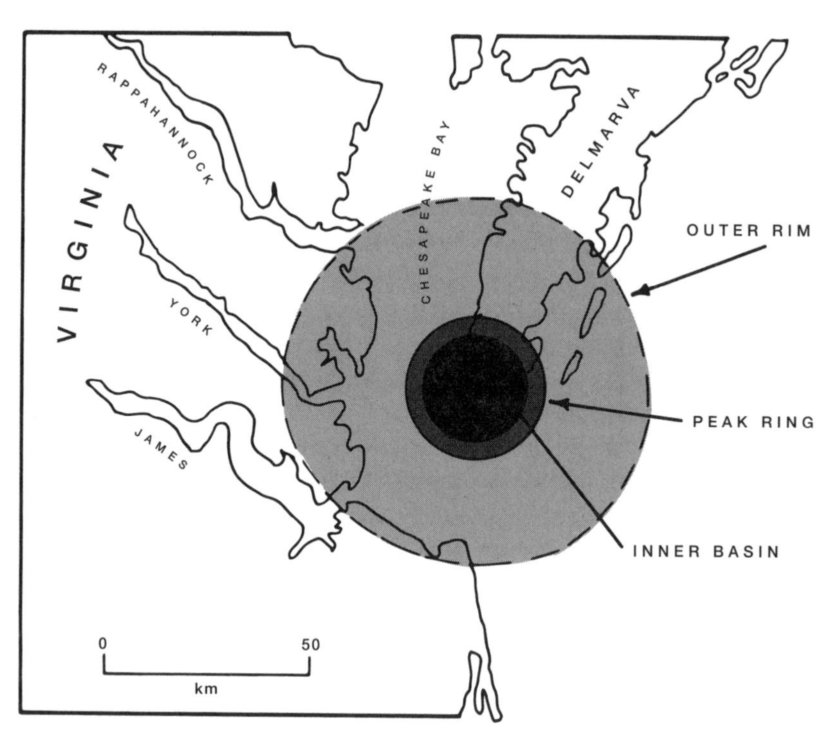

Figure 4.3 Chesapeake Bay Bolide Crater

How did the bolide affect the formation of the bay millions of years later? We know that only 18,000 years ago this area was dry land. About 10,000 years ago the Pleistocene ice began to melt rapidly, and gorged rivers carried meltwater to the sea, which rose and flooded the low-lying coastal areas that became the Chesapeake and Delaware Bays. Evidence indicates that the rivers of the region converged directly over the buried crater when it was dry land. It is also known that the ancient Susquehanna River had several successive ice-buried channels from 450,000 to 20,000 years ago. The ancient channels all changed course after crossing the rim of the crater. Because the surface of the crater was much lower than the surrounding surface, having created a topographic depression, the crater had a major effect on the courses of the ancient rivers, the valleys they cut, and the eventual formation of the Chesapeake Bay.

The ancient Chesapeake Bay crater is located 124 miles (200 kilometers) southeast of Washington, DC. It is buried 984 to 1,640 feet (300 to 500 meters) beneath the lower bay, surrounding peninsulas, and continental shelf. In 1983, the drill ship *Glomar Challenger* was part of the National Science Foundation's Deep Sea Drilling Project. Offshore 75 miles (120 kilometers) east of Atlantic City, New Jersey, the project's scientists, led by Wylie Poag, brought up a core sample with debris that indicated a bolide impact. It contained certain minerals whose physical properties had been altered by a tremendous impact. Glass beads measuring 0.04 to 0.4 inches (1 millimeter to 1 centimeter) in size, known as *tektites*, are derived from sediment melted by the impact. Shocked minerals, especially quartz, showed sets of closely spaced intersecting dark stripes. No mechanism other than a bolide produces tektites and shocked quartz. Also present in the sediment were fossilized remains of microorganisms that lived when the bolide debris was deposited. Laboratory analysis dated this layer of sediment to the late Eocene Epoch, about thirty-five million years ago.

In 1986, the US Geological Survey was drilling in southeastern Virginia in search of sources of fresh groundwater when they found a sandy rubble bed. Fragments from various geologic ages were mixed together in the rubble layer that covered an area about twice the size of Rhode Island. A terrific, unexplained force had torn apart the normal stacked layers of substrata in this part of Virginia at about the same time the bolide impact deposited the tektites near New Jersey. Scientists saw a link between the two sites. Texaco and Exxon Exploration Company found the final piece to the puzzle in 1993 while exploring beneath the floor of the bay for oil and gas. Their seismic reflection profiles revealed a huge peak ring impact crater buried beneath the bay, centered near the small town of Cape Charles on Virginia's lower Eastern Shore. The crater is 54 miles (90 kilometers) in diameter and 0.78 miles (1.3 kilometers) deep, covering an area twice the size of Rhode Island and deeper than the Grand Canyon. The rubble bed filled the crater and formed a thin halo around it called an ejecta blanket. Over the next thirty-five million years, the crater was buried under sedimentary layers.

Scientist Wylie Poag has since identified and mapped over one hundred faults around and over the crater. Cities since built over or along the rim of the crater—including Hampton Roads, Newport News, Norfolk, Portsmouth, Virginia Beach, Yorktown, and towns of the lower Virginia Eastern Shore—are at risk from fault-related ground instability. There have been four recorded earthquakes aligned with the rim of the crater; the last was in 1995 in York County.

Another result of the bolide impact is that the orderly stack of underground aquifers found in most of Virginia is not present inside the crater. The crater truncates the freshwater aquifers, and in their place is a huge reservoir whose water is 1.5 times saltier than seawater and not suitable for drinking. This disruption of the aquifers limits the availability of freshwater for further development in this part of Virginia. The current-day course of rivers in the area is also a result of the bolide's impact. The normal course of rivers flowing from the Appalachians to the Atlantic is a gentle incline to the sea. Yet the James, Rappahannock, and York Rivers do something different when they reach the coast; they bend abruptly, turning their mouths north and east to face the village of Cape Charles. Millions of year ago, these rivers flowed into the low-lying ancient crater, and that pattern continues today.

The Kudzu Menace

Kudzu is a lush, fast-growing vine that has been engulfing much of the southern United States for several decades. A hairy vine with broad, pointed leaves, kudzu arrived when the Japanese used it as an ornament in their exhibit at the 1876 Centennial Exhibition in Philadelphia. Today kudzu covers approximately seven million acres (2,832,746.55 hectares) from Massachusetts to Maryland to Florida, blanketing hillsides, entangling power lines, and destroying trees. In the South, kudzu is "as familiar as black-eyed peas and okra" (LoLordo 2000). First planted to provide ground cover and nourish the soil, kudzu vines spread rapaciously into field and forest. Some vines can grow up to 1 foot (0.3048 meters) in one day, and it is not unusual for them to grow 100 feet (30.48 meters) or longer in one summer. Kudzu grown in the United States is a poor seed producer, but it can produce a new plant from each joint, about every 12 inches (30.48 centimeters). To most people, kudzu looks like a verdant and beautiful cover plant. Environmentalists and farmers, however, are quick to point out that where kudzu grows, nothing else does; this vine swallows everything in its path. James Miller, an Auburn University ecologist who has been studying kudzu for twenty years, states that kudzu now covers every southern national park, forest, and conservation area and is responsible for $340 million in lost forest productivity (LoLordo 2000). The invasive species reduces biodiversity by swamping native plants and their associated ecosystems.

At first kudzu was welcomed as a ground cover to mitigate soil erosion. The US government even paid southern farmers as much as $8 an acre to plant it. Chickens, cows, and goats like to eat the plant, and county agricultural agents soon began to promote kudzu as a supplemental forage crop (it has a nutritive value comparable to alfalfa). Its 7-foot (2.1-meter) roots aerate the soil and reach deep for water during dry periods. As a member of the legume family, kudzu restores nitrogen to the soil. Promoters felt that the plant would grow on the hard red-clay soil, stop erosion, and provide good grazing and an attractive ground cover.

Kudzu was strongly promoted during the New Deal period in the 1930s. The Agricultural Stabilization and Conservation Service, created to reclaim land, began a massive kudzu promotion campaign, and the Soil Conservation Service grew millions of kudzu plants in its nurseries from Maryland to Texas. The Civilian Conservation Corps planted them along highways in eroded gullies. Seeds were also imported from Japan prior to World War II. By the end the war, kudzu clubs had sprung up. Auburn, Alabama, even chose a kudzu queen. In the mid-1950s, however, a number of problems with kudzu became apparent. Kudzu could not stand up to intensive grazing. The tangled vines could not be harvested as easily as hay, and the vines freely jumped fences and ran wild. The US Forest Service and the lumber industry opposed the spread of kudzu, as it kills timber by cutting off sunlight and smothering trees. Farmers increasingly began to see kudzu as a scourge that damages field crops and fruit and nut trees. Telephone companies have reported that kudzu pulls down telephone lines and poles. Power companies must spray herbicides yearly to keep the vines off high-voltage towers because the vines can cause an electrical arc, causing extensive damage. These powerful herbicides penetrate the soil and flow across the watershed and into the Chesapeake Bay, especially during strong storm events.

The kudzu menace is not going away. Chemical controls can be effective, but they are not applied over widespread areas because of their adverse environmental effects and the financial cost. The use of bulldozers to remove roots is not a practical control method, either; kudzu roots penetrate deep into the soil and can weigh up to 400 pounds (181.4 kilograms). There are some promising new treatments being explored; for example, David Orr, an entomologist at North Carolina State University, has done research using an experimental mutated version of the soybean looper, an insect that eats twice as much kudzu as its unmutated relatives (LoLordo 2000).

As kudzu spreads unchecked, the acreage in Maryland covered by the vine will increase and the state will be faced with a major ecological problem. Kudzu is but one example of the environmental challenges presented by nonnative invasive species of plants and animals. Maryland's environment is under attack from a whole host of invasive species, including northern snakehead fish, various thistles, Johnson grass, devil's tong weed, and many more. The ramifications for native species can be fatal.

Urbanization on the Ocean City Shoreline

Attraction of the Coast

The US coastline is becoming a victim of its own magnetism, attracting development that fouls its beauty and undermines its ecological balance. As environmentalist Anne Simon stated in her book *The Thin Edge*, "In the last ten years, the coast's magnetic pull has become stronger than ever—more industry, more oil, more people, hotels, motels, boatels, more sewage, more waste. The coast is informing us that there is a saturation point beyond which its natural functions no longer flourish, often diminish, or simply cease."

Many environmentalists believe that much of the Atlantic shore seems destined to become a stretch of boardwalk and pizza parlor tackiness. Powerful economic arguments advocate "needed growth" to bring tourists and along with them new jobs, profits, and tax revenues to coastal communities. As incomes and available leisure time rose, people could afford to buy vacation homes, which led to rapid development on Fenwick Island, Maryland, starting in the 1960s. In trying to regulate such development, state and local authorities have faced challenges in planning and directing coastal development, which have been compounded by property ownership patterns. Over 90 percent of the US coastline is privately owned, 5 percent is controlled by the armed forces, and less than 5 percent is open to the public. Yet over 50 percent of the US population lives in coastal counties.

Natural hazards occur within urban development all along the Atlantic and Gulf coasts. The resort center of Ocean City on Fenwick Island presents a prime example of a people-environment conflict. This important narrow coastal margin is heavily used for recreation, industrial and commercial activities, waste disposal, and food production. Those who choose to live on the immediate coast face risks associated with hurricanes, other major storms, and geomorphic changes. The tremendous growth in development and population at Ocean City intensifies these ecological and economic risks. Worcester County has a permanent population of about 51,000 today and it is projected to grow to 72,000 by 2020. Ocean City itself has a permanent population of about 7,100. The population of the Coastal Bay region, including Fenwick Island and the adjacent mainland, swells to about 300,000 during the summer. Between Memorial Day and Labor Day, about four million people visit Ocean City. As development continues unabated and the population in these areas grows, there is an ever-higher risk of environmental disaster for more and more people. Decisions about development of the shoreline should be made on the basis of information about rates of natural processes; planners could use this information to guide future development to less hazardous areas. Unfortunately, this has not been the case in Ocean City.

The Nature of Barrier Islands

Barrier islands are found along the coast of the United States from Maine to Texas, where they are the primary terrestrial-marine interface. Barrier islands are attractive places to live on or visit, but they have the potential of being hazardous. Gilbert White, an environmental geographer who on December 1, 2000, received the National Medal of Science from President Bill Clinton, has stated that "the most rapidly growing site for catastrophic events in the United States is the Gulf and Atlantic coast of the country."

The most unstable coastal lands utilized by people in the United States, barrier island beaches and dunes are only temporary in form—they continually shift and move in response to natural forces. These islands adjust to natural forces by building new beach, destroying old beach, and shifting position (fig. 4.4). Misunderstanding the true nature of barrier islands, people have tried to "improve" these natural processes with various engineering constructions intended to stabilize these areas. Development of this rapidly changing landscape

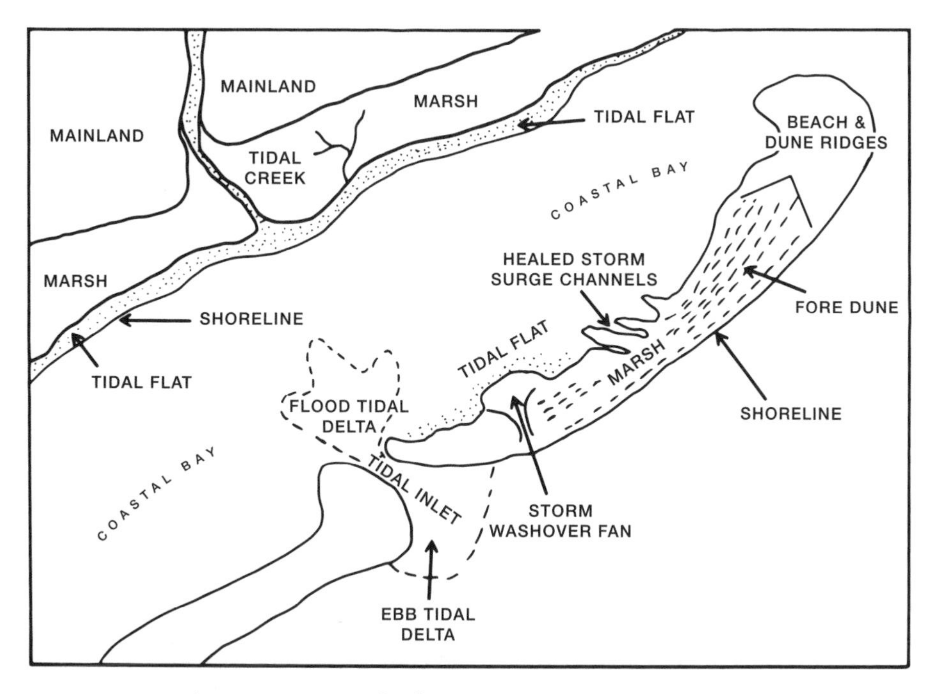

Figure 4.4 Barrier Island System

is based on the belief that the landscape can and should be engineered to be stable.

But barrier islands are naturally unstable because the constant movement of sand by wind and currents, the occurrence of catastrophic storms, and the general trend of rising sea level due to global warming are exacerbating coastal recession. Sandy barrier island beaches are stable only if sand arrives at the same rate that it is carried off. A surplus of sand causes seaward beach building, and a deficiency of sand results in coastline retreat. Recently, the amount of sand leaving the mid-Atlantic coast has generally exceeded the amount of sand being carried in. The net result of these forces is the natural migration of barrier islands closer to the mainland. This natural coastal retreat process, which is occurring along the beaches of barrier islands such as the one on which Ocean City sits, causes concern to property owners, planners, scientists, and community leaders.

The two principal causes of major damage on Fenwick Island have been hurricanes and extratropical storms (northeasters). Records going back to the seventeenth century show that although numerous storms have passed nearby, between 1672 and 1970 only three hurricanes with winds in excess of 75 miles per hour (121 kilometers per hour) have struck Fenwick Island. The hurricane of August 23, 1933, had wind velocities averaging 60 miles per hour (96.5 kilometers per hour) with 10-foot (3-meter) waves, and the tide reached a record level of 7.5 feet (2.3 meters) above mean low water. Total damage from the storm amounted to $5 million (equivalent to $86 million in 2011). The hurricane opened the Ocean City Inlet, a breach 10 feet (3 meters) deep and 250 feet (76.2 meters) wide, which would have closed naturally had it not been for subsequent "maintenance" by the US Army Corp of Engineers. Each year between thirty and forty northeasters cause surge waves of at least 5 feet (1.5 meters) above normal along the Ocean City shoreline. Another storm, the northeaster of Ash Wednesday, March 1962, caused major damage to the Ocean City area. It had winds averaging 50 miles per hour (80.5 kilometers per hour) that persisted over several days. Waves ranged from 10 to 15 feet (3 to 4.6 meters), and the tides were 9 feet (2.7 m) above mean low water. Much of Fenwick Island was submerged for two days at depths up to 8 feet (2.4 meters). Damage from the storm totaled $7 million (equivalent to $52 million in 2011). Although this storm ravaged Ocean City, it set the stage for the vast reconstruction and development that occurred immediately thereafter.

Coastal scientists use statistics from studies of storm and wave conditions to construct storm recurrence interval graphs. A storm producing a maximum deepwater wave height of 25 feet (7.6 meters) offshore from Ocean City can be expected to occur at least once every twenty-five years. Many ocean-side resorts such as Ocean City, which have large numbers of new high-rise apartments and condominiums, have yet to be tested by major storms because the hurricane cycle was unusually lull during the 1970s, when many of these buildings were constructed.

Varying gravitational attractions associated with the moon, Earth, and sun can also affect the sea level along the coast. The highest spring tides at Ocean City occur twice each month when the moon, Earth, and sun are aligned, which increases the tidal range by 20 percent. Although tides have their greatest adverse effect if coastal flooding occurs, not all high tides or coastal storms cause flooding. Under certain circumstances, uncommonly high tides (perigean spring tides) coincide with strong on-shore winds and flood the island; this is what happened on Ash Wednesday in March 1962.

The Growth of Ocean City

Ocean City is a model of beach overdevelopment, but this problematic situation developed only recently. During the colonial period, the island was used as pastureland. The growing market for oysters, clams, diamondback terrapin, and waterfowl in the 1830s encouraged more settlement, and by 1872 several residents of Worcester County joined with some Baltimore businessmen to form the Sinepuxent Beach Corporation (the original name for Ocean City was Sinepuxent Beach) to promote the construction of a resort on the island. The Atlantic Hotel opened its doors in 1875 and welcomed visitors to the shore. Visitors came by rail from Berlin, Maryland, to the coast of the mainland across from Ocean City. In 1878 the railroad was extended across Sinepuxent Bay to Ocean City.

Ocean City remained a small resort community until after World War II. Starting from the small town center on the southern end of Fenwick Island, it began to spread north to the Delaware border. The most recent growth spurt started in the early 1950s and intensified in the 1960s and 1970s. From 1950 to 1972, urban development on Fenwick Island more than doubled, mostly at the expense of wetlands. An important factor in this rapid growth was the opening of the Chesapeake Bay Bridge connecting Sandy Point on the Western Shore to Kent Island and then the Eastern Shore. Before the bridge opened in 1952, the trip to Ocean City from Baltimore or Washington, DC, took three to four and a half hours via ferry and railroad. After the bridge opened, one could reach Ocean City in less than three hours. The reduced travel time, coupled with more leisure time and affluence among visitors, led to an explosion in development and property values in the 1960s and 1970s. The 1970s became the decade of condominiums, many of which were sited on filled wetlands (fig. 4.5).

Since 1970, many apartments and condominiums have been built close to the beach. Some of these buildings in "high-rise row," an area of densely developed beachfront, are over twenty-two stories high. Although development stretches along 10 miles (16.1 kilometers) of Ocean City shoreline, only two blocks are open for public recreation. The vast

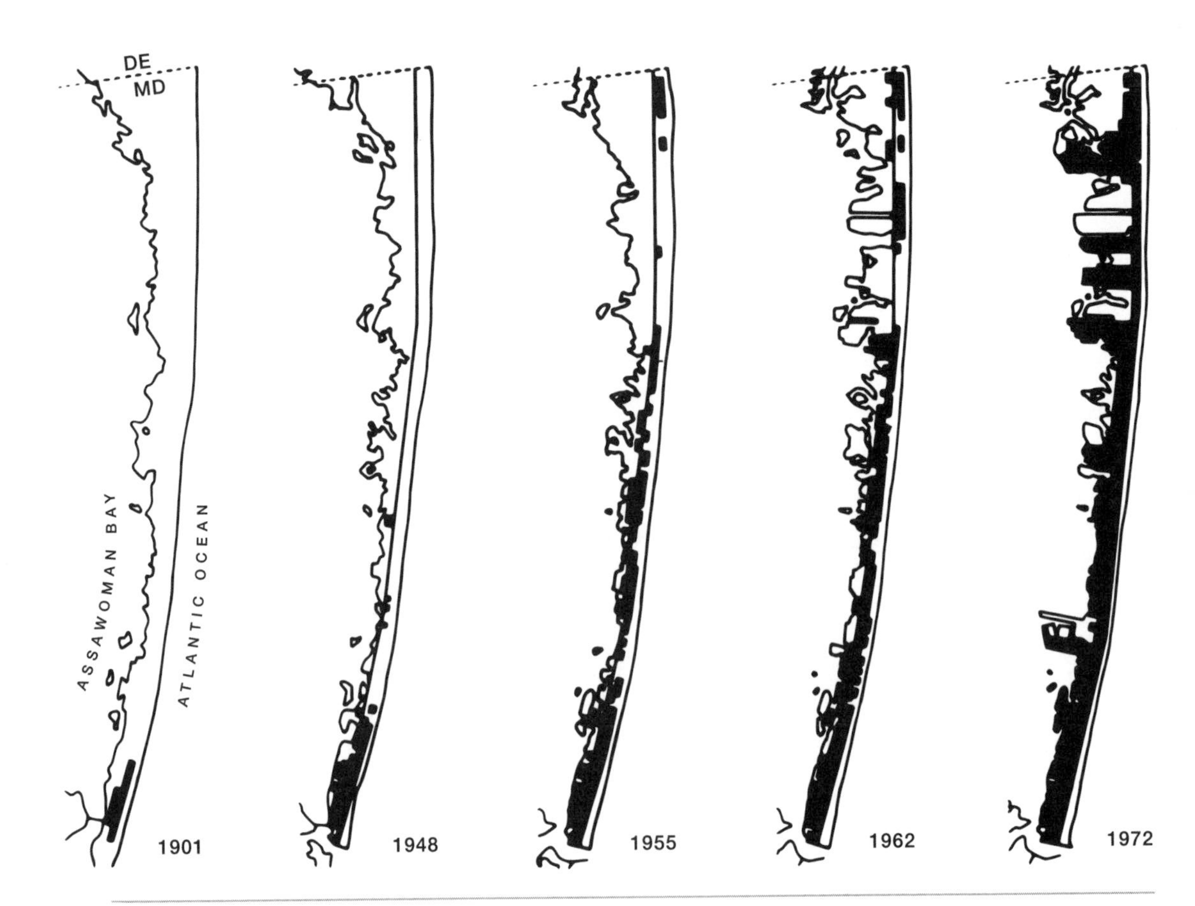

Figure 4.5 Ocean City Urbanization

majority of housing units in Ocean City are hotels, motels, apartments, and condominium units. Although Ocean City's permanent population has always been small (about 7,100), the peak weekend summer visitor population at times exceeds fifty times the permanent population.

Beach Stabilization Measures

Along the Maryland coast from Delaware to Virginia, beach sand is being carried southward by littoral drift. Sand is also being lost by winds blowing westward (deflation) and by seaward losses during violent storms. Erosion has caused the beach along the shoreline at Ocean City to recede over the past fifty years at an average rate of 2.4 feet (0.73 m) per year. Between 1850 and 1965, the mean high-water contour shifted landward 270 feet (82.3 meters). Of this movement, 86 feet (26.2 meters) occurred since 1930, coinciding with the period of greatest growth at Ocean City (fig. 4.6).

There are three general types of shoreline protection measures: (1) seawalls, (2) groins and jetties, and (3) artificial nourishment of beaches. Seawalls are meant to absorb and reflect wave energy; they protect the elevated land behind them. Because the beach in front of a seawall often loses sand at an accelerated rate, seawalls have not been used at Ocean City. A groin is a structure built perpendicular to the shoreline to impede littoral drift of sand. At Ocean City, sand is entrapped on the north side of the groins while a shortage of sand is experienced down-beach on the south side. To use groins effectively, littoral drift must be significant in volume and the land down-beach must be expendable. Groins and jetties (a jetty is a type of groin used also as a retaining

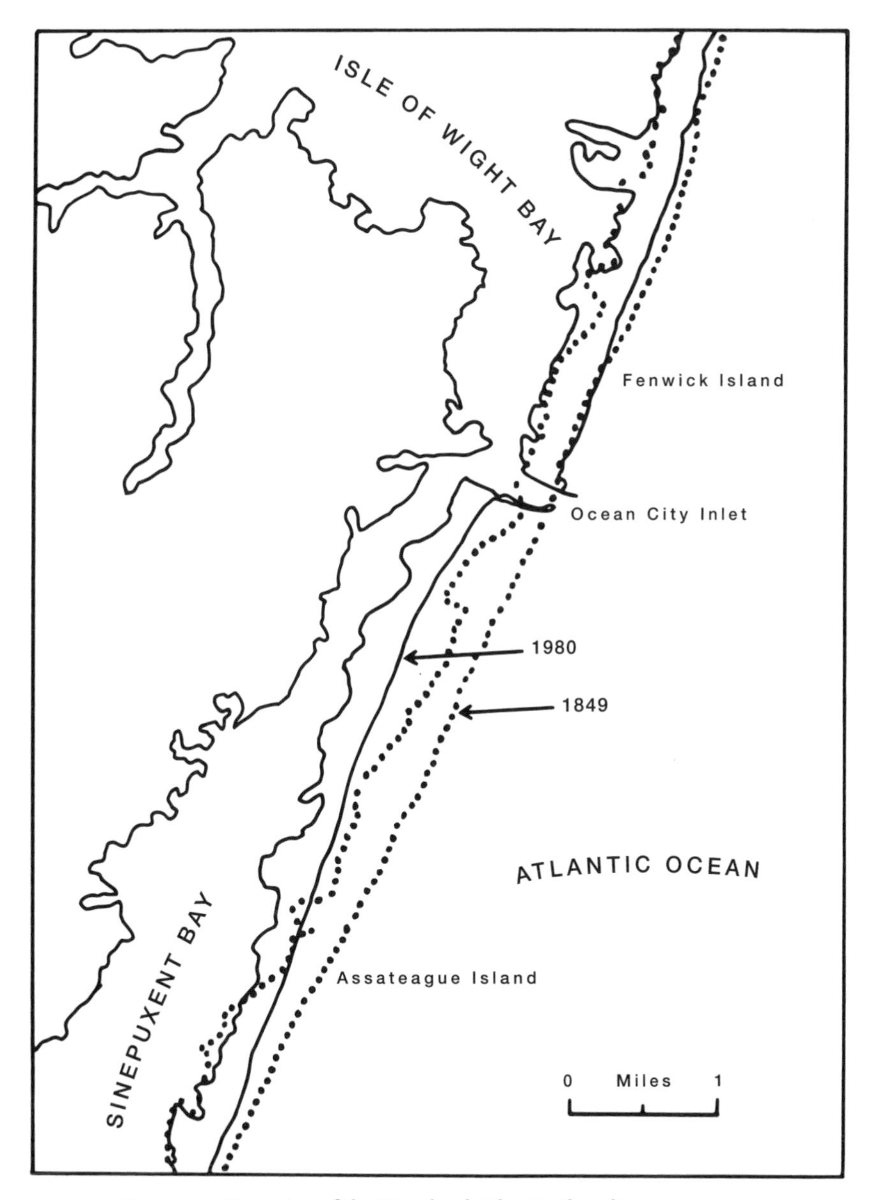

Figure 4.6 Recession of the Maryland Atlantic Shoreline

wall) have long been used at Ocean City to trap sand and protect the beaches. In a study of beach loss at Ocean City, the US Geological Survey concluded that groins and jetties have not been generally successful and have instead exacerbated erosion. The location of groins and jetties must create a system (one that works together) to be successful; those at Ocean City were not. Instead, the mayor of Ocean City subjectively selected sites along the 9 miles (14.5 kilometers) of coastline considered to be problem areas.

The north jetty, built in 1934 and located on the northern side of the Ocean City Inlet, has created a massive impoundment of sand that blankets the most southerly mile of Ocean City's shore. This impoundment has benefited the beaches to the north and has kept some of the sand out of Ocean City Inlet. Still, the inlet must be dredged from time to time. Before the south jetty was built on the southern side of the inlet in 1935, the southward moving sand moved onto what is now Assateague Island. Since the breach in 1933 and the construction of the jetties, little sand has reached Assateague from the north, which has resulted in the massive erosion of the shore face of the northern 3 to 4 miles (4.8 to 6.4 kilometers) of Assateague. Before the breach, there was one continuous barrier island. By 1980, there was a noticeable offset: Fenwick Island lies more to the east and is wider than Assateague, which has been starved of sand deposition. From 1943 to 1963, the beach on Assateague retreated 800 feet (244 meters). The Ash Wednesday storm of 1962 ripped a space through the decimated northern end of Assateague Island, later repaired by the US Army Corps of Engineers. The northern end of Assateague Island has severely eroded, resulting from the jetties built north of the Ocean City Inlet. The beach on Assateague just south of the inlet is the object of the North End Restoration Project undertaken by the US Army Corps of Engineers, National Park Service, and Maryland Department of Natural Resources (National Park Service 2004).

A 1985 study by the Maryland Department of Community and Economic Development justified the extension of the groin system by concluding that public funds should be used to protect the beaches with a system of groins from Ocean City north to the Delaware border. Their rationale was based on the economic terms of employment, income, tax revenue, and tourist attraction provided by the beach attraction. Over 95 percent of the travelers to Ocean City are tourists who stay only a couple of days. The report concluded that the use of public funds to stabilize and protect the Ocean City beaches and the tourism industry founded on them appears to be quantitatively justified. The US Geological Survey and Army Corps of Engineers have traditionally favored a beach nourishment program. They consider this approach of moving sand onto the beaches from offshore borrow areas as the most desirable method of protection. Efforts to stabilize this barrier island landscape have had serious ecological effects. Inhibiting overwash and

AN AGREEMENT FOR THE BAY

A recent major step toward coordinating efforts to mitigate bay degradation is the 2000 Chesapeake Bay Agreement. This third such agreement (previous agreements were made in 1983 and 1987), signed by the Chesapeake Bay Commission, Commonwealth of Pennsylvania, Commonwealth of Virginia, District of Columbia, state of Maryland, and US Environmental Protection Agency (noticeably missing is Delaware, whose southwestern subwatersheds also feed freshwater into the Chesapeake Bay), was a declaration of signatories to:

> rededicate ourselves to the restoration and protection of the ecological integrity, productivity and beneficial uses of the Chesapeake Bay system. We reaffirm our commitment to previously-adopted Chesapeake Bay Agreements and their supporting policies. We agree to report annually to the citizens on the state of the Bay and consider any additional actions necessary.

The agreement was submitted for public comment prior to its release, signifying an orientation toward "stakeholder-based" approaches. It outlines commitments in areas including ecology, public health, and livelihoods in the Chesapeake Bay watershed. Complementing the top-down state approaches have been grants by nongovernmental organizations to support bottom-up work to protect and restore the bay. The agreement also addresses evolving pollution and development concerns. The language speaks of managing for the present to accommodate the needs of the future in ways that are reminiscent of "sustainable development" discourse, which developed over the last several decades and made its biggest splash in the now-famous 1997 report of the World Commission on Environment and Development, "Our Common Future." Sustainable development revolves around a strategic decision-making system that incorporates both economic and ecological concerns while attempting not to surpass nature's limits. For many years, in the Chesapeake Bay and elsewhere, people often assumed that nature had no limits, and that it would absorb unlimited human impacts and cleanse itself. The 2000 Chesapeake Bay Agreement and others place the onus on the diverse range of stakeholders rather than nature to improve environmental conditions. The signatories to the agreement made a commitment to:

- living resource protection and restoration,
- vital habitat protection and restoration,
- water quality protection and restoration,
- sound land use, and
- stewardship and community engagement.

An implication of these commitments is that nature's limits will be exceeded if we do not plan and conserve wisely. One of the agreement's byproducts, the potential to label growth as inappropriate at times, is another theme in harmony with sustainable development. Most importantly, this approach seeks prediction and regulation for the purposes of finding a necessary equilibrium. Others who have not accepted the underlying premise that environmental limits exist, however, consider such perspectives to be alarmist and reactionary, resulting in negative impacts for big business and industry. Ironically, they hope that nature's limits can be extended dynamically through efficiency and institutional improvements led by science and technology. But blind faith in the belief that science will create new technologies that will solve the problems of the Chesapeake Bay reflects a mentality that has led to this great estuary's poor current state of health.

concentrating windblown sands in the form of man-made dunes alters overwash channels, changes vegetative communities, and interferes with the natural landward migration of the island. Inhibiting longshore currents with groins causes shore accretion on the up-current side and erosion on the down-current side. The construction of roads, parking lots, and campgrounds alters sediment processes, waste attenuation, and freshwater runoff.

There is a clear conflict between environmental and economic priorities. Yet the tremendous growth on Fenwick Island demands that threats to the environmental be constantly monitored and that actions be taken.

Where Do We Go from Here?

To combat the negative effects of development, erosion, and storm surge, projections and data must be continually updated; a systematic monitoring program—including use of remote sensing, geographic techniques, and fieldwork—is key. Such efforts, like most environmental preservation and restoration work, require popular and political support to secure funding. Economic and environmental interests need not be mutually exclusive. Nevertheless, identifying hazards, assessing risks, estimating costs, and ensuring that mitigation measures are effective should always come before development. Over the past twenty-five years, various federal, state, and local agencies have cooperated to study and monitor the situation in and along other ocean barrier islands of Maryland. They must not only assess the risks of storms, natural migration of sand, and various beach stabilization measures, but now they must also address the short- and long-term ramifications for the low-lying coastal areas of Maryland threatened by global warming and its resultant rising sea level.

Throughout their history, Marylanders have created an interactive and diverse economic landscape that has evolved through various stages. Although the physical landscape and geographic location have presented resources and opportunities that have been used to create what is today a dynamic and prosperous knowledge-based economy, this progress has not been achieved without environmental costs, as rapidly developing cities like Ocean City reflect.

PART II

Old Economy, New Economy

5 Sustaining a Farming Heritage

In 2005, farms covered about two million acres (825,559 hectares), or over one-third of the state of Maryland. Nonfarmland accounts for 65.5 percent of its area, and farmland accounts for 34.5 percent. Although farming in Maryland is diversified, there are some clear patterns and concentrations of various crop and livestock activities. Eastern Shore farms are noted for their poultry, vegetables, grains, and cattle. Divided by the Choptank River, the Eastern Shore constitutes upper and lower parts. When traveling south on Route 50, the upper Eastern Shore is left behind on the northern bank of the river. On the southern bank of the river sits the city of Cambridge in Dorchester County, the gateway to the lower Eastern Shore. The lower Eastern Shore accounts for 40 percent of total state agricultural receipts, primarily because it is one of the nation's leading producers of broiler chickens. Croplands and forests mix in the four southernmost counties of the Eastern Shore, where the light, sandy loam soils are excellent for raising vegetables. Grains are also important on the Eastern Shore and are used mainly as feed for poultry. On the upper Eastern Shore, the percentage of land used for agriculture reaches the highest level in the state, 75 percent in Kent County. The heavier soils of the upper Eastern Shore support crops such as wheat, corn, and hay.

Although southern Maryland is still considered tobacco country by many, tobacco acreage has been declining over the past few decades. In 2007, it accounted for only 0.2 percent of the state's total agricultural receipts (table 5.1). In addition to tobacco, the light loam soils in this part of Maryland are used to grow some wheat, alfalfa, and an increasing amount of soybeans.

North-central Maryland is an area of general farming, livestock production, and dairy farming. Carroll, Frederick, and Washington Counties account for nearly 20 percent of Maryland's total gross receipts from agriculture. Farming in western Maryland is more restricted because of the higher elevation, thinner rocky soils (except in the valleys), and shorter growing season. This area has a concentration of fruit orchards along with some production of potatoes, livestock, and maple syrup. Dairying is locally important in Garrett County to

Table 5.1. Cash receipts from Maryland farms

	2005		2007	
Farming activity	*Cash receipts (thousands)*	*Percentage of total*	*Cash receipts (thousands)*	*Percentage of total*
Livestock and products	**$843,102**	100	**$1,205,787**	100
Broilers, poultry, and eggs	$583,343	69.2	$903,531	74.9
Dairy products	$169,458	20.1	$192,426	16.0
Cattle and calves	$50,570	6.0	$58,293	4.8
Hogs	$8,268	1.0	D	0
Others (including, farm chickens, turkeys, sheep and lambs, honey, and wool)	$31,463	3.7	$51,537	4.3
All crops	**$450,202**	100	**$630,303**	100
Grains	$167,555	37.2	$307,944	48.9
Tobacco	$2,713	0.6	$1,367	0.2
Vegetables	$60,488	13.4	$56,394	8.9
Fruits, nuts, and berries	$12,967	2.9	$19,393	3.1
Greenhouse and nursery	$188,484	41.9	$208,692	33.2
Other	$17,995	4.0	$36,513	5.7
Total receipts	**$1,293,304**		**$1,836,090**	

Source: Data compiled from US Department of Agriculture, *Maryland Agriculture Overview, 2005 and 2007.* Washington, DC: US Department of Agriculture, 2012

meet the needs of the western Maryland milkshed (a milkshed is an area supplied with fresh milk products).

Greenhouse and nursery farming are increasing dramatically, especially around the population centers of Maryland. In 1977, greenhouse and nursery farming accounted for only 3.8 percent of Maryland's total gross sales from all agriculture (crops and livestock), increasing to 11 percent by 2007. Greenhouse and nursery farming accounted for 33 percent of gross crop sales in 2007. In 2002, the five outstanding counties in total receipts from farming were Caroline, Somerset, Wicomico, and Worcester on the Eastern Shore and Frederick on the Piedmont. The largest component of agriculture on the Eastern Shore is poultry, while on the Piedmont it is dairying and other livestock.

Like most other states, especially those in megalopolis, Maryland experienced a shift in employment from rural agricultural jobs to service-oriented jobs usually concentrated in urban and suburban areas. As a result, the total land in agriculture, the number of farms, and the number of farm workers has been in a long decline (tables 5.2 and 5.3).

The farm employment figures displayed in table 5.2 represent mean

Table 5.2. Workers on Maryland farms

Year	*Number employed on farms*	*Percentage of total employment in Maryland*
1970	28,100	1.7
1980	29,600	1.4
1990	22,100	0.8
1995	19,900	0.7
1997	17,400	0.6
2005	17,614	0.6
2010	17,300	0.6

Source: Maryland Department of Agriculture

Table 5.3. Maryland farm size

	1992	*1997*	*2005*	*2007*	*Trend*
Total number of farms	13,037	12,084	12,198	12,834	decrease
Total acres	2,223,476	2,154,875	2,077,630	2,051,756	decrease
Average farm size (acres)	171	178	170	160	decrease
Market value of agricultural products (thousands)	$1,169,331	$1,312,086	$1,293,304	$1,836,090	increase
Market value of crops (thousands)	$388,143	$458,719	$450,202	$629,303	increase
Market value of animal products (thousands)	$781,188	$853,367	$843,102	$1,205,787	increase
Farms of 1 to 999 acres	12,686	11,715	11,808	12,468	stable
Farms of 1,000 acres and larger	351	369	390	390	increase

Source: US Census of Agriculture

employment for the months of April, July, October, and January. These employment figures vary significantly by season, especially for hired labor. Migrant workers come to Maryland, particularly to the Eastern Shore, from as far south as Florida. Most of these migrant workers, the majority of whom are African and Hispanic Americans, come north by bus and stay about five months to pick vegetables. Many live in labor camps owned by contractors who supply workers to local farmers. The camps are usually fenced in and consist of cabins that provide poor living conditions. Contractors deduct from the workers' income fees for the use of cabins, utilities, and gasoline for travel to and from the fields. The state of Maryland provides limited medical care, including health professionals who make visits to the camps, but the effort is limited. Farm laborers often live apart from the local society, almost as aliens, and yet many return year after year.

There also are a number of part-time farmers in Maryland. Part-time farming supplements an income from another job, is an avocation, or is done by retired individuals. Some coal miners in Allegany County own and operate part-time farms, as do watermen in Wicomico County. In southern Maryland, some people grow tobacco on small plots to supplement their income.

From 1970 to 2010, the number of farm workers in Maryland fell from 28,100 to 17,300, a decline from 1.7 percent to only 0.6 percent of the state's total labor force (table 5.2). While the acreage in farmland and the number of farms decreased, the average size of farms decreased only slightly. The numbers of both smaller and larger farms (1,000 acres—404.7 hectares—or more) have remained stable, although there are fewer small farms and more large farms. The trends reflected in tables 5.2 and 5.3 are clear. The number of farms and total farmland in Maryland are decreasing, the overall value of agricultural products in Maryland has increased, and the value of animal products continues to exceed crops by a factor of nearly 2 to 1.

The long-term decline in the number of farms, average size of farms, and acreage in farmland is dramatic. In 1900, over 80 percent of Maryland's land was in agriculture. By 1951, there were over 38,000 farms in Maryland constituting about 3,800,000 acres (1,537,805 hectares), representing about 61 percent of the land. By 2007, those figures had dropped to only 12,834 farms with a total of only 2,051,756 acres (830,316 hectares), representing only 31 percent of the land. Similar trends can be seen throughout many other states.

The American Farm Bureau examined historical data on how much time it took a farmer to produce 100 bushels of corn:

- In 1850, about seventy-five to ninety hours of labor were required to walk a plow, harrow, and plant seed by hand. Yields were about 40 bushels per acre.
- In 1900, labor time is reduced to thirty-five to forty hours using a two-bottom gangplow, disk, peg-tooth harrow, and two-row planter. Yields remain at about 40 bushels per acre.
- In 1950, farmers work ten to fourteen hours using a tractor, three-bottom plow, disk, harrow, four-row planter, and two-row picker. Corn yields about 50 bushels per acre. Farmers make up about 12.2 percent of the population, and one farmer feeds twenty-seven people.
- In 2000, only two and a half hours and 1 acre of land are needed to produce 100 bushels of corn. Farmers use a tractor, five-bottom

plow, 25-foot tandem disk, planter, 25-foot herbicide applicator, 15-foot self-propelled combine, and trucks. Farm population is only about 1.9 percent of the total US population, but one farmer feeds 129 people. (Maryland Conservation Partnership Press 1999)

Today's agricultural system produces more with less labor and in less time, thanks to advances in technology and increased consumption of fossil fuels. The modern system of agriculture in both Maryland and the United States requires a lot of energy. Fossil fuels are used to manufacture chemical fertilizers, pesticides, and herbicides. Energy is necessary to manufacture machinery, and then more is needed to run those machines. At the same time, farmland in Maryland is in decline. Just during the ten years from 1990 to 2000 there was a net loss of 120,000 farmland acres, though a few counties did gain farmland acreage (i.e., Allegany, Carroll, Cecil, Queen Anne's, Talbot, Washington, and Worcester). Yet the total statewide addition of farmland was more than offset by the statewide loss. A particularly strong pattern of declining farmland is evident in the urban web stretching outward from Baltimore and Washington, DC. This region, which accounted for 62 percent of farm acreage lost between 1992 and 1997, includes Anne Arundel, Baltimore, Calvert, Charles, Frederick, Harford, Howard, Montgomery, Prince Georges, and St. Mary's Counties. The growth and rapid development of this urban region are directly related to the loss of valuable farmland. The urban sprawl of Washington southeast toward St. Mary's County and northwest toward Frederick County is particularly noticeable. The loss of farmland in this urban region continues to increase as residential and commercial development consumes land at an alarming rate.

There also has been a noticeable loss of farmland in Garrett County in the west, and in Kent and Queen Anne's Counties in the east. Garrett County continues to grow and develop as more people are discovering the amenities of this mountainous county. Winter skiing, boating, and other outdoor activities have attracted significant development, but that development has come at the cost of farmland. The central and upper parts of the Eastern Shore are also losing farmland to the rapid development of new housing, shopping malls, and other commercial projects. The disappearance of farmland in Queen Anne's County is partly due to the sprawling growth of Dover, the capital of nearby Delaware. Statewide, developed land increased by 40 percent between 1982 and 1997 (Alliance for the Chesapeake Bay 2000).

Crops

In the colonial period and for a long time thereafter, tobacco was the main commercial crop in Maryland. Today, tobacco has all but disappeared, as corn used for grain has come to dominate a diverse cropland. Although the future of agricultural crop production in the suburban counties of the Washington–Baltimore urban corridor is in doubt, in most other areas of Maryland, agriculture is competitive and farming continues to be a viable way of life for many. By value of crops sold, Maryland ranked thirty-sixth among the fifty states in 2011, an impressive number given the relatively small cropland area of the state.

Maryland farms are quite heterogeneous and face different challenges depending on the region in which they reside. The major challenges to the state's croplands are economic and demographic in nature. Some segments of agriculture are at risk as the state's population continues to grow and affluent residents spread out onto its farmlands. Still, agriculture remains a major economic engine in Maryland, and the single biggest factor in the economy of some areas.

Truck Farming

Truck farming, raising small fruits and vegetables in large quantities for nearby large urban markets to be sold fresh or processed and canned, is an important component of Maryland agriculture. The Eastern Shore is Maryland's leading truck-farming area (table 5.4); it is also one of the outstanding truck-farming regions in the nation. In 2002, the Eastern Shore produced about 60 percent of Maryland's commercial vegetables and melons for the fresh market and processing. The four counties of the lower Eastern Shore accounted for 32 percent of that production, while the upper Eastern Shore counties grew 28 percent of these crops.

In addition to having a long growing season and good soils, the Eastern Shore is readily accessible to the large city markets of Baltimore, Dover, Philadelphia, Richmond, Washington, Wilmington, and the rest of megalopolis. West of the Chesapeake Bay, Carroll County was the leading truck-farming county in 2002, followed by Baltimore County. In 1978, this order was reversed, reflecting the greater loss of agricultural land to development in Baltimore County during the 1980s and 1990s. Much of the produce from these counties goes to the metropolitan Baltimore market. Most labor-intensive truck farms are small to medium in size. Table 5.4 shows that the harvested acres in the various regions of the state do not add up to the state total, because in some counties there is little truck farming, and reporting the acreage would disclose information about specific farms (a practice prohibited in the US Census of Agriculture). Harvesting some crops is more labor intensive because these crops must still be picked by hand (e.g., melons, strawberries, and beans). It is to these areas that seasonal laborers travel for the harvest.

Although the Eastern Shore is one of the nation's leading tomato-producing areas, tomato crops have declined over the past twenty-five years as farmers have turned to more profitable crops requiring less labor (Shelsby 1980). Early in the twentieth century, Maryland was the nation's leading producer of tomatoes; today it accounts for only a small

Table 5.4. Acres devoted to growing major commercial vegetables and melons

Crop	*West of Chesapeake Bay*	*Upper Eastern Shore*	*Lower Eastern Shore*	*State total*
Beans (lima)	12	283	13	308
Beans (snap)	2,555	930	1,084	4,569
Cantaloupes	255	171	153	684
Cucumbers	67	1,063	2,089	3,552
Kale	118	18	14	164
Peas (green)	2	1,086	1,059	2,842
Peppers (bell)	175	49	21	246
Peppers (chili, etc.)	52	8	317	378
Pumpkins	1,422	154	160	1,770
Squash	233	33	80	907
Corn (sweet)	2,910	3,944	2,432	10,096
Tomatoes	472	112	216	811
Watermelons	289	217	1,510	2,122
Berries	340	35	19	461
Total	8,902	8,103	9,167	28,910
Percentage	31	28	32	100

Source: Compiled from the 2007 Census of Agriculture, Maryland County Data
Note: The number of acres per region do not add up to state totals because small farms were not disclosed.

share of the market. In the 1930s, Maryland farmers planted an annual average of 51,200 acres (20,736 hectares) of tomatoes for processing, while California farmers planted 53,400 acres (21,627 hectares). During the 1940s and 1950s, Maryland's acreage declined while California's increased. From 1945 to 1955, Maryland farmers planted an annual average of 33,900 acres (13,730 hectares) of tomatoes, compared to 111,300 acres (45,077 hectares) for California. These figures diverge even more from 1959 to 1963, when 9,100 acres (3,636 hectares) were planted in Maryland and 142,500 acres (57,713 hectares) in California. In 2002, Maryland farmers harvested only 811 acres (312 hectares) of tomatoes.

The cannery industry started on the East Coast in the 1930s. Most of the Maryland canneries were small, family-owned businesses. Since 1940, the number of tomato canneries in Maryland has dropped from 450 to fewer than forty. As the industry moved west, the size of canneries became larger, and it became increasingly difficult for Maryland canneries to compete with the huge California operations. West Coast canneries have other advantages in addition to the economies of scale associated with larger operations, including longer growing season, higher yields per acre, and greater mechanization on larger farms.

Picking Strawberries

On the site of what is now Baltimore-Washington International Thurgood Marshall Airport, there were once farms that grew strawberries for customers in the Baltimore area. Strawberries were grown in that region until 1937, when refrigerated trucks and better roads allowed cheaper strawberries from Georgia and the Carolinas to take over the Baltimore market. The heyday for growing strawberries in the region was in the 1930s, when Baltimore writer and storyteller Gilbert Sandler wrote about Baltimore's Polish American strawberry pickers. Sandler became intrigued when strange-looking coins began to turn up in the Annapolis area in antique shops or as jewelry. As it turned out, the coins were "picker checks," a special currency used to pay crop pickers in Maryland at the time. The row bosses, hired to supervise, laid out empty boxes in the rows. When filled with strawberries, the picker was given a "picker check" (a coin made of brass, aluminum, or sometimes cardboard) with an amount written on it. Families could earn about $300 in a season. The strawberries were sent off immediately to the Baltimore market to be sold in grocery stores.

The process of recruiting new strawberry pickers for the next picking season, which ran from May through late June, began at Christmastime, when a "row boss" (a boss of the strawberry crop) visited "Little Poland," the Polish community in Baltimore. Many Polish American families—including children—piled into wagons, taking with them clothes and food to last through the picking season. They left the city for six weeks to live in the fresh open air of the farm. Husbands usually stayed in Baltimore to work, but they often came to the farm on weekends. The men traveled from Baltimore to Anne Arundel County by train, stopping at various stations—Harman, Marley, Pumphrey—named for the farms near which they were located. The *Baltimore Sun*'s poet-columnist Folger McKinsey wrote a poem about the strawberry-picking era:

> Down to Anne Arundel the pickers go
> Women and children, row on row
> Dew in the morning and birds in rhyme
> Chattering, chattering, strawberry time

In late June or early July, the farmers' wagons, loaded with Polish pickers, headed back to southeast Baltimore. After a few days' rest, many of these pickers would travel across Delaware, Maryland, and Pennsylvania to pick other crops.

Pickers lived in wood shanties with gabled roofs about 100 feet (30.5 meters) by 30 feet (9 meters) in size. A support wall ran down the middle, and there were two floors. About fifty people lived on each side of each floor, or about 200 people per shanty. Sheets and blankets were hung to create some privacy for families. There was no indoor plumbing. Pickers awoke at the crack of dawn, around 4:30 a.m. They ate a hearty breakfast and then picked all morning. After a noon lunch break, it was back to picking until around 5:30 p.m., when pickers washed up and ate dinner. Then the pickers relaxed, often singing Polish songs and dancing polkas. For six weeks, farm owners with names such as Clark, Cunningham, Marley, and Smith employed pickers with names like Borowski, Dobrowski, Tyszkiewicza, and Ziemski. It was a symbiotic relationship—the farmers needed pickers for these six weeks, and the Polish immigrants of Baltimore needed work.

Field Crops

Field crops are the leading types of crops grown in Maryland (table 5.5). In the past twenty years, soybean acreage has surpassed crops of corn intended for grain to become the most widely grown field crop. Soybeans have increased in importance in Maryland farming, as they have in the rest of the nation. This versatile crop is used to produce oil, meat substitutes, dairy product substitutes, and plastics, among other things. It is also used as hay, silage, and a cover crop. In 1925, only 6,000 acres (2,430 hectares) of soybeans were raised in Maryland. By 1951, this figure had increased to 77,000 acres (31,185 hectares), and by 2002 there were 465,780 acres (186,312 hectares) of soybeans, making it Maryland's top field crop by acreage. The leading counties in soybean production are clustered on the Eastern Shore (constitut-

Table 5.5. Major field crops

	Crop production			
	West of bay	*East of bay*	*State total*	*Acreage harvested*
Corn for grain (bushels)	12,857,182 (43%)	17,184,714 (57%)	30,041,896	406,841
Barley (bushels)	1,431,462 (49%)	1,476,261 (51%)	2,907,723	36,241
All hay (tons dry)	509,006 (92%)	44,817 (8%)	553,823	227,727
Soybeans (bushels)	3,721,629 (35%)	6,974,244 (65%)	10,695,873	465,780
Tobacco (pounds)	1,739,926 (100%)	0 (0%)	1,739,926	1,162
Wheat (bushels)	3,396,366 (32%)	7,228,767 (68%)	10,625,133	162,062

Source: Compiled from the 2007 Census of Agriculture, Maryland (various tables)
Note: Percentages in parentheses are the share of the state total.

ing 65% of state production): in order, Queen Anne's, Worcester, Caroline, Dorchester, Kent, Talbot, Wicomico, and Somerset Counties. The smaller, yet still significant, production west of Chesapeake Bay is concentrated in a band of counties stretching from Cecil County in the northeast to Montgomery and Frederick Counties in the southwest. Soybeans are commercially produced in all counties west of the bay except in Allegany and Garrett (US Department of Agriculture 2002).

Corn for grain is the second-largest field crop by acreage grown in Maryland, with significant production both east and west of the Chesapeake Bay. Most of the field corn grown on the Eastern Shore is used as feed for chickens and other livestock, while much of that grown west of the bay is used to feed dairy and beef cattle. Although other grains are used in livestock feeds, corn is usually the main ingredient. Grown in Maryland by the indigenous Indians before the European colonists arrived in the early seventeenth century, corn was a good crop for early settlers because it could be grown among the rocks and stumps on rough land. Some corn is raised in all twenty-three counties (53% east of the Chesapeake Bay and 47% west of it), but several are particularly outstanding producers: Queen Anne's and Worcester Counties lead the state, although the entire Eastern Shore is a strong corn-producing region. The leading corn-growing counties west of the bay are in a belt from Cecil in the northeast and westward to Washington County: in order, Carroll, Frederick, Harford, Cecil, Baltimore, Montgomery, and Washington. The rolling Piedmont is physically well suited for growing corn because of the long, hot, and usually wet summers as well as the clay loam soil. A noticeable increase in corn production in Washington County has been seen over the past few decades thanks to the rich soil of the Hagerstown Valley in the Appalachian physiographic province. By 2011, Maryland acreage in corn was increasing because of higher gasoline prices and increased demands for biofuel, as corn is the crop of choice for biofuel production in the United States. Yet corn may not be the best solution. Scientists at the University of Maryland and College Park and Bowie State University have been working together to find an alternative crop that can be used exclusively for biofuel and not food.

Hay is another important field crop in the state; the bulk of the hay crop is grown west of the Chesapeake Bay (92%), where the thriving dairy and equine industries of central Maryland require hay as feed. More of this valuable roughage feed is produced in Frederick County than anywhere else in the state. On the Piedmont, alfalfa, timothy, and clover are important types of hay. On the Eastern Shore and in southern Maryland, soybeans and cowpeas are increasingly being fed as hay.

Winter wheat was not grown extensively in Maryland before the American Revolution. Soon after the war, the three Ellicott brothers from Pennsylvania established a mill on the fall line at what is today Ellicott City (Blood 1961). Soon much of the local land, exhausted from tobacco production, was planted in wheat. With the help of fertilizer, these Piedmont soils began to produce good wheat crops. At one time wheat was such an important crop that the tax levy was valued in bushels of wheat, and wheat exports from this area helped establish Baltimore as a leading grain-handling port. Winter wheat is sown in the early fall so that the plants are well established before the winter freeze. In early spring, growth resumes, and farmers harvest the wheat in July. Wheat growing is not labor intensive; rather, this labor-extensive type of agriculture relies heavily on the use of machinery such as combines. Although acreage planted in wheat dropped in Maryland, yields increased into the 1990s (except for drought years such as 1999) because of improved wheat varieties, fertilizers, and farming methods. As the Midwest was settled and extensive wheat fields were planted there, Maryland wheat farmers found it increasingly difficult to compete despite their increased yields and closer proximity to large East Coast markets. By 2002, the leading county in wheat production was Queen Anne's (18% of the state's crop), followed closely by Caroline County (10% of the state's crop). Both counties are sited on the Eastern Shore, which overall produced 68 percent of the Maryland crop, representing a significant geographical change in Maryland's wheat production. Twenty-four years earlier, the production figures for wheat crops were 53 percent west of the bay and 47 percent east of it. The leading counties in wheat production until the 1990s were always west of the bay, especially the Piedmont counties of Frederick and Carroll. Today the Eastern Shore dominates in wheat production.

Barley is a popular grain feed that can be substituted in part for corn, and it requires less labor to produce the same amount of feed in value. Barley is grown both east and west of the Chesapeake Bay. Production is slightly higher on the Eastern Shore, which produced 51 percent of the barley crop in 2002. After World War II, the production of oats in Maryland declined drastically as the large-scale replacement of horses and mules by machines reduced demand. Barley has replaced much of the acreage formerly in oats. In 2002, only 3,700 acres (1,480 hectares) were planted in oats; that figure had been as high as 23,000 acres (9,315 hectares) in 1978. Oats grow better in a climate cooler than most of Maryland.

Tobacco

Tobacco deserves special attention because of its historic importance to Maryland. Since colonial days, a distinct way of life or culture has developed in Maryland's tobacco region. Grown in the state for over 300 years, tobacco is still grown today in areas originally planted in colonial times, and Maryland farmers until recent times utilized many of the same growing, curing, and marketing processes as previous generations. Other regions today dwarf Maryland's contribution to the tobacco industry, but the state still produces a distinct type of to-

WESTERN MARYLAND

Because most of this book's discussion of agriculture in Maryland focuses on the Coastal Plain and Piedmont, you might think that there is little beyond Eastern Shore chicken broilers and Piedmont dairying. But in western Maryland resides a different pattern of agriculture. In the highland areas of western Maryland, agricultural production drops off dramatically. The rough mountainous terrain and shorter growing season make it difficult to grow many crops. Freezes occur later in the spring, winters are colder, and summers are cooler, too. Allegany County ranks lowest in the state in most every agricultural category. Yet some livestock and feeds are raised in these areas, and in Garrett and Allegany Counties, maple syrup is an important local forest product.

By far the most important crops of western Maryland are fruits. Fruit production in western Maryland is centered in the Hagerstown Valley in Washington County (part of the Great Appalachian Valley), a rich limestone-soil valley that is one of the finest apple-producing regions of the United States. West of the Hagerstown Valley, extensive orchards grow on the hillsides, especially around Hancock. Commercial orchards are generally sited on sloping ground, which allows for two types of drainage: water and air. Fruit trees need good under drainage; they do not do well on sites with standing water, a high water table, or poorly drained, heavy clay soils. Air drainage is equally important, especially in the fall and spring (Blood 1961). During the day, solar insolation heats the Earth, which in turn reradiates heat to the atmosphere. At night, the thin air containing less moisture on the hilltops loses heat faster than air in the valleys. The heavy cold air flowing down the slopes collects in the swales and valley bottoms, pushing the warmer valley air upward. The rising warmer air spreads to the valley slopes, creating a thermal belt that farmers take advantage of by planting fruit trees in these areas. While frosts may occur on the colder valley floor, they do not often affect the relatively warmed slopes protected by air drainage and temperature inversion. Early budding trees on slopes have some protection, allowing the fruit to set. Mechanical air circulators that look like large propellers mounted on high towers above the trees assist with air circulation in some orchards.

In 2005, Maryland orchardists grew 15,000 tons (13,608 tonnes) of apples and 4,200 tons (3,810 tonnes) of peaches. The apple and peach crops of Maryland have declined considerably in recent decades, however. In 1978, Maryland orchardists grew 42,500 tons (38,548 tonnes) of apples and 12,000 tons (10,884 tonnes) of peaches, representing a decline of 65 percent in both apple production and peach production between 1978 and 2005. From 1997 to 2002, the market value of agricultural products sold from the three counties of western Maryland dropped by 0.3 percent overall. Washington County experienced a small gain of 1 percent, Allegany County had a large decline, of -28.4 percent, and Garrett County declined slightly, -0.1 percent.

Allegany County's modest agricultural sector produces various crops, including corn, wheat, oats, forage (hays), a small vegetable crop, apples, cherries, grapes, and peaches. Other items are produced in small amounts.

Garrett County has been developing and growing, bringing some changes in agriculture. The growing population in the Deep Creek Lake area has supported growing sales in nursery and greenhouse products, although other crops—most significantly corn, barley, wheat, oats, rye, forages, apples, and tame blueberries—are still produced. Of special note is the production of maple syrup from tapping the plentiful maple trees of this cooler highland region of Maryland. Agriculture held nearly steady in Garrett County between the 1997 and 2002, according to the US Censuses of Agriculture. This county is also a source for garden peat from bog lands.

Although it ranks thirteenth in livestock product sales overall, Garrett County ranks fourth in cattle and calves, and sixth in sheep and goat products. The high rankings in hay forage crops, corn for silage, and oats relate to the county's livestock activities. The county ranks second in cut Christmas trees and woody crops.

Washington County has the largest agricultural sector in western Maryland, with its major crops including corn, wheat, barley, sorghum, oats, and rye. It is the only county in the western part of the state with significant soybean acreage. Forage crops, mainly hays and corn for silage, are grown to feed hogs, cattle, and dairy herds. The county ranks seventh in the state for overall livestock sales, and second among the counties in value of cattle and calves as well as milk and dairy products. Washington County's modest vegetable production is intended mostly for local markets, but its fruit crops (mainly apples, cherries, grapes, nectarines, peaches, blackberries, tame blueberries, raspberries, and some strawberries) rank first among Maryland counties in value. A growing segment of agriculture in Washington County is the nursery and greenhouse business.

bacco leaf not found in many other states. The unique characteristics of Maryland tobacco make it highly desirable among cigarette manufacturers in the United States and Western Europe.

As early as 1637, the Maryland economy was based on tobacco. During the colonial period, the British purposely limited the amount of currency in America, and tobacco became so important that it became the prime medium of exchange for paying debts, taxes, and salaries. By 1750, tobacco production had reached 13,500 tons (12,247 tonnes) and accounted for 75 percent of the value of all Maryland exports (Baker 1957). No large tobacco settlements developed in the tobacco country of southern Maryland. Rather than building extensive roads, plantation owners depended on water transportation. Ships came up the rivers to the doorsteps of plantations, where hogsheads (large barrels holding 600 to 700 pounds) of tobacco were loaded for export and imported goods from England were unloaded. Because the plantations were well connected by water, overland roads were few and poor. The lack of central settlements in southern Maryland is still evident today.

The Maryland tobacco region encompasses five contiguous counties in southern Maryland: Anne Arundel, Calvert, Charles, Prince George's, and St. Mary's. All of Maryland's remaining commercial tobacco is grown in these counties, which have experienced a remarkable locational geographical persistence for this crop. Because they are closer to large urban centers, Anne Arundel and Prince George's Counties urbanized earlier than the others. But the effects of urban sprawl and the resultant loss of agricultural land are being experienced throughout southern Maryland today. The climate, topography, and soils of southern Maryland are ideal for growing tobacco. The mild climate's annual average is 56°F (13°C), with precipitation normally averaging around 40 inches (102 centimeters) per year and the growing season lasting approximately 189 days. Lying between the Chesapeake Bay and the Potomac River, southern Maryland with its humid and warm climate is ideal for growing tobacco. Tobacco is grown on sandy loam soils of the Sassafras series, which have loose surfaces but do not dry out or cake (Coddington and Derr 1939). Even so, today tobacco is a rapidly dwindling cash crop for a declining number of southern Marylanders.

The Patuxent River basin occupies a central position in southern Maryland between the Chesapeake Bay to the east and the Potomac River to the south and west. Along the middle course of the Patuxent, between Benedict and Queen Anne's Counties, lies the single greatest concentration of rich tobacco-producing land in Maryland. From 1978 to 2004, the total tobacco acreage in Maryland declined from 23,000 acres (9,200 hectares) to 1,100 acres (440 hectares). This remarkable 95 percent decline in tobacco acreage was a result of powerful pressures on Maryland tobacco growers. An interesting characteristic of tobacco growing in southern in Maryland is that many people grow it as a side business. It is not unusual for an auto mechanic and his teacher wife to raise small 8- to 10-acre (3.2- to 4-hectare) plots of tobacco to supplement their income. Tobacco is both a side crop and a traditional way of life for many southern Marylanders.

Tobacco quickly wears out the soil. As earlier farmers abandoned old fields for new land, the old land grew over with weeds and trees. Much of this wasted land has given way to commercial and residential development, but some still remains unused today. In recent years some of this land has been reclaimed, especially by Amish people who moved into southern Maryland from Pennsylvania. Eight Amish families led the way when they moved into southern Maryland from Lancaster County, Pennsylvania, in 1940. More Amish families followed, settling on the swampy area southeast of Washington, DC, that they soon turned into productive tobacco acreage. Amish-grown tobacco typically commands a higher price. The Amish tend be more careful in stripping the leaves and separating the grades into similar lengths and colors. They also do not use heavy doses of chemical fertilizers; instead they use horse manure. Overall, it costs the Amish less to produce good-quality tobacco.

Much of the work on Maryland tobacco fields is still done by hand. Machines are used mainly to prepare the fields and to set out the plants. In October, seedbeds of fertile soil are prepared; the following March, the seedbeds are sown and packed firmly. White muslin cloth or plastic covers are draped over the delicate plants to keep them warm and to protect them from insects. In May, the plants are transferred to well-drained, light, sandy loam fields. While maturing, sprouts on the sides of the plants, called suckers, are removed to preserve the quality of the large leaves. After sixteen to twenty leaves have grown, the top of the plant is broken off below the seed head to allow the lower leaves to develop (a process called *topping*).

After topping, the plants grow another two to four weeks before being harvested in August. At harvest time, the stalks are cut close to the ground and left in the field to wilt. Once they are easy to handle, each stalk is pierced and strung on a tobacco stick; the leaves are then removed to a curing barn. Tobacco-curing barns are a common—but dwindling—sight in southern Maryland. These tall and usually unpainted barns have hinged side boards to allow for air flow during the four- to six-week process of air curing, which takes place in late winter or early spring and without the use of artificial heat. The dry, brittle tobacco softens up with the spring humidity, called "down weather" by the locals. The hanging leaves are then taken down from the rafters and stripped from the stalks, separated by quality and bunched into "hands" of twelve to fifteen leaves each. Each hand is placed into a large pile called a burden. This labor-intensive process of separating leaves requires great skill. Although a machine for stripping was once developed, it was not successful because it shook the leaves, damaging them.

Tobacco is a relatively tolerant crop in terms of temperature requirements. Some varieties prefer warm, humid conditions such as in the Carolinas, while others thrive in more vigorous climates such as Connecticut. In Maryland, high humidity produces a mild tobacco with thin leaves, low nicotine, and a steady burning quality (Alexander and Gibson 1979). There are six classes of tobacco: flue cured, fire cured, air cured, cigar filler, cigar binder, and cigar wrapper. Classes are further broken down into types based on quality and color. Maryland tobacco is Class 3 light air-cured Types 31 and 32. Type 31, called burley, is the type most widely grown in the United States, accounting for 25 percent of production. Very little burley was traditionally raised in Maryland; most of the state's crop is Type 32, often called "Southern Maryland." Type 32 is unlike any other tobacco, and it is difficult to duplicate because of the particular soil and climatic conditions necessary to grow it. Air-cured Type 32 is reddish brown, thin, and very dry. Unlike other tobacco, Type 32 continues to burn evenly after lighting until totally consumed. Because it also has no distinctive odor, cigarette manufacturers prefer Type 32. Europeans especially like cigarettes with high quantities of Type 32 in the blend; Europe has historically been a major market for Maryland tobacco.

Hogshead Markets

For over 200 years, Maryland tobacco farmers sold their crops exclusively at hogshead markets, named for the vessel in which tobacco was stored and subsequently sold. Tobacco was packed tightly into hogsheads—casks approximately 44 inches (112 centimeters) in diameter and 54 inches (137 centimeters) long using a mechanized screw. At Gieske and Niemann, a Marlboro packing plant (named for the town, not the cigarette brand) in business since 1858, coopers built as many as 300 hogshead barrels a day. Many tobacco farmers felt that selling their tobacco in hogsheads afforded them some advantages. The tobacco was not vulnerable in a barrel. If farmers were dissatisfied with the auction price, they could store their tobacco for up to four years in hogsheads at the warehouse. But loose-leaf markets eventually replaced hogshead markets; the last hogshead market in the United States was the Cheltenham market in southern Maryland. After World War I, high-quality Maryland tobacco became popular in Belgium, France, Germany, Luxembourg, and Switzerland. In the early 1930s, the American Tobacco Company and the French government were the largest purchasers of Maryland tobacco. But by 1933, the company had dropped its purchases dramatically, and the French government, which operated a tobacco monopoly, bought none. Purchasers objected to the method of buying only from hogshead samples, as they could not inspect the tobacco throughout the casks. A Maryland visitor to a French cigarette factory was shown a yard littered with rocks, rusty plowshares, and other trash found in Maryland tobacco. After World War II, however, domestic and European demand for Maryland tobacco increased once again. Starting in 1950, the Europeans heavily subsidized the Maryland Tobacco Improvement Foundation at the University of Maryland's experimental tobacco farm near Upper Marlboro. In 2000, about one-third of the dwindling Maryland crop still went overseas (Meyer 2003).

By the late 1970s, the hogshead coopers were gone, and tobacco was being packed in bundles by a "dump packer" machine that could handle up to 250,000 pounds (113,400 kilograms) of tobacco in a single day. But even though the hogshead markets were history by 2000, Maryland laws regulating the handling of hogsheads remained on the books. The carefully prescribed procedure was performed almost as a ritual. First, the full cask was weighed. Next, three state tobacco inspectors removed the barrel, except for the bottom, leaving a compressed cylinder of tobacco. An inspector then poked a breaker bar into the tobacco at five different locations. At each location, an official of the Maryland Tobacco Growers' Association extracted some leaves. One bunch of leaves per cask was marked and sealed for later inspection by prospective buyers. The barrel was weighed to determine the net weight of the tobacco. Finally, the tobacco was placed back into the hogshead cask (Meyer 2003).

Loose-Leaf Markets

Loose-leaf markets have operated in Maryland since 1938. Starting in that year, the tobacco marketing system was revolutionized when Frank M. Hall and R. L. Hall built a large shed in Upper Marlboro and asked tobacco growers to bring their loose tobacco to sell. Buyers could now see the exact quality of all the tobacco leaves (Dozer 1976). Bundles of tobacco were placed in baskets on the auction floor, and state inspectors graded each basket. The baskets were lined up in rows so that people could inspect both sides of each basket. In one line, an auctioneer called out the prices bid in mesmerizing, rhythmic chants. Behind the auctioneer walked an assistant who recorded the highest bid. Both domestic and foreign buyers shuffled along the rows to procure Maryland tobacco for their companies. These tobacco auctions—held at Hughesville, LaPlata, Upper Marlboro, and Waldorf—were indeed auctions in motion.

When the 2005 Maryland Type 32 tobacco crop came up for auction in March 2006, it was the smallest crop since records began in 1866. The only auction warehouse remaining at the time was in Hughesville, Charles County; this warehouse had operated every year since being built in 1938. Only about fifty farmers and three buyers showed up to deal for the 300,000 pounds (136,080 kilograms) of tobacco on the floor. Some predicted it would be the last Maryland tobacco auction. To give some perspective, in the 1950s Maryland warehouses auctioned about 40 million pounds a year. In 2007, only 843,000 pounds were sold. When March 2007 rolled around, there was no tobacco auction held in Hughesville. The old tobacco warehouse is now the site of a flea market.

The Future of Tobacco Growing in Maryland

Farming of all types is in retreat in southern Maryland, but there are a number of serious threats to the tobacco industry and culture of southern Maryland. The encroachment of the Baltimore-Washington urban corridor continues to consume more tobacco land for commercial and residential use each year. Attractive jobs in nearby urban areas lure labor away from the farms. Starting in the early 1980s, Type 32 tobacco was being raised in the flue-curing areas of Virginia and the Carolinas, where farmers harvested nearly as much as Maryland farmers. Owing to these pressures, the number of Maryland tobacco growers and the acreage in tobacco declined significantly.

One major problem for tobacco growers is finding the laborers needed for the grueling harvesting, curing, and stripping of tobacco (Wheeler 2000). The booming economy of the 1990s with its low unemployment made it difficult for tobacco growers to compete for workers in this labor-intensive form of agriculture. While tobacco still remains a strong tradition in the minds of many southern Marylanders, few young people opt to become growers.

A much more serious threat to the Maryland tobacco industry has come from the medical and legal issues associated with the effects of tobacco consumption on health. In November 1998, forty-six states, the District of Columbia, and five territories agreed to settle pending litigation with tobacco manufacturers. According to this landmark settlement, payments amounting to about $198.5 billion will be made to states between 1998 and 2025. The remaining four states (Florida, Minnesota, Mississippi, and Texas) individually settled their lawsuits. Medicaid claims were the focus of the litigation initiated by the states; a substantial portion of the payments will be used to reimburse states for Medicaid costs generated by illnesses associated with tobacco consumption (Mann, Scheider, and Thom 1999). Some of the more than $4 billion that the state of Maryland is scheduled to collect from the tobacco settlement was designated to fund a buyout program. Beginning in 2000, the state offered to pay tobacco farmers to switch to other crops. The dramatic effect of the buyout program can be seen in figure 5.1. Tobacco acreage and total sales dropped off precipitously. By 2007, only 423 acres were planted in tobacco on only seventy Maryland farms, harvesting about 843,000 pounds (over 1.7 million pounds had been harvested as recently as 2002).

Pressured by farm labor shortages, legal and political assaults on smoking, competition from other areas, and declining prices for tobacco, many Maryland tobacco growers saw the buyout as a way to switch gracefully to other endeavors. Because fewer younger people are going into tobacco growing, the average age of tobacco growers is now sixty-two. The buyout plan appeals to those nearing retirement. By late 2000, Maryland officials were surprised by the overwhelmingly positive response from tobacco growers to the buyout. By the spring of 2006, about 90 percent of eligible farmers—854 in all—had opted for the buyout. According to the Maryland Tobacco Authority, there

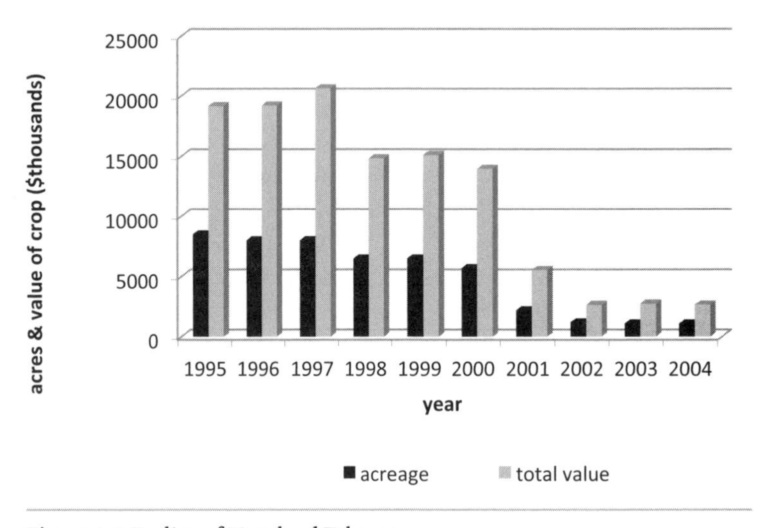

Figure 5.1 Decline of Maryland Tobacco

were only 150 remaining tobacco farmers by the middle of 2006; about 90 percent of those were Amish or Mennonite farmers. Both of these groups have traditionally been wary of government programs.

To receive buyout money, tobacco farmers must sign a ten-year contract promising to keep their land in agriculture other than tobacco; such promises are an attempt to fend off pressures from developers. Growers were paid $1 per pound of tobacco they produced in 1998, and the payments are set to continue for ten years. The overall objectives of the Maryland buyout program are to end tobacco growing in Maryland, to move farmers to other crops, to preserve open space, and to maintain farming as a way of life (Wheeler 2000). But if farmers commit to the buyout program, they must continue farming for another ten years. They therefore need solid alternatives. Farmers are switching to growing corn, soybeans, Christmas trees, cut flowers, nursery plants, and herbs and are raising cattle, hogs, poultry, and even llamas as alternatives. Interestingly, more and more farmers are growing grapes in southern Maryland for the state's developing wine industry. Grape vines love sandy loam soil, and southern Maryland has just that. Maryland's fifty-four wineries (as of mid-2014) regularly experience a shortage of grapes and often must buy them out of state. A stronger Maryland grape crop could help combat these shortages and build the local economy. But, like tobacco, growing grapes is labor intensive and might not appeal to older farmers. Those who choose to grow grapes still face the problem of finding the needed labor, especially at harvest time.

The beginning of the twenty-first century is a time of great uncertainty for tobacco growers in Maryland. Long-term trends and recent events all point to a further decline in tobacco growing in the state. But the story of tobacco farming in southern Maryland is more than just its demise. It is also the story of an economic and cultural transformation of a landscape. The over 370-year history of tobacco growing in Maryland is drawing to a close. Southern Maryland is adjusting to its new place in the Maryland economy.

Livestock

Livestock activities in Maryland accounted for 66 percent of total receipts from agriculture in 2007, as compared to 34 percent for all crops (table 5.1). The two largest components of livestock production in the state are broilers (49% of all receipts from agriculture) and dairy products (11%). Recently there has been significant growth in the broiler industry, which accounted for only 28.1 percent of all agricultural receipts in 1977. Meanwhile, dairy receipts have dropped well over one half from 24.6 percent in 1977. The information shown in figure 5.2 relates to the livestock side of Maryland agriculture only.

MARYLAND'S WINE COUNTRY

Although not a major agricultural sector, Maryland's developing wine industry offers a surprising variety of good-quality wines and contributes to the culture and economy of the state. Nationally, wine consumption and sales have grown tremendously over the past thirty years. In 1975, about 600 wineries existed in thirty-four states. By 2013, there were over 8,000 wineries in all fifty states. Grape vineyards now grow in twenty-one Maryland counties. The greatest concentration is on the Piedmont in Baltimore, Carroll, Frederick, and Harford Counties. A second concentration is found on the middle Eastern Shore (Maryland Grape Growers Association 2013).

Regionally, the Maryland wine industry is smaller than that of Virginia and Pennsylvania, but it is growing rapidly. In 1979, Maryland and Virginia had the same number of wineries. By 2006, Virginia had ninety-four wineries to Maryland's twenty-six. Virginia does not have a physical geographical advantage over Maryland, but its greater growth is due in large part to Virginia's state laws, regulations, and state support for the industry. Virginia has a full-time enologist (winemaking scientist) at Virginia Polytechnic Institute and State University and a full-time viticulturist (expert on cultivation of grapes) at Virginia Tech's Agricultural Experiment Station. The Maryland wine industry likewise needs the support of both a full-time viticulturist and plant pathologist, perhaps located at the University of Maryland, to provide grape growers with on-the-ground consultation, diagnosis, education, and research. Despite the lack of such expert support, the Maryland wine industry continues to grow, reaching fifty-four wineries in 2014.

Growing Grapes and Making Wine

Of the fifty-four operating wineries in the state, the heaviest concentration is on the Piedmont of central Maryland, forming an arc from Montgomery and Frederick Counties in the west to Harford County in the northeast. Cabernet Franc, Cabernet Sauvignon, Chardonnay, Merlot, and Pinot Gris constitute the major grape varieties grown and used on the Piedmont. Southern Maryland (from Anne Arundel County south) is one of the newest wine regions of the state, with ten wineries in the region. This part of the Coastal Plain gets hot and stays hot for much of the summer. Southern Italian and other heat-loving Mediterranean varieties of grapes, such as Sangiovese, grow well here. There are ten wineries on the Coastal Plain of the Eastern Shore; vineyards in that region are generally planted in Chardonnay, Pinot Noir, Sauvignon Blanc, and Vidal. Far western Maryland (Garrett County) has only one winery. The grapes grown there must be able to withstand the colder and longer winters and shorter growing season: Cabernet Franc, DeChaunac, Foch, Norton/Cynthiana, Seyval, and Vidal.

Overall, Maryland's midlatitude climate is mild. The Chesapeake Bay also moderates temperatures by keeping the summer and winter temperatures from reaching extremes. The growing season averages 195 days, except in western Maryland, where it is shorter; this is long enough to grow many varieties of grapes. Although Maryland averages 42 inches (106.7 centimeters) of rain annually, it tends to fall mainly in the spring and fall. The often hot and drier summers are good for grape growth and the buildup of sugar content in the fruit necessary for good vintages. In 2005, grape growers and winemakers in Maryland contributed about $8 million to the Maryland economy, and production has been steadily increasing. Between 1995 and 2000, Maryland wineries increased production by 183 percent.

Viticulture is another vital segment of the Maryland wine industry. In 2010, the Maryland Grape Growers Association reported that there were about 150 vineyards across the state growing 225 to 300 acres (91.1 to 121.4 hectares) of wine grapes. The stock of vines in 2010 numbered about 525,000 vines of sixty-seven different varieties. The top four (Cabernet Franc, Cabernet Sauvignon, Chardonnay, and Merlot) account for 47 percent of Maryland vines.

Some Tough Challenges

Maryland vineyards do not currently grow enough wine grapes to meet the demands of the state's wineries. Most Maryland wineries have their own vineyards, but there is still an excess of demand for quality wine grapes, forcing some Maryland wineries to buy grapes from out-of-state vineyards. The major barriers to increasing vineyard acreage and production in Maryland are economics and lack of accurate information about disease and access to disease-free vines, geography, and incentives. The estimated cost of planting a vineyard is $9,000 to $15,000 per acre (Maryland Wine and Grape Advisory Committee 2005). Like orchardists, grape growers must wait several years before plants are fully productive.

Maryland's growers, as well as prospective new investors, need sound economic, biologic, and geographic information on which to base their decisions. Presently, quality data of those kinds are not easily available. Economic data on the costs to establish a vineyard at a specific site is needed, including the costs of land, taxes, machinery, trellis material, vines, labor, and more. Market prices and trends in supply and demand are also critical pieces of information. Biologic information is needed about grape vines and diseases, as well as where to acquire disease-free vines. Diseased vines have posed serious challenges for some Maryland wine grape growers, as they can lead to major losses of both time and money. Removing diseased vines and preparing the ground for new plants can cost up to $1,000 per acre (Maryland Wine and Grape Advisory Committee 2005). California and Washington are two states with "clean vine" laboratories that screen for and eliminate diseases in plant tissue (see the National Clonal Germplasm Repository; http://www.ars.usda.gov/Main/site_main.htm?modecode=53-06-20-00). They then grow clean vines to sell to nurseries. No such facility exists in the eastern United States. Viticulturists from agricultural cooperative extension services in Maryland, Pennsylvania, and Virginia are currently negotiating to coordinate efforts to create a laboratory facility. Maryland is in a good position to lead this effort because the Maryland Department of Agriculture already operates a similar laboratory to create clean strawberries and brambles.

Maryland grape growers also need more detailed geographic information. The general environmental conditions of Maryland make much of the state suitable for growing quality wine grapes, but growers need much more specific information about the suitability of a piece of land, including elevation, microclimate, slope, orientation of the slope relative to the sun, soil quality, and weather. Virginia has developed site-suitability maps for most of its counties. A pilot project conducted by the University of Maryland Cooperative Extension Service with mapping assistance from the Maryland Department of Planning has produced a draft suitability map for Washington County based on needed geographic criteria. After scoring each variable, scores are standardized, and then the composite scores are used to create classes of suitability for viticulture that are then mapped.

The wine industry is becoming more and more technical. Maryland's grape growers and wineries face various economic market forces, including demand and supply, changing cultural preferences, biologic issues related to suitable grape varieties and disease, and geographical-environmental limits. The economic-geographic landscape is the playing field on which grape growers and wineries must compete. A few bad decisions and the game can be over, ending in a failed business and bankruptcy. Quality information is the key to success. Growers and producers also need the expert assistance and advice on how to best use that information.

An important legal issue concerning the sale of wine by Maryland wineries that had previously been a hindrance to the state's wine industry was settled in May 2011. As of July 1, 2011, consumers may order wine for delivery to their doorstep from wineries both within Maryland and throughout the United States that hold a new direct shipper's permit. Maryland has some of the tightest alcohol controls in the country, including a three-tier system requiring a separation of wholesalers, distributors, and retailers. The new wine law reflects a new attitude toward regulating wine sales and is expected to boost sales and have an overall positive effect on the growth of the wine industry in Maryland.

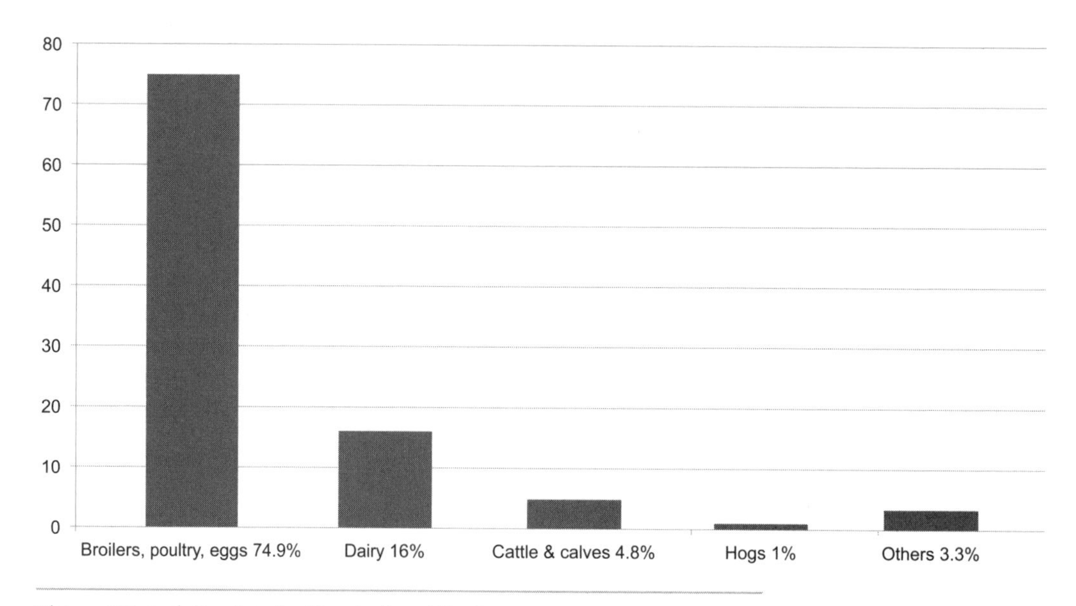

Figure 5.2 Cash Receipts for Livestock and Products

Poultry

If the blue crab is an icon of the Chesapeake Bay, then the broiler chicken—a chicken raised for meat rather than for eggs—is the premier icon of the lower Eastern Shore. In this part of the state, the broiler industry reigns supreme, dominating agriculture and the economy. Prior to the development of the broiler industry, chicken meat, usually eaten only on special occasions, was a byproduct of egg production. Today, Americans eat about 80 pounds (36.3 kilograms) of chicken meat per person annually (AgrAbility Project 2005). In the early 1900s, millions of small, backyard flocks of chickens existed across the United States. Today there are less than fifty specialized, vertically integrated broiler firms. The large companies, called "integrators," have hatcheries, feed mills, grain elevators, processing plants, and distribution networks. Contract growers (often family farms) sign a contract with integrators to raise chickens. These contract growers agree to follow the integrator's specific standards, and the integrator delivers the chicks from the company hatcheries to the growers. The grower's job is to water and feed the chicks, and to maintain the environment of the chicken house (humidity, temperature, air quality, etc.). About fifty days later, the integrator returns to pick up the adult chickens. Then it's off to the processing plant.

The US broiler industry began in the 1920s on the Delmarva Peninsula in Delaware, when the region was economically depressed and desperately searching for a cash crop. By 1935, Maryland produced two million commercial broilers, and two of every three broilers produced in the United States were from Delmarva. As broiler production grew, it stimulated growth in hatcheries, feed companies, and processing plants. By 1951, Maryland production reached 58.4 million broilers, and by 2011 it reached 300 million (fig. 5.3). In 2005, Maryland ranked seventh among the states in the number of broilers produced, and eighth in value of production and pounds of broilers produced. Americans are eating twice as much chicken as they did in the 1960s, and Maryland growers have no trouble finding markets. The more than $1 billion poultry industry on Maryland's Eastern Shore is a massive operation dominating its agricultural landscape.

The Maryland broiler industry is concentrated on the lower Eastern Shore in Dorchester, Somerset, Wicomico, and Worcester Counties. There is also significant production in the middle and upper Eastern

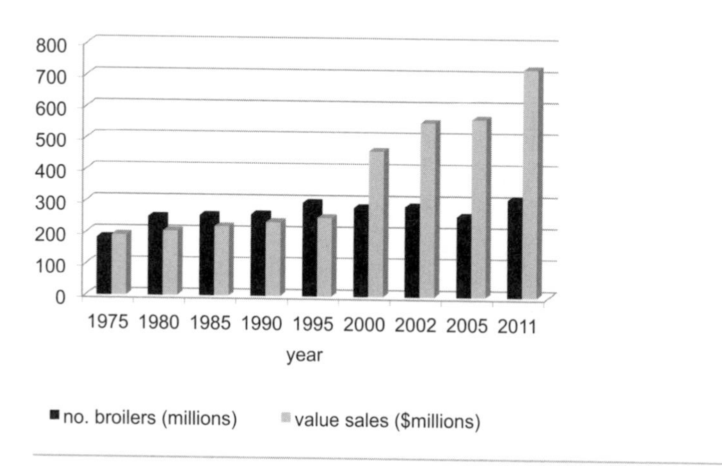

Figure 5.3 Maryland Broiler Production and Sales, 1970–2005

Shore in Caroline, Kent, Queen Anne's, and Talbot Counties. Few broilers are raised on the Piedmont. Production on the entire Eastern Shore accounts for 99.8 percent of all Maryland production of broilers. The Virginia and Delaware parts of the Delmarva Peninsula also produce a significant number of broilers, making it one of the leading chicken-producing regions of the country. Because chickens cannot tolerate extreme heat, the mild climate of the Eastern Shore makes it a good place to raise them. Even so, climate control systems are still needed in these huge, modern chicken houses. The location of the Eastern Shore relative to the large markets of megalopolis is ideal.

Up to the early 1950s, most chickens were sold live in big cities, but today there are large-scale dressing plants on the Eastern Shore. About 300 million broilers were sold in 2011 with a gross sales amount of $700 million, compared to $426 million in 1986. The many components of the industry include hatcheries, chicken farms, corn feed and other grain production, dressing plants, crate and package production, and refrigerated truck transportation. The broiler industry of the Eastern Shore creates many jobs other than direct farming. Seventeen poultry slaughtering-dressing-processing plants in Maryland employing 3,199 people existed in 2008. Of this total, 3,181 were employed on the Eastern Shore, and only eighteen in the Piedmont region (table 5.6).

On the Eastern Shore over the past fifty years, the trend has been to consolidate poultry firms. As recently as 1966, there were still twenty firms. In 2000 there were twelve, and by 2008 there were only eight. Today, the company with the largest number of employees is Perdue Farms, which owns several plants in Salisbury, one in Berlin, and one in Church Hill. The next-largest company is Allen Family Foods. Perdue is currently the dominant chicken marketer in the region as well as in major markets outside the region, including New York City and Boston. Several Maryland poultry companies are also exporters, including Allen Family Foods, Chestertown Foods, and Perdue Farms. Another trend in recent decades has been to increase automation in poultry plants. In 1979 there were 7,617 employees in sixteen plants. In 2000 there were 4,199 employees in the same number of plants, and in 2008 there were 3,199 employees in just seventeen plants. While white American workers dominated the broiler industry at the beginning, they were followed by an influx of African Americans. Today, Latino immigrant workers make up as much as 40 percent of the workforce in some plants (Nalewajko 2004). Some of these Maryland workers are undocumented, and the US Immigration and Naturalization Service conducts periodical raids to identify them.

The Downside of Poultry: Runoff and Pollution

The more than 6,000 chicken houses that raise more than 600 million birds a year also turn out more than 750,000 tons (680,385 tonnes) of manure, which is raked by tractors into the soil as fertilizer. The effect

Table 5.6. Maryland poultry plants

Poultry dressing plant	*Number employed*	*Location*	*Region*
*Allen Family Foods	500	Cordova, Talbot County	lower Eastern Shore
*Allen Family Foods	600	Hurlock, Dorchester County	lower Eastern Shore
Campbell Soup	355	Chestertown, Kent County	upper Eastern Shore
Case Foods	2	Salisbury, Wicomico County	lower Eastern Shore
Chestertown Foods	200	Chestertown, Kent County	upper Eastern Shore
Coast to Coast Poultry	3	Ridgely, Caroline County	upper Eastern Shore
Ise America	10	Warwick, Cecil County	upper Eastern Shore
Paradise Farms	2	Willards, Wicomico County	lower Eastern Shore
*Perdue Farms	400	Salisbury, Wicomico County	lower Eastern Shore
*Perdue Farms	188	Salisbury, Wicomico County	lower Eastern Shore
*Perdue Farms	7	Berlin, Worcester County	lower Eastern Shore
*Perdue Farms	4	Salisbury, Wicomico County	lower Eastern Shore
*Perdue Farms	300	Salisbury, Wicomico County	lower Eastern Shore
*Perdue Farms	5	Church Hill, Queen Anne's County	upper Eastern Shore
*Perdue Farms	5	Salisbury, Wicomico County	lower Eastern Shore
*Perdue Farms	600	Salisbury, Wicomico County	lower Eastern Shore
Rocko Meats	18	Thurmont, Frederick County	Piedmont

Source: Compiled from Maryland Department of Business and Economic Development, *Maryland Manufacturerst Directory*. Twinsburg, OH: Harris Info Services, 2008. www.HarrisInfo.com

Note: Each plant is listed separately; for example, Perdue has four plants in Salisbury, one in Berlin, and one in Church Hill. There were a total of seventeen plants employing 3,199 people.

*Companies that export.

of all this waste is potent as it runs off fields during rains and settles in the groundwater, rivers, and eventually the Chesapeake Bay. The poultry industry is the number one source of pollution for sections of the Chesapeake Bay and for coastal bays of Delaware, Maryland, and Virginia (Goodman 1999).

Poultry manure has been spread as fertilizer on the Eastern Shore for years, but the sheer volume of waste is far too great for crops to absorb. As acreage in farmland shrinks owing to more development and sprawl from Baltimore, Dover, Washington, and even Philadelphia, the result is a higher concentration of chickens raised on less land. The US Environmental Protection Agency's Chesapeake Bay Program has identified poultry manure as the largest source of excess nitrogen and phosphorous reaching the Chesapeake Bay from the lower Eastern Shore. Both nitrogen and phosphorous stimulate algae growth. As algae dies and decomposes, it consumes oxygen in the water, choking fish, plants, and other aquatic life. Chicken manure contains much more phosphorous than nitrogen relative to what a biologic plant needs. When a farmer spreads enough chicken manure to satisfy the nitrogen needs of a crop, excess phosphorous remains. So much phosphorous is washed into water bodies that the Chesapeake region has been referred to by ecologists as a "phosphorous sink."

A controversial hypothesis proposed by some scientists is that a link exists between the toxic microbe *Pfiesteria piscicida* (see chap. 4) and excess nutrients and algae blooms. Although links between fertilizing with heavy amounts of chicken manure, nutrient runoff, algae growth, and *Pfiesteria piscicida* seem evident, the complex biochemical nature of the link has proven difficult to identify. A number of major academic and government studies have detailed the environmental effects of growing millions of chickens on an increasingly smaller area of land. The magnitude of the poultry operations is greater than most Marylanders realize. Every working day, a dozen Eastern Shore slaughterhouses slice the necks of more than two million birds, using more than twelve million gallons of water to flush away more than 1,600 tons of guts, chicken heads, fat globules, feathers, and blood. This flush water goes directly into tributaries. Slaughterhouses treat the water before they release it to creeks, but it still contains some pollution (Goodman 1999).

Water treatment plants used by slaughterhouses develop a mud-like sludge that is scraped off and taken in large tanker trucks to be used at farms as fertilizer. Tractors with special injectors squeeze the sludge into the soil. When there is more sludge than the soil and crops can digest, the excess nutrients percolate to the groundwater. Another method of applying waste from poultry plants is through irrigation nozzles that on a daily basis spray more than three million gallons of treated wastewater onto fields. Maryland has led the nation with strict rules to limit these kinds of farm pollution. In response, the major poultry corporations and growers associations have hired lobbyists to prevent any additional regulations. The poultry industry is also waging a vigorous campaign to improve its public image and to counter the charges of pollution.

The poultry pollution issue has many components, but the environmental and economic factors are central. The industry is trying to meet the nation's demand for cheap chicken, which requires large-scale operations in order to operate more efficiently to market chickens at a low price (i.e., economies of scale). As a result, more chickens are being grown on less acreage. Meanwhile, the demand for pristine waterways and statewide environmental regulations must also be addressed. Many people on the Eastern Shore benefit directly from employment in the poultry industry; they have jobs as construction crews that build chicken houses, grain farmers, plant workers, truck drivers, advertising and publicity firms, bankers who finance the industry, and others. It has been estimated that a drop of just 4 percent in Maryland's poultry production would eliminate 1,000 jobs and $74 million in economic output.

A number of recent encouraging developments seem poised to help address the problems associated with poultry pollution. In an experiment conducted by the University of Delaware, flocks of chickens fed a hybrid strain of corn produced manure that had less phosphorous than normal. Some of the chickens were also fed an enzyme to aid in the digestion of phosphorous, so that less phosphorous passed through the chicken and into its manure. The manure from chickens fed both hybrid corn and the enzyme had 41 percent less phosphorous. The hybrid corn is being tested to determine if it produces desirable yields and is resistant to pests and diseases. Another development also provides some hope for helping to alleviate the poultry pollution problem. Two promising ventures would use some 400,000 tons (362,872 tonnes) of poultry manure a year (about 40% of what comes out of chicken houses across the Delmarva Peninsula). Fibrowatt, a British firm that has successfully operated manure-to-energy plants in Britain since 1991, proposed in August 2000 to build a 40-megawatt power plant in Dorchester County near Vienna and the Nanticoke River. The plant would use a closed-cycle cooling system that is like a giant air-cooled radiator turning water into steam and back to water. It will not need to draw water from nor discharge any into the Nanticoke River, an important spawning area for shad, herring, and rockfish. The company would pay farmers for their chicken manure; currently farmers usually pay to have it removed. Interest and support for the proposal have come from a diverse group of politicians, farmers, and environmental groups.

One of the strongest opponents to the Fibrowatt proposal has been Perdue Farms, the largest poultry company in Maryland. Perdue is building a $12 million plant near Seaford, Delaware, that will use chicken manure to manufacture fertilizer pellets that would be marketed and sold locally, as well as to outside areas such as the Midwest. Perdue has offered to clean out chicken houses for free, but they have not offered to pay for the manure. Estimates from both Perdue and Fibrowatt of their needs do not point to a pending shortage of chicken manure, which is now being redefined as a raw material (Horton 2000).

Dairying

To understand the geographical pattern of milk production in Maryland, one must understand the market areas for dairy products. Maryland's dairy industry began with the growth of Maryland's cities and towns. Today the Maryland dairy belt is centered on Frederick County, but it stretches from Harford County west to Washington County. This area is also Maryland's center of hay and field-corn production; those crops are stored and used during the winter to feed dairy herds. Most of the milk produced in Maryland's primary dairy area on the Piedmont goes to the Washington–Baltimore metropolitan area, which demands fresh milk daily. Milk from the middle and upper Eastern Shore goes into local consumption or to the Philadelphia and Baltimore fluid-milk markets. Washington County's milk is sold around Hagerstown and in nearby areas of Pennsylvania, while milk from Allegany and Garrett Counties is consumed locally or sold to towns in nearby West Virginia and Pennsylvania.

Dairy products are the second-largest segment of Maryland's livestock sector of agriculture, representing 16 percent of all livestock receipts and 10 percent of total agricultural receipts. Maryland west of the Chesapeake Bay accounts for 92 percent of Maryland's milk production. Frederick County is by far the leading producer of milk and other dairy products in Maryland, accounting for 30 percent of total state production ($50,042,000 in sales). Neighboring Carroll County ($21,631,000 in sales; 13% of state total) ranks second in dairy production. The third ranking dairy producer is Kent County on the Eastern Shore ($12,644,000; 7% of state total). Other significant dairy product production is found in Baltimore, Harford, Howard, and Washington Counties.

There is a noticeable lack of dairy farming in the ridges and valley region of western Maryland, southern Maryland, and the lower Eastern Shore. It was expected that as tobacco farming declined in southern Maryland, dairy farming would increase to supply the sprawling Washington urban area, but that has not happened. The urban sprawl into southern Maryland has consumed large tracts of farmland, and the urban area continues to be supplied from the strong dairy area centered on Frederick County to the north. Many of the southern Maryland farmers opting out of tobacco farming are older, and dairy farming is not attractive to them. Dairying is a hard, full-time job with no off-season. Cows must be milked morning and evening all year long. Despite the long hours and hard labor, income from a dairy farm is steadier than from some other types of farms.

Other Livestock

Other livestock raised in Maryland are considerably less important than chickens and dairy cows. Some hogs are raised, but it is difficult for Maryland hog farmers to compete with farmers in the Midwest, where corn is plentiful and converted quickly into pork on a huge scale. The number of hogs in Maryland dropped from 335,000 in 1867 to 270,000 in 1951; to 215,000 in 1978; and to a mere 65,000 in 2002. In 2002, only 23,000 head of sheep and lambs were recorded in Maryland. Beef cattle made up 23 percent of the state's cattle in 2002.

From 1940 to 1950, as farms turned to mechanization, the number of horses in Maryland decreased by 50 percent. Still, the business of breeding horses is important to fox hunting, racing, and other activities that are part of Maryland's equestrian culture. A number of farms throughout Maryland are devoted to raising thoroughbreds; hunters, show horses, and ponies are also bred in the state. Horse racing in Maryland contributes to the well-being of agriculture through the State Farm Board, which receives a share of levies imposed by law on race tracks. Breeding operations are likewise important customers of farm products such as hay, oats, and other grains. Although the horse remains an important icon or symbol of the Maryland landscape, attendance at Maryland racetracks has steadily declined over the last decade.

6 The Watermen's Domain

Although only a mere 0.3 percent of Maryland's employment is in fishing, it remains an important economic activity, especially on the Eastern Shore and parts of the Western Shore. Seafood is not only an important resource, but also a major component of the cultural identity of Maryland. One can hardly think of Maryland without associating it with oysters and crabs. In 2004, there were 7,313 licensed watermen in Maryland who altogether accounted for a dockside commercial catch of 45,655,294 pounds (20,832,510 kilograms) of fish. In reality, the catch was higher because this figure does not include the catches of a large number of sport fishermen. The major share of Maryland's catch is credited to blue crabs, 33,495,212 pounds (15,193,428 kilograms), or 73.3 percent by weight (table 6.1).

Crabbing

The Chesapeake Bay has provided more crabs for human consumption than any other estuary in the world. The bay's blue crab has by far the greatest commercial value of the Maryland fishery. As recently as 1980, oysters were the most valuable component of Maryland's Chesapeake Bay fishery, but by 2000 the blue crab catch exceeded the value of oysters more than five times over. In 2004, the value of Maryland's blue crab catch was over $31 million; the oyster catch was valued at only $181,000. Soon after the oyster season ends in April, Maryland watermen pursue the blue crab, an icon in Maryland just as cod is in Massachusetts. The Atlantic blue crab, *Callinectes sapidus* (*Callinectes* is the Greek word for beautiful, *sapidus* Latin for tasty or savory), is sought after, caught, and more eagerly consumed than any other crab (Warner 1976). Chesapeake Bay watermen have been developing new techniques and equipment to pursue and catch crabs for over 150 years. Many families in Maryland, and especially the bay region, have long considered the blue crab to be not only an important food source, but also a source of economic livelihood, social and cultural identity, and enjoyment.

The blue crab, one of the fastest swimming of all crabs, is found from Canada to South America. It is most abundant along the US Atlantic coast from New Jersey to Texas. In the mid-1990s, over half of the US commercial catch came from the Chesapeake Bay. But the Maryland blue crab industry is in trouble. In 2004, the bay accounted for a diminished 19 percent of the US blue crab catch. Maryland crab watermen are facing strong competition from crab production in other parts of the United States as well as imports, mainly from Asia. The picture is brighter for other crab species. In 2004, Maryland alone accounted for 93 percent of production of soft and peeler blue crabs in the United States. The United States is still the top crab-consuming nation in the world, with an annual catch in 2004 of 169 million "whole crab" pounds—equivalent to $127 million in sales.

Although it takes the amateur's bait readily, the blue crab is difficult to catch in commercially valuable quantities, as it migrates with

Table 6.1. Commercial fishery harvest

Fish type	*Pounds landed*	*Percentage of the total Maryland catch*
American eel	258,176	0.57
Black sea bass	191,474	0.42
Blue crab	33,495,212	73.3
Bluefish	52,682	0.12
Butterfish	8,950	0.02
Carp	38,444	0.08
Catfish	819,277	1.8
Croaker	1,354,982	3
Drum (red)	1,051	0.002
Flounder	314,419	0.67
Grey sea trout	50,519	0.11
Menhaden	5,369,952	11.8
Oyster	63,057	0.14
Perch (white)	1,502,201	3.3
Perch (yellow)	67,354	0.15
River herring	86,938	0.2
Shad	3,944	0.009
Shark	47,159	0.1
Striped Bass	1,924,470	4.2
Tautog	3,814	0.008
Tuna	1,173	0.003
Whiting	46	0.0001
Total	45,655,294	100

Source: Maryland Department of Natural Resources, 2013. Data are for 2004

the seasons. In late summer and early fall a layer of freshly oxygenated seawater moves slowly northward along the bottom of the Chesapeake Bay and its tributaries. Above it lies the tired, biologically exhausted water that has become oxygen starved after the long, hot summer. In late fall and early winter, the now-cooling surface water sinks as the deeper, oxygenated waters rise. The change in temperature and the churning action is a cue for the blue crabs. The big "jimmies" (mature males) and the younger crabs head for the deep central part of the bay, where they bury themselves in the sediment, slow down their metabolism, and hibernate for the winter. The "sooks" (mature females) move down the bay to winter in the more saline waters of Virginia, where most of the bay's blue crabs are hatched. Until recently, it was believed that nearly all of the sooks migrated to Virginia waters. Recent winter crab samples taken in dredges by the Maryland Department of Natural Resources (DNR), however, have revealed that large numbers of sooks do not finish their migration before winter hits and instead hibernate in the middle or upper bay. The DNR found in 1993 that more than 40 percent of the crabs found in its winter survey were sooks (Fincham 1994b). Come spring, the up-bay sooks once again head for the spawning grounds at the mouth of the bay. To get there, they now have to run the gauntlet of the watermen's spring and summer crab pots.

Maryland watermen have long faulted Virginia watermen for dredging crabs in the winter months and taking many of the fertile sooks (winter dredging is illegal in Maryland). New evidence of the failure of many sooks to reach Virginia waters before winter has led many Virginia watermen to fault Marylanders, who run as many as 800 to 1,500 crab pots per boat, taking many sooks during the fall migration south as well as during the spring run. These differing viewpoints neglect to recognize that the Chesapeake Bay blue crab fishery is part of one ecosystem and that it should be managed as one ecosystem. Both Maryland and Virginia scientists have long realized this fact and have been cooperating for years. In 1989, a unified management approach for the bay was initiated with the Chesapeake Bay Blue Crab Management Plan, whose goal is to manage blue crabs in the Chesapeake Bay, conserve the bay wide stock, protect its ecological value, and optimize the long-term utilization of the resource. The plan is part of the Chesapeake Bay Program, which includes the Chesapeake Bay Commission (a tristate legislative body), District of Columbia, Maryland, Pennsylvania, US Environmental Protection Agency, Virginia, and various citizen groups. In 2001, the Chesapeake Bay Commission's Bi-State Blue Crab Advisory Committee (Maryland, the Potomac River Fisheries Commission, and Virginia) reaffirmed the goals of achieving a crab fishery that is biologically, economically, and socially sustainable.

In the spring, vertical mixing occurs again, though in reverse of the fall pattern. As the waters warm, the sooks that have come south the previous fall head up the bay, and the jimmies come out of the deep channel. Both head for the warm shoal waters and eelgrass of the lower creeks and tributaries. In order to make a living by catching blue crabs, a waterman needs to understand their life cycle, which requires him to be somewhat of a biologist. Most of the successful Maryland crabbers have learned their skills from their fathers. Generations of Maryland watermen wake up at 3:30 a.m. and work hard under a blazing sun or driving rain to follow the blue crab's life cycle. Making a living on the water requires not only the strength of an athlete but also the skills of a mechanic, the mind of an accountant, and the spirit of a stoic able to adapt to the vicissitudes of time and tide (Meyer 2003).

Hard Crabs

Most hard crabs are caught in crab pots. These steel-mesh baited traps allow the crab to enter through a funnel, but they cannot escape once they pass through it. The crabbing season runs from the first of April to the last of November. The size limit on crabs is 5 inches (12.82 centimeters); anything smaller must be legally returned to the bay. Crab pots need to be checked regularly, because after a couple of days these cannibalistic critters will feed on each other.

Hard crabs are also caught on trotlines, the traditional method for catching crabs for over one hundred years before the introduction of the pot. Trotlines are up to one-third of a mile (0.53 kilometers) long, with bait (preferably eel) attached every 3 feet (0.914 meters). The line is anchored at the ends and marked with buoys. The crabber visits the trotline daily to pull it in, being careful to scoop the crabs with a deep net before they hit the surface when they let go of the bait. Running trotlines is hard, messy work. Not much trotlining is done in Virginia; it is most commonly practiced in the eastern part of the Chesapeake Bay south of Kent Island and in the marshlands of Dorchester County. Maryland restricts the use of the more efficient crab pots to the open bay. Creeks, tributaries, and subestuaries are the domain of the trotliners.

Although soft-shell crabs are gaining popularity, the hard crab is still the heart of the industry. Crabs are sold dockside to crab houses, where they are cooked whole to firm up the flesh. Large cylindrical pots can boil up to 1,200 pounds (544.3 kilograms) of crabs in twelve minutes. The boiled crabs are cooled overnight, and pickers then split, quarter, and dissect the crabs by hand to obtain every last gram of meat. Pickers, mostly women, are paid by the pound or hourly (at minimum wage), whichever is higher. Picking is a difficult skill, but a good picker can pick 40 to 50 pounds (18.2 to 22.7 kilograms) of meat in each seven- or eight-hour workday. There was an attempt to manufacture an automatic crab picker, but it was a failure. The machine shook the crab and was unable to distinguish between the delicate, prized lump meat and the rest of the crab; it still takes the nimble fingers of a picker to properly do the job.

Soft-Shell Crabs

About 93 percent of the catch of soft crabs in the United States comes from the Chesapeake Bay. Crabs grow by shedding, or molting, and forming a new shell soon after. Because they do not eat while shedding, soft crabs cannot usually be caught in baited pots or on trotlines. Most are caught from late May to September by the use of crab scrapers in the shallow grass bottoms. A specially designed shallow draft boat, called a Jenkins Creeper, is used in scraping. In 1870, L. Cooper Dize of Somers Cove in Somerset County patented the toothed crab scrape. The start of the soft-shell crabbing industry is credited to former US congressman John Woodland Crisfield, who used his influence to have the rail line pushed south to Somers Cove in 1873. It was now possible to ship live soft-shell crabs on ice to relatively distant Baltimore and Philadelphia. The town of Somers Cove was renamed Crisfield, and it became the crab capital of the Eastern Shore (Mountford 1999).

Most soft-shell crabs are caught while still hard or about to shed, which crabbers call "peelers." They can tell the crab is about to shed by the color rimming the last two links of the back paddle fin. Pink and red coloring means that the crabs are "rank," ready to shed in about two or three days. White coloring indicates that the crab is five to seven days from shedding and will still feed on other crabs; watermen crack their claws and then separate them from the others. The crabs are placed in a shedding float, a rectangular floating enclosure made of strips of lathing about 0.25 inches (0.64 centimeters) apart that allow water to circulate while preventing schools of hungry minnows or other predators from devouring the helpless soft crabs. Moored with stakes inside protected waters, acres of floats can be observed on the way into Tylerton on Smith Island. Many shedding floats can also be seen at Crisfield, Deal Island, and Dorchester County marsh towns. Skill, hard work, and infinite patience are needed to hold crabs in float pens. Pound operators try to keep the peelers healthy and happy. Their claws are nicked to discourage damaging combat, and the floats are constantly monitored for "stills," or dead crabs. It is not unusual for as many as 50 percent of peelers to die in float pens.

Soft-shelled crabs are removed from the pens before they begin to form a hard shell. They are packed in trays and covered with sea grass and ice. If their gills are kept wet, the crabs can live for several days. Each tray holds three-dozen jumbo or six-dozen medium crabs, and three trays make a 60-pound (27.2-kilogram) box. Specialized seafood-trucking firms pick up the boxes and deliver them overnight to markets in Baltimore, Philadelphia, and New York's famous Fulton Fish Market. The reputation of Maryland's soft-shell crabs has spread far. In 1980, soft-shell crabs were sold mostly in the mid-Atlantic region. Today, the demand is high and growing in places such as England, Japan, and the western United States. Both frozen and iced soft-shell crabs are flown daily from Baltimore–Washington International Thurgood Marshall Airport. It is now possible to order Maryland soft-shell crabs in restaurants in Chicago, Detroit, Kansas City, Louisville, San Francisco, and many other cities. Many Marylanders believe that the soft-shell crab is to crustaceans what a filet mignon is to steak—a delicacy. The soft-shell crab has even taken a larger share of the seafood market in New England, land of the lobster.

Maryland's leading crabbing area is around Smith Island and its three communities of Ewell, Rhodes Point, and Tylerton. This somewhat isolated island can be reached by ferry from Crisfield. People with the name of Evans constitute about half the population of Ewell, Tylers reign in Rhodes Point, and Bradshaws dominate Tylerton. This quaint island has no local government (the Methodist church is the closest thing to a ruling institution). Children go to school in Crisfield by boat. The Virginia watermen on nearby Tangier Island catch more crabs than Smith Islanders only because they are allowed to join the winter crab dredging fleet in the lower bay. Winter on Smith Island used to mean relying almost solely on oyster fishing; in summer, Smith Island is the champion of the bay's soft-shell crab production.

The Future of the Crab Fishery

There is some controversy about whether the Chesapeake Bay crab fishery is in danger. Some watermen and scientists believe that the blue crab population is healthy, numbers will once again increase, and that crabbers should not worry about overfishing (Cahill 1996). But the majority of bay scientists, who are much more cautious and support bay-wide management and study, do not agree. The historical pattern of the blue crab catch is uneven (fig. 6.1). From the mid-1980s to the mid-1990s, the blue crab catch was high. Since 1997, however, it has dropped to a prolonged period of low abundance.

Other bay fisheries—those harvesting shad, menhaden, oysters, and rockfish—have diminished to an alarming degree. All of these experienced declines in harvests, which sent prices higher, spurring more watermen to work longer hours and harder days, depleting these dwindling resources at an even faster rate (Fincham 1994a). Although the blue crab seems to be holding its own, there are several important factors working against it:

- Since the 1970s, eelgrass beds used by crabs for hiding and nursing their young reduced by 70 to 90 percent.
- Many watermen have moved away from oystering and other kinds of fishing as certain types of seafood have become depleted; many now fish for crabs as their sole means of support.
- Demand for crabmeat is rising as the population of the bay region grows.

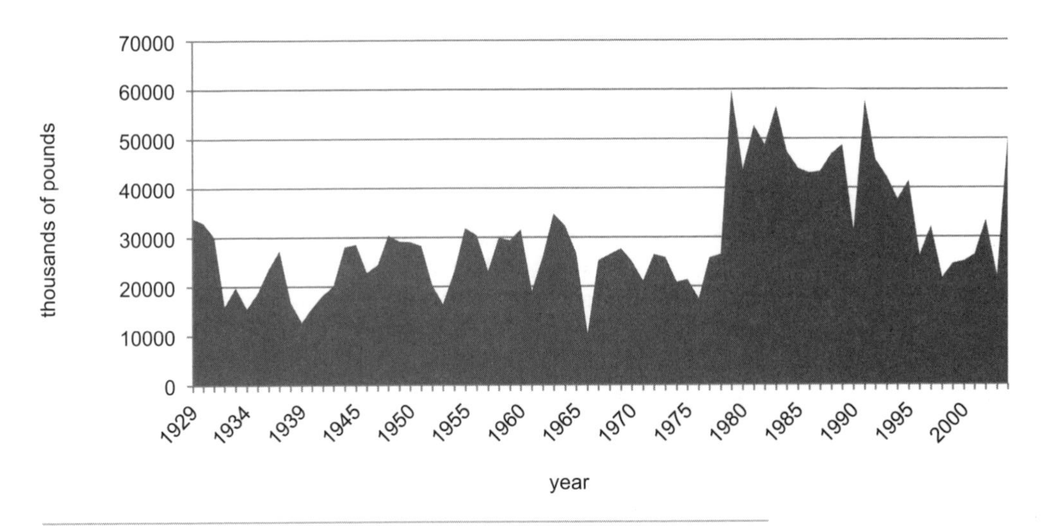

Figure 6.1 Maryland Blue Crab Catch, 1929–2012

- Many more watermen are pursuing the blue crab, but the catch has not increased significantly; it now takes a lot more work to catch a crab.
- Habitat encroachment through urban sprawl and nitrate- and phosphorous-rich runoff from farms robs oxygen from the bay.
- Crabs live very near those noisy, dirty, and ill-tempered predators known as humans.

In the late 1980s, Chesapeake Bay crabbers once removed 40 to 50 percent of the adult crab population in a season. Today that figure is over two-thirds. Although scientists believed that the egg-bearing sooks were in the greatest danger, evidence now suggests that adult jimmies are under stress, too. Estimates are that the number of legal-size males in the bay's blue crab population went from nearly 60 percent in the late 1960s to just 10 percent by 1997. Young males used to scramble up the tributaries to seek shelter among the woody debris along the banks, but housing and farm developments have deprived them of much of this habitat. A serious predator eats many of the young crabs: cannibalistic elder crabs. The diminishing catch of blue crabs from the Chesapeake Bay is directly related to the population of blue crabs themselves (fig. 6.2). Here, too, there is a pattern of severe decline since 1997. The Chesapeake Bay blue crab population today hovers near historic lows.

Scientists such as Bill Goldsborough of the Chesapeake Bay Foundation, a nonprofit conservation organization, believe that the Chesapeake Bay blue crab fishery is being fully exploited and perhaps overexploited. They believe that the only reason the blue crab population has not collapsed is because it is biologically very resilient. Over the past fifteen years, crabbing has increased fivefold in the bay without a corresponding increase in yield. Crabbers are already working harder and harder for less and less (Lerner 2000). The predicament of the blue crab, oyster, rockfish, and other species in the bay can be tied to what is happening on land. Governmental agencies responsible for water quality improvements need to implement changes on the land to reduce nutrients, sediment, and other pollutants that are harming the wildlife of the Chesapeake Bay. The other part of the equation is fishery management, which must complement land-use management to ensure the survival of a sustainable Chesapeake fishery.

The challenges facing the commercial fishing industry of the Chesapeake Bay are severe and demanding. There is a need for greater investment in biological monitoring of bay fishery populations to learn more about life cycles, populations, trends, and reactions to habitat alterations. There are severe economic challenges, too. The commercial fishery faces highly variable harvests. Competition from other states and overseas suppliers is growing. Maryland seafood companies such as Phillip's now import huge quantities of crabs from India, Indonesia, Thailand, and other Asian countries. Low wages in these countries in labor-intensive jobs such as crab picking make it difficult for Maryland processors to compete. The effects on some locales are already being felt; for example, in 2005 there were about thirty crab processors left in Maryland. Of those, twenty-one were in Dorchester County, representing 2 percent of the county's manufacturing base and 3.3 percent of the labor force.

The future of Maryland's fishery will be bleak unless all participants in the industry make strong commitments. Fisheries are interrelated with other economic sectors, including resource management,

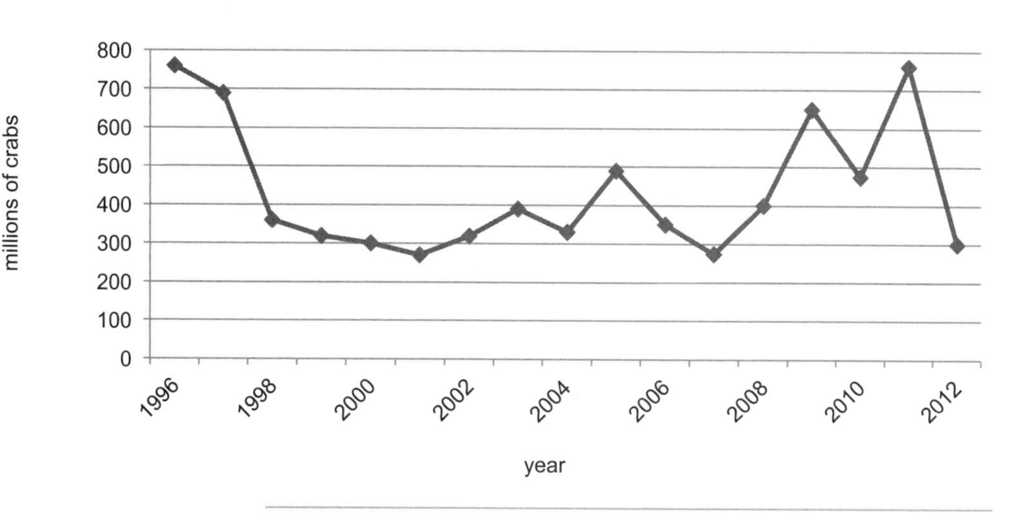

Figure 6.2 Chesapeake Bay Blue Crab Population, 1990–2012

harvesting, processing, seafood safety, marketing, and distribution. A failure to sustain the fishery will have multiplier effects on these related economic factors. When a processor plant closes, it no longer buys cans for packing, insurance, fuel, or services from accountants and lawyers. It no longer pays salaries to workers who would be spending money in the local economy. The Maryland fishery is economically and socially too important to let it fail.

Fishing

An estimated 295 species of fish live in the Chesapeake Bay. About thirty-two species live in the bay year round, while others migrate to the bay from the Atlantic Ocean to feed and reproduce. Finfish play important roles in the bay's health and food chain. Anchovy and menhaden are critical links between the lower food web and predators higher up the food chain. They live on plankton and are themselves a vital food source for striped bass and bluefish.

Recent decades have witnessed poor water quality and overfishing in the bay, leading to declines in many finfish species. The much sought-after striped bass, also known as rockfish, has struggled to return to previously larger populations even after aggressive management and a five-year moratorium on fishing for striped bass. In the past, fishery management usually addressed one species at a time with little consideration as to how all species are linked to their common ecosystem. Starting in 1996, the Magnuson-Stevens Fishery Conservation and Management Act established principles and guidelines for ecosystem-based fishery management. In 2004, the Chesapeake Bay Office of the National Oceanic and Atmospheric Administration released its Fisheries Ecosystem Plan for the Chesapeake Bay. This approach to managing the fisheries of the bay recognizes that fisheries are sustainable only if the relationships between predator and prey are accounted for along with efforts to restore water quality and habitat.

Rockfish

The rockfish (*Morone saxatilis*) is a local favorite, better known in the Chesapeake Bay region as the striped bass. These migratory fish can swim from Florida all the way up the East Coast. A Chesapeake-tagged rockfish was once found in Canada, 1,000 miles (1,609 kilometers) away. Although mature rockfish can weigh 100 pounds (45.4 kilograms) and measure 5 feet (1.5 meters) in length, the largest rockfish ever found in the bay weighed 67.5 pounds (30.6 kilograms). Rockfish spawn in estuaries or at the edge of tidal freshwater in rivers. The Chesapeake Bay is the primary spawning area for 60 to 70 percent of Atlantic rockfish. Their migratory behavior is more complex than many other anadromous fish (i.e., fish that migrate from the sea into rivers to spawn). Mature rockfish move into the warmer tidal freshwater areas of the bay in late winter and spring. After spawning, the mature rockfish return to the Atlantic Ocean. Summer and early fall are spent in New England water. In late fall and early winter, rockfish migrate south to the waters off Virginia and the Carolinas. The smaller, immature rockfish spend their first three to five years in the Chesapeake Bay.

Rockfish are the most sought-after commercial and recreational finfish in the Chesapeake Bay. Their population in the bay has experienced a varying up-and-down pattern over the years. The Maryland Juvenile Rockfish Index tracks the average number of newly hatched fish caught in seine nets during summer-long monitoring programs (Alliance for the Chesapeake Bay 1999). Recent indices suggest that recent management of this resource after ten years of steep decline in the 1980s has been a success. The Chesapeake Bay Striped Bass Fishery Management Plan of 1989 was passed to manage and enhance the fishery in the bay as well as along the entire Atlantic coast. From 1998 to 1999, the commission set harvest quotas for the entire Chesapeake Bay, limiting the yearly catch to 11.2 million pounds (5.1 million kilograms). It then divided the quota, with Maryland at 52 percent, Virginia at 33 percent, and the Potomac River Fisheries Commission at 15 percent. Maryland has a separate quota of 91,000 pounds (41,278 kilograms) for its Atlantic coastal waters (also restricted to fish with minimum total length of 28 inches (71.1 centimeters). The rockfish population rebounded, starting in the early 1990s. The federal and state governments of Maryland and Virginia will continue to closely monitor and manage this important fishery.

Eels

Of increasing importance to the Chesapeake Bay fishing industry is the American eel. Eels are caught in pots made of fine mesh wire with nylon funnels. The pots are baited with razor clams, horseshoe crabs, or fish. After finding their way into the pot, the eels cannot get out. Although eels are good to eat and are considered a delicacy in some countries, they are not yet popular in the United States. Most Maryland eels are shipped by air to the European market. Spring, summer, and fall are the eel seasons; spring is the most important because it is when eels proliferate.

Shad

Shad were once the Chesapeake Bay's most valuable commercial species of finfish. They were a staple for the Indians of the region, and early European settlers depended on the shad runs for survival. It was common for most colonial homes to have a barrel of salted shad (Chesapeake Bay Foundation 2010). Shad are an anadromous fish, spending most of their life migrating along the Atlantic coast. Then at about age four they return to spawn in the streams where they were born. Historically, shad swam hundreds of miles up the tributaries of the Chesapeake Bay. But pollution, development, and other environmental challenges in recent years have severely hindered shad migration. Restoration projects have focused on stocking shad in tributaries and removing obstacles to their migration. By 2000, there was a baywide moratorium on catching shad. Maryland banned bay shad fishing in 1980, but many observers believed that it was already too late. In 1979, only fifty shad returned to spawn in the Susquehanna, which had once witnessed a shad population in the millions. Since 1980, millions of young shad (also called fry) have been released into the rivers with the hopes that they will return to these upstream release points to spawn when they are four to five years old (Blankenship 2004). A record thirty-six million fry were released in 2000. The construction of a fish ladder at York Haven dam along the Susquehanna in Pennsylvania reopened the Susquehanna to shad for the first time in a century.

The effects of pollution, sediment, and overall environmental degradation on the shad population are representative of many other species of finfish in the bay. In 1890, the all-time high catch of 3,500 tons (3,175 tonnes) of shad was landed. By 1909, the catch had dropped to 1,500 tons (1,361 tonnes). From 1940 to 1970, the annual commercial catches of shad in Maryland waters ranged from 500 to 1,000 tons (453 to 907 tonnes). After 1970, the shad catch went into a precipitous decline. Maryland responded by limiting fishing for shad, promoting hatcheries, and building fish ladders and elevators around dams along rivers that are shad-spawning runs. By 2000, the shad restoration program had spread bay wide, as Delaware joined the District of Columbia, Maryland, Pennsylvania, Virginia, and several nonprofit groups in the effort.

Menhaden

Called "pogy" in New England, menhaden are known as "moss bunker" in the Middle Atlantic states and "fatback" in North Carolina. Chesapeake Bay watermen often call them "bunker" or, on the Eastern Shore, "alewife" or "oldwife" (Warner 1976). In precolonial days, eastern North American Algonquin Indians called menhaden *mannawhatteaug*, or "fertilizer." Menhaden are likely the fish that the Indians urged the Pilgrims, and also early settlers in Maryland, to plant with their corn. Menhaden are an important finfish catch in Maryland, accounting for 12 percent by weight (although only 0.6% of the total catch value). Menhaden have a relatively low value per pound in comparison to some other species. Most menhaden are not sold for human consumption; rather, they are reduced to oil, animal feed, and solubles used to manufacture cosmetics, linoleum, and steel. Menhaden are also important for several other reasons. Along with oysters, menhaden are the Chesapeake Bay's major filter feeders. Filter feeders graze on large amounts of algae from the water, thus removing nitrogen and helping to improve the clarity and quality of bay water. Clearer water allows sunlight to reach the bottom, which enables valuable grasses to grow. Together, both menhaden and oysters at one time probably removed as much as 20 million pounds (9,071,940.4 kilograms) of nitrogen from the bay annually; today they remove less than 5 percent of that amount (Wheeler 2000). Being at the bottom of the food chain, menhaden are also a major source of food for rockfish and others. The Atlantic coast menhaden population has dropped 79 percent since 1991.

Watermen

"Watermen" is a curious term that was likely first used around 1400 in Sir Thomas Malory's *Le Morte d'Arthur*. From the Elizabethan period on, frequent references to watermen were confined to the Thames River in England to describe waterborne taxi men. Today, the term has only limited and archaic use in England; it is used mainly in August when the London Fishmongers' Company sponsors the annual Watermen's Race. It is not known exactly why in America the term took root only along the shores of the Chesapeake Bay. It may be that early Chesapeake watermen plied for hire on the rivers. The word came to differentiate those who had the resources to acquire land from those who did not and were forced to turn to the water for subsistence (Wheeler 2000).

Clams

A 1970s entry into the Maryland seafood industry was the soft-shell clam, *Mya arenia*, the succulent steamer of New England clambake fame. For years, Chesapeake Bay watermen neglected and even despised these clams, locally called "manoes," as being unfit for human consumption. In New England, *Mya arenia* lie in the mud at low tide. In Maryland, they remain under 10 to 20 feet (3.1 to 6.2 meters) of water and must be dredged. In 1950, Fletcher Hanks of Hanks Seafood Company in Easton learned of a serious decline of New England clams. The next year he patented a hydroescalator clam harvester that became an instant success and is still used today. The front end is lowered to the bottom. High-pressure water jets on the front end of the conveyor wash and stir up the clams from the bottom, pushing them onto a mesh chain conveyor belt that takes them up to the boat. Because of this churning cultivating process, clam dredges are restricted to certain areas of the bay so as to avoid disturbing valuable eel grass areas and oyster beds.

Bob Orth, professor of biology at the Virginia Institute of Maritime Science, has been taking aerial surveys of grass beds in the Chesapeake Bay since the mid-1980s and generally tracking the condition and extent of submerged aquatic vegetation (SAV). He estimates that there are only about 60,000 acres (24,281.67 hectares) of grasses in existence today; decades ago, the bay was home to about ten times that amount. SAV is one of the bay's most important resources because it helps to improve water quality and provide food and habitat for waterfowl, blue crabs, and other species. Extensive damage to grass beds was detected in the late 1990s. Although there are numerous causes of the depletion of grass beds, a significant amount of the damage is apparently being caused by hydraulic clam rigs that cultivate and churn the bottom. The process also leaves a large sediment plume in the water. When the sediment settles on grasses, it can smother them.

Although most of Maryland's soft-shell clam catch goes to New England, other regions of the United States are beginning to demand them. Hard-shell clams are marketed all over the country. Razor clams are used locally as crab and eel bait. In 1952, nearly 95 percent of the US catch of soft-shell clams was taken in New England. By 1960, the Chesapeake Bay soft-shell clam fleet was supplying over 600,000 bushels (211,440,000 liters) annually, representing 70 percent of US production. The peak harvest year for soft-shell clams was in 1969, when 659,000 bushels (232,232,000 liters) were brought in. In 1998, Maryland watermen caught only 12,300 bushels (4,335,000 liters) of soft-shell clams, valued at $1.3 million. In recent years, coastal New England has experienced frequent "red tides" caused by marine algal blooms associated with polluted waters. During red tides, New England waters are declared closed for clamming, and Maryland's soft-clam fishers sell all they can catch to New England.

Problems may be developing for the soft-shell clamming industry in the form of a parasite called *Perkinsus Chesapeaki*. It is closely related to *Perkinsus marinus* (also called "Dermo"), which has devastated the Chesapeake Bay oyster population. Between 1990 and 2000, marine biologists found both *Perkinsus Chesapeaki* and Dermo in soft-shell clams taken in the upper bay. Finding both of these parasites in soft-shell clams both surprised and challenged marine biologists. Scientists are not yet certain if these parasites are harming clams, but they are trying to determine whether there is any relationship between the parasites and the dying of clams in hot, dry summers as well as the decline in harvest by nearly 93 percent during the 1990s (Shelsby 2000). Scientists know that clams prefer the colder waters of the upper bay and that they die at a higher rate when temperature and salinity are high, but they do not know why. Perhaps higher temperatures encourage the parasites to multiply and spread.

Oysters

From the rocky coast of Maine to the sandy beaches of California, when Americans think of Maryland, they usually envision the Chesapeake Bay and its oysters. "Those succulent, tender, juicy, plump, delicious, magnificent Chesapeake oysters which Marylanders thrive upon, brag about, and even lie about are certainly the crown jewels of the world of gastronomy" (*Maryland Business Journal* 1980). The bay and the oyster are both significant components of Maryland culture. Oystering (or "arstering," as it is often pronounced on the Eastern Shore) is more

MARYLAND AND VIRGINIA DISPUTE THEIR BORDER

Since the eighteenth century, Maryland and Virginia have had a number of conflicts over fishing rights in the water boundary separating the two states. In 1785, a new compact was signed at George Washington's Mount Vernon estate recognizing the boundary established in 1668. It allowed residents of both states to fish the Potomac River, although Virginia conceded that the river was entirely within Maryland. Although George Washington was the arbitrator in this dispute, he himself had operated a shad and herring fishery in the Maryland waters of the Potomac River from his estate at Mount Vernon; he had neither sought permission from Maryland nor paid any fees.

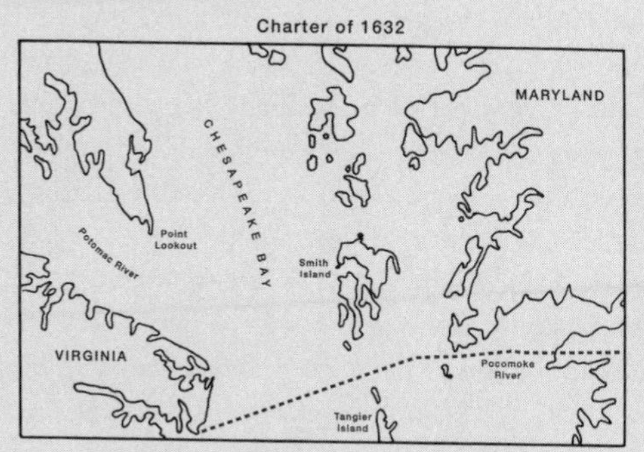

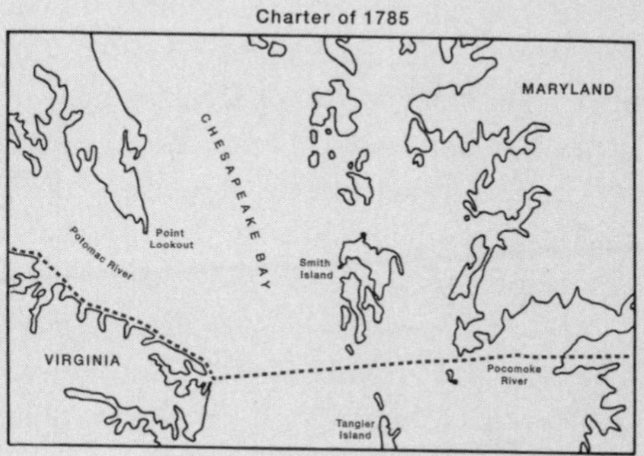

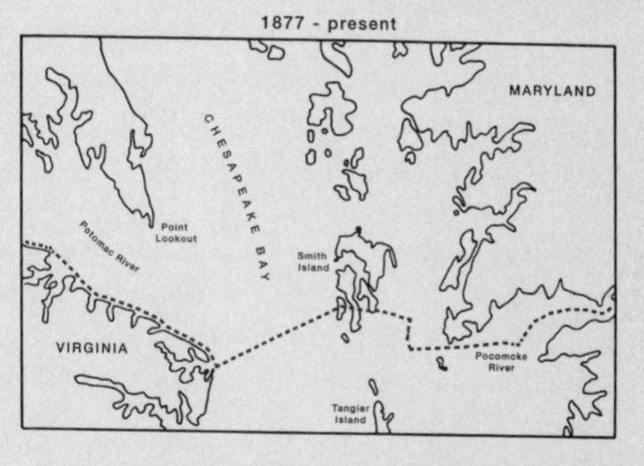

Maryland–Virginia Border Dispute

As oyster supplies diminished, Maryland passed laws that Virginia refused to honor. In 1957, the Maryland legislature finally abrogated the Compact of 1785, and in 1958 the two states drew up a new compact that recognized the de facto boundary that had existed since arbitration between the states in 1877. Under the terms of this compact, a joint commission regulates the Potomac River fishery, and patrol boats from both states enforce the laws. Until 1958, casinos housing slot machines operated on the ends of piers stretching out into the Potomac from Virginia. Although illegal in Virginia, gambling was legally occurring in Maryland territory. Finally, in 1958, Maryland in essence outlawed these casinos by mandating that they had to be accessed by land from the Maryland side of the Potomac (Papenfuse 1976).

Although fishing laws vary significantly from the Maryland to the Virginia regions of the Chesapeake Bay, both states actively patrol the border across the bay. Although watermen are not allowed to cross the state boundary in search of a catch, incursions and incidents are common. One notable incident occurred in July 1949, when a Virginia Fisheries Commission inspector boarded the boat of Earl Lee "Pete" Nelson, a Maryland crabber out of Smith Island. When Nelson refused to go to a Virginia port for impoundment, a gunshot rang out, and Nelson was killed. Open conflict on the waters between Maryland and Virginia raged into the 1950s. Smith Island has been at the heart of the Maryland–Virginia border dispute because the state line cuts across the southern part of the island. Virginia officials actively patrolled the border to prevent Smith Islanders from pursuing crabs in the Virginia waters, where rich eel grass harbors "peelers" (soft-shelled crabs). Maryland crab boats caught over the line below South Point were often towed into Onancock, Virginia, where a stiff fine plus court costs were levied. Virginia police once pursued some teenaged Smith Islanders, allegedly accused of illegal oystering, to Tylerton on Smith Island. Junior Evans recalled, "there were twenty five or thirty Smith Islanders with rifles behind every pole . . . they came to when they saw the rifles; they then turned tail and left" (Meyer 2003).

In 1971, Maryland watermen were successful in getting Maryland courts to abolish the extension of Maryland county borders into the Chesapeake Bay. Today, Maryland watermen can work up and down the bay within Maryland. But negotiations as delicate as any international treaty talks failed to achieve an informal settlement with Virginia for a neutral zone where Smith Islanders could crab (Meyer 2003). The watermen took their case to the federal courts, and in June 1982 the US District Court in Richmond, Virginia, ruled that Virginia had no right to ban any Maryland waterman from crossing the line to pursue their livelihood. In October 1982, a federal judge in Baltimore ruled that Virginians could crab in Maryland waters, lifting the ban on commercial crabbing across the state line. Yet controversies remain; for example, dredging for crabs is illegal in Maryland but legal in Virginia during the winter months.

The common border of the Potomac River is another longstanding disagreement between Maryland and Virginia. Growth and environmental concerns have led to competing claims to the river's waters. Virginia wanted to build a 725-foot (220.98-meter) pipeline into the Potomac River to draw drinking water for its growing population in Northern Virginia. Maryland maintained its longtime historic right to control the waterway because the boundary is on the southern (Virginia) side of the river. Maryland asserted its ownership rights by tracing them to its royal charter in 1632, insisting that its territorial rights are at stake. England's King Charles I gave Lord Baltimore, Cecilius Calvert, a grant of land that included the entire river to the high watermark on the Virginia shore. In 1877, arbitration changed the Maryland boundary to the low watermark, thus giving Virginia control of its own shoreline. Virginia later argued that it had a right to an adequate supply of drinking water for the approximately two million residents of Northern Virginia. Eventually reaching the US Supreme Court in May 2000, the case was decided in Virginia's favor in 2003 (Clemons 2003). The Supreme Court ruled that Virginia had a right to access water from the Potomac, while maintaining that the river was within the boundaries of Maryland.

Virginia may now access the cleaner water flowing in the deeper parts of the Potomac River's central channel. Although the long-standing state boundary runs along the Virginia side of the river, Virginia maintains that it has a right to decide its own fate without interference from Maryland. Maryland is claiming to protect the environment by regulating the Potomac River, a main feeder river of the Chesapeake Bay. Maryland officials cite the rapidly rising population of Northern Virginia as a threat to the environment. New Maryland legislation enacted in May 2000 forbade new water intake pipes in the Potomac until the state and federal governments completed studies on the impact on the water quality and the general environment. The Supreme Court has ruled on Potomac River disputes between Maryland and Virginia four times, in 1894, 1910, 1958, and 2003. This will not likely be the last dispute.

than a job—it is a way of life. Yet recent decades have brought troubled times to the Chesapeake oyster industry. Environmental degradation, pollution, diseases, overfishing, habitat destruction, and competition have all combined to reduce annual catches and drive away many watermen from their traditional occupations. "Still, the Chesapeake Bay for all of its size and problems remains an amazingly intimate place where land and water intertwine in infinite varieties of mood and pattern" (Warner 1976).

In terms of value, the oyster is the second-leading resource of the bay, but the oyster fishery is in serious decline. Maryland ranks fourteenth of thirty-three coastal and Great Lakes states for its value of landed fish. Blue crabs, oysters, and striped bass dominate the Maryland catch. In 2010, only 82,958 bushels (2.9 million liters) of oysters were landed, valued at $4.3 million. In 1885, Maryland watermen harvested 15 million bushels (528.6 million liters) of oysters (fig. 6.3). Since that peak year in the late nineteenth century, the number of oysters harvested in the Chesapeake Bay has steadily declined. Data on oyster catches are reported in pounds or bushels, but different states use different conversion factors. In Maryland, 6.48 pounds (2.94 kilograms) of landed oyster meat is equal to one bushel of oysters. Virginia uses 6.59 pounds (2.99 kilograms) to the bushel.

The harvest ranged between two and three million bushels from the 1960s through the 1980s. During the 1990s, the Maryland oyster harvest averaged only 150,000 bushels, representing a 90 percent decline from the previous decade. This decline is reflected in the relatively small number (fewer than 1,000) of watermen licensed to oyster in 2010 as well as the loss of jobs in shucking houses, processing plants, and transport. The way of life that has long embodied social and work patterns with deep regional historical roots has also significantly eroded. The long-term decline of the Maryland oyster fishery and near-collapse in the early twentieth century are vividly shown in figure 6.3. The Virginia oyster fishery has declined even more drastically (table 6.2). Both Maryland and Virginia share the Chesapeake estuary's oyster fishery, but they have significant differences in how they manage their oyster beds. Virginia's share of the oyster catch for the entire bay declined from about 34 percent in 1980 to a mere 6 percent in 2001. Although the Maryland share of the catch stood at 94 percent in 2001, it was a larger percentage of a severely diminished total catch.

Oyster harvests from the bay have not fallen evenly. From 1980 to 2008, the Maryland harvest fell by 88 percent from 1980 levels, and the Virginia harvest fell by 64 percent. As the oyster harvest has declined, the value per unit of oysters has increased. The price of a pound of oysters in 2001 was over three times the 1975 value, an increase to be expected as supply declines. But between 1970 and 2001, inflation eroded the value of the dollar. After adjusting for inflation, price increases have not been enough to offset the decline in harvest. In nominal terms

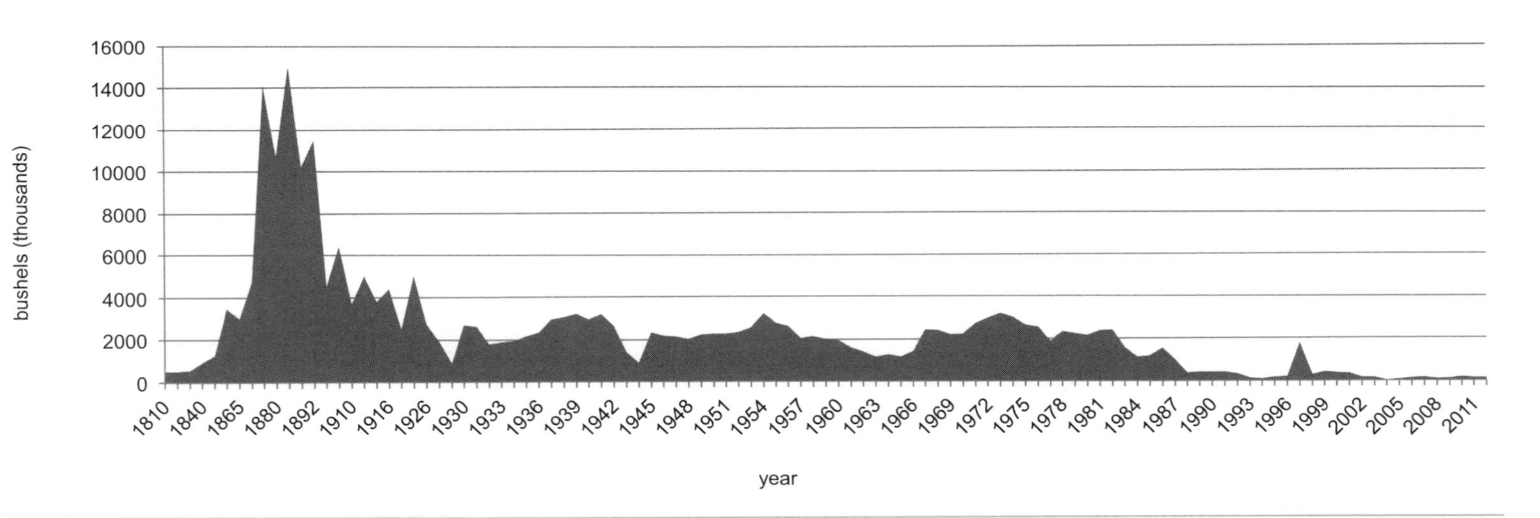

Figure 6.3 Maryland Oyster Harvest, 1810–2012

Table 6.2. Chesapeake Bay oyster catch: Maryland and Virginia

	1980	*1985*	*1990*	*1995*	*2000*	*2001*	*2008*
				Bushels landed			
Maryland	2,111,080	1,142,493	414,445	164,641	380,675	347,968	249,000
Virginia	1,074,776	627,052	162,618	55,414	22,623	22,573	382,000
Maryland share (%)	66	65	72	75	94	94	40
Virginia share (%)	34	35	28	25	6	6	60
				Oyster license holders			
Maryland	2,246	1,807	1,156	603	1,773	1,293	587
Virginia	2,622	2,081	1,782	399	295	320	420

Source: Data extracted from Nonnative Oysters in the Chesapeake Bay. Washington, DC: Ocean Studies Board, National Academy of Sciences, 2004. Data for 2008 added from the Maryland Department of Natural Resources and Virginia Department of Conservation and Recreation

(using the actual price for each year), the total value of the 2001 oyster harvest was less than the value of the 1980 harvest. The value of the 1980 harvest was $29.3 million, and in 2001 it was $4.3 million, a drop of about 85 percent. Expressing the total values of the harvests in real terms (2001 dollars), the drop was closer to 93 percent. In 2001 dollars, the harvest in 1980 was valued at $60.1 million, and the 2001 harvest $4.3 million. As the supply of oysters diminishes, it takes more work hours to harvest a pound of oyster meat. Oyster watermen are caught in a bind of falling revenues and rising costs.

There are some significant differences between the Maryland and Virginia portions of the Chesapeake Bay, including physical environment, state laws regulating the oyster fishery, and ownership of the oyster beds. Virginia's lower bay is plagued by more predators and diseases that prefer higher salinity. The lower salinity of Maryland's upper bay means fewer incidences of devastating diseases. Natural oyster bars are restricted by the 30-foot (9.14-meter) isobath, a line connecting points of equal water depth. The deeper waters of the middle bay have too little oxygen and too much unstable silt and clay sediment that can smother oysters.

Oysters reproduce by spawning eggs between June and September. The eggs hatch into larvae, which then develop in several stages. The larvae swim, eventually falling to the bottom, where they look for a place to attach, a process called spat setting. Spat are particular about where they attach, selecting only clean oyster shells or other clean objects. Growth is slow, taking about three years to reach 3 inches (7.69 centimeters). Oysters live only in salty water. They begin life as males and change to females at about three years of age.

Salinity is a major environmental factor affecting oysters. The Chesapeake oyster, *Crassostrea virginica*, is restricted to waters with salinity ranging from 5 to 30 parts per thousand (ppt). Bay salinity ranges from nearly 0 ppt at the northern head to 25 ppt at the mouth; it also varies by season and with climatic patterns. The eastern shoreline of Chesapeake Bay is more saline than the western shoreline, partly owing to the Coriolis force, which deflects freshwater moving down the bay from the north and west toward the Western Shore. Dense, salty water moving up the bay is deflected toward the Eastern Shore. At the same time, less saline water from the upper bay flows southward. The result is a slow churning action. Another reason for lower salinity in the upper bay and along the Western Shore is that the rivers flowing into the bay from the west carry much greater volumes than the rivers feeding the bay from the Eastern Shore. In the lower bay, the higher salinity allows a number of predators and diseases to flourish.

Oyster Diseases

Major predators and diseases have attacked the Chesapeake Bay oyster population over the past sixty years. Small oyster-killing snails, known as oyster drills *Urosalpinx cinerea* and *Eupleura caudate*, are prevalent in the Virginia and southern Maryland portions of the bay and have inflicted severe damage on the oyster crop. The wild oyster fishery in the mid-Atlantic region began to experience a serious decline in the mid-1950s. A mysterious parasite appeared in 1956, killing over 90 percent of the oysters in Delaware Bay. Because of the parasite's multispherical appearance, it was called MSX. By 1957 MSX, or *Minchinia nelson*, was spreading up the Chesapeake Bay, and by 1959 it had reached all the way into Maryland waters. MSX, which invades the connective tissue around the oyster's intestines, eventually resulting in death, has been a greater problem for Virginia than for Maryland. Much of the Maryland waters have been spared from MSX because the parasite cannot tolerate salinities less than 15 ppt. During dry periods when salinity in the upper bay increases, MSX moves northward. In Virginia, especially at the mouth of the James River and immediately to the north, the oyster fishery fatality level reached 60 percent.

Starting in the 1940s, oysters in the Gulf Coast of the United States began succumbing to a protozoan parasite, *Perkinsus marinus*, often called "Dermo" because it was originally classified as the fungal species *Dermocystidium*. This parasite invades the oyster through its digestive tract and kills it. Dermo first appeared in the Chesapeake Bay in 1949. By the late 1980s, the Dermo parasite had infested nearly every major oyster bed in the bay. This resilient parasite continued its march up the East Coast to Delaware Bay and has even appeared in the oyster grounds of Maine. Dermo extends up the Chesapeake Bay to the 15-ppt fall isohaline, restricting it to the extreme southern portion of Maryland waters. Dermo infects most of the oysters that have been on the floor of the southern half of the bay for over two years.

Oysters are recognized as a keystone species in the Chesapeake Bay ecosystem, so efforts are being made to save them. One adult oyster can filter about 40 to 50 gallons of water (151.5 to 189.4 liters) per day in warm weather. As filter feeders, oysters ingest algae (phytoplankton). Although algae form the base of the food chain and are vital to organisms higher up the chain, excessive amounts of algae, as are found in many parts of the bay, decay and deplete oxygen from the water. As a natural water purification system, oysters could help to remove more algae and improve water quality, which in turn would help sustain a larger oyster fishery.

Roger Newell, a researcher at the University of Maryland Center for Environmental Science at Horn Point, estimates that the bay's oyster population was once so vast it could filter the entire volume of the Chesapeake Bay (18 trillion gallons) in a few days during summer. Today's diminished oyster population, only about 1 percent of historic high levels, may require a full year to accomplish the same task (Mertz 1999). Many Maryland schoolchildren are familiar with a science project that illustrates the filtering power of oysters. Take two large jars of murky bay water. Place a few oysters in one jar but none in the other. After one or two hours, the jar with the oysters will be clear. The ability of mollusks to clean up a water body has been witnessed in Lake Champlain and the Great Lakes, where the inadvertent introduction of the zebra mussel led to an improvement in water clarity. Yet no responsible scientist would recommend introducing the zebra mussel to the Chesapeake Bay. The zebra mussel population grows rapidly and clogs intake pipes of power and water treatment plants, causing great damage. The bay already has a great filter feeder in the oyster; its restoration should be a high priority.

Until the mid-1990s, prospects for the restoration of the oyster fishery were not good. MSX and Dermo decimated nearly all of the major oyster grounds in the bay, and scientists still did not have a complete understanding of these parasites. In 1989, the US Congress enacted legislation that led to the creation of the Oyster Disease Research Program, which has made considerable progress since 1995. Oysters surviving research at the Oyster Disease Research Program facilities are now being put back into the water. Research on breeding oysters that are more tolerant to both MSX and Dermo is progressing at several universities in the region, including Rutgers University, the University of Delaware, the University of Maryland, and the Virginia Institute of Marine Science. Cross-breeding native oysters has produced an oyster that is resistant to MSX and more tolerant of Dermo. Called "CROSBreed," this oyster is being bred in hatcheries; field trial plantings of CROSBreed on oyster bars began in the fall of 1999. It has taken over a

DOES THE BAY NEED A NEW OYSTER?

Oysters serve several functions in the state of Maryland. In economic terms, they provide jobs and incomes for watermen, processors, food outlets, and businesses that service the oyster industry. Oysters also have important ecological functions. A large enough oyster population can better filter algae and sediment from the Chesapeake Bay, resulting in better water quality for all species. Oysters also build reefs that are important habitats for many species. There is a cultural function, too; the oyster is an icon of the bay as well as the entire region. For centuries, Marylanders have included the oyster fishery as an important part of their cultural identity. Eating succulent oyster during oyster season is part of the annual cycle of life in most of Maryland. With the Chesapeake commercial oyster fishery severely depleted, various interest groups have pressured lawmakers to take faster action. There have been long-running efforts to restore the oyster fishery based on the native eastern oyster, *Crassostrea virginica*. To date, these efforts have been unsuccessful. Because many watermen and processors want immediate relief for the oyster fishery and their economic interests, they have begun looking at an alternative solution—bringing a new type of oyster to the Chesapeake Bay.

The long-term historic decline of the Chesapeake Bay oyster population is attributed to overharvesting and habitat degradation. But neither of these causes will be ameliorated by the introduction of a new nonnative oyster. The major modern cause of high mortality in the native *C. virginica* population is disease, especially Dermo and MSX. Both were discovered in *C. virginica* in the 1950s and are more prevalent in the higher-salinity waters of the lower bay, primarily in Virginia waters. These diseases have also expanded northward into less saline Maryland waters.

In 1995, the Virginia legislature authorized the Virginia Institute of Marine Sciences (VIMS) to begin research on introducing nonnative oysters to the Chesapeake Bay. The first experiment was with the previously introduced *Crassostrea gigas*, also known as the Pacific or Japanese oyster (the most widely cultivated oyster in the world). The results were disappointing. Compared to the native *C. virginica*, *C. gigas* had both slower growth and survival in lower salinities. Scientists believe that the initial introduction of the nonnative *C. gigas* in the 1930s also inadvertently introduced MSX as the oyster spread from Delaware Bay into the Chesapeake Bay via the Chesapeake and Delaware Canal. This fact is foremost on the minds of scientists today as they study the introduction of yet another nonnative oyster. Once *C. gigas* was determined to be unsuitable, research efforts turned to the Asian Suminoe oyster, *Crassostrea ariakensis*. VIMS conducted some preliminary trials and found that *C. ariakensis* grew more rapidly than *C. virginica*, had less mortality, and an acceptable taste. While *C. virginica* takes two to three years to grow to market size (three inches in Maryland), *C. ariakensis* took only nine months. These test results excited the seafood industry, which called for a rapid adoption of *C. ariakensis* for the bay.

By July 2004, about 860,000 *C. ariakensis* oysters had been tested in the Chesapeake Bay. These test oysters were triploids, meaning that they were sterile, or not able to reproduce. Federal and state permits to experiment with these caged triploids require strict biosecurity and monitoring measures. Scientists know that there could be some diploid (fertile) oysters among the triploids, and some triploids could biologically revert to diploids. The result could be a free-living *C. ariakensis* population in the bay. The introduction of this nonnative species could be irreversible and could severely harm the health of the bay ecosystem, including the native *C. virginica*. There is also the possibility that private individuals could illegally introduce

C. ariakensis to the bay before any conclusive research is completed. Hurricanes and storms could also displace them into the wild from test sites and hatcheries.

At this point, scientists do not fully understand the biology of *C. ariakensis*. They do know that in its native Chinese waters, heavy mortalities from disease (80–90%) of *C. ariakensis* have occurred since the early 1990s. It is possible that *C. ariakensis* may introduce even more pathogens to the bay. *C. ariakensis* have also contracted Dermo (*Perkinsus mannus*) in Chesapeake waters. In one trial, 90 percent of *C. ariakensis* were infected, but only 20 percent of *C. virginica* were. *C. ariakensis* is also susceptible to herpes (Committee on Non-Native Oysters in the Chesapeake Bay 2003). Tests showing faster growth rates of *C. ariakensis* over *C. virginica* compared triploid *C. ariakensis* to diploid *C. virginica*. Yet sterile organisms often grow faster than fertile organisms because they do not direct resources to reproduction. Further tests demonstrated that in crowded conditions *C. ariakensis* grew more slowly than *C. virginica*. The growth advantage of *C. ariakensis* disappears under space-limiting conditions. When grown together, *C. virginica* will in the long run outgrow *C. ariakensis*, although *C. ariakensis* grows faster in the early stages (Luckenbach 2004).

The debate continues between those advocating immediate introduction of *C. ariakensis* and those calling for caution and more study. Various organizations and agencies have conducted studies and issued positions, including the Chesapeake Bay Foundation, Chesapeake Bay Program, National Research Council of the National Academy of Sciences, University of Maryland Center for Environmental Science, US Army Corps of Engineers, and Virginia Institute of Marine Sciences. Introducing a new species of oyster to the Chesapeake Bay is not just a scientific issue—it has also become a socioeconomic and political issue. There is still considerable uncertainty about the outcomes of introducing the nonnative *C. ariakensis* into the bay. Current research suggests that *C. ariakensis* is not a panacea for the woes of the Chesapeake Bay oyster fishery. Research to save the native *C. virginica* should continue. Long-term historic overharvesting and habitat destruction related to human activities (pollution, sedimentation, etc.) will not be overcome just by introducing a nonnative oyster. Policymakers face the challenges of incomplete knowledge of the biology of *C. ariakensis*, a lack of consensus among scientists about the impacts of a nonnative species on the bay, and pressures from commercial interests to revive the bay shellfish industry (Richards and Ticco 2002).

century of overfishing and pollution to decimate the oyster population, and it will take decades to restore it. Computer simulation models predict that increasing the oyster population to only 10 percent of its historic high would lead to better water clarity and the growth of vital sea grasses (Zimmerman 1997). Oysters are quite flexible and can tolerate a range of changes in temperature, salinity, and dissolved oxygen, but they are not as flexible about where they attach. They seem to live and thrive best on oyster shell reefs that rise off the bottom away from the bottom surface, where sediment can smother them. Restoring oyster habitats is another critical part of the overall Chesapeake Bay restoration effort.

Oyster Bars

In the Chesapeake Bay, the oyster beds with the best production lie between the spring 5-ppt and summer 15-ppt isohalines. This area, which lies entirely within Maryland, is protected from frequent severe floods from the north and predators from the south. Some scientists believe that good oyster years are those with dry, climatically stable summers; summertime is when oysters spawn and when spat (young oysters) attach to objects. Scientists from the University of Maryland regularly sample and count the spat in the bay's oyster bars; they have found the pattern to be erratic.

There are two kinds of oyster bars—growth bars and seed bars—found in different areas of the bay. Growth bars, which sustain rapid development because of the availability of food, are found in waters with greater circulation. Seed bars lie in quiet waters, which are needed to minimize larvae dispersal. Most of the seed bars are in Virginia, and most of the growth bars in Maryland. In 1970, over 90 percent of seed oyster production in the United States came from the Virginia's James River, which was then reputed to be the greatest natural seed-producing area in the world (Alford 1975). Because Virginia's waters are infested with various predators, disease, and pollution, it would be reasonable to try and move the seed oysters to Maryland's growth bars. Oysters will grow about 1 inch (2.54 centimeters) per year; it takes three years to reach the 3-inch (7.6-centimeter) size that is legal to harvest. If moved to cleaner waters soon enough, oysters are often able to flush out pollution and become fit for consumption. Unfortunately, however, most oysters that have lain on Virginia bottoms for more than two years are infected with parasites. The Chesapeake Bay oyster fishery would be more productive if Virginia watermen concentrated on seed oysters and Maryland watermen on growth and fattening. A major challenge has been that Virginia laws restrict the movement of large quantities of seed oysters to Maryland.

Virginia Oyster Bars. Oyster diseases such as Dermo and MSX devastated the Virginia oyster bars during drought years in the 1980s, when higher salinity allowed these diseases to flourish. Virginia's oyster bar recovery strategy has been to use "strong oysters" to help improve the oyster fishery. Large oysters that have survived diseases are brought in from Tangier and Pocomoke Sounds and then planted on sanctuary oyster reefs in Virginia waters. Scientists hope that these oysters will pass on their disease resistance to future generations. Virginia is taking a long-term ecological approach by constructing reefs, protecting them, and encouraging natural reproduction (Mertz 1999, 12). Shells are being dredged from upper bay areas that no longer support oysters owing to low salinity; they are then used in Virginia to construct three-dimensional reefs in intertidal areas that are ideal for seed oysters. About one-third of Virginia's oyster bars are leased to private individuals, who are also encouraged to put down shell. There are about 100,000 acres (40,469 hectares) of Virginia bars in private leases and about 240,000 acres (97,127 hectares) in public bars.

Maryland Oyster Bars. At first traditionally, but now legally, most Maryland oyster bars were considered public lands. About 10,000 acres (4,047 hectares) are leased to private individuals, while 285,000 acres (115,338 hectares) of oyster bars are public. The state of Maryland is working to restore its public bars, and it remains dedicated to the public oyster bar principle. In Maryland, oyster diseases are only a serious problem during dry years when salinity increases farther up the bay. Maryland now produces millions of "certified seed" oysters at several hatcheries. The spat are stocked in sanctuary oyster bars in areas where diseases are not likely to be severe. Maryland's public oyster bars are replenished through another program providing 200 million to one billion spat. A $300 surcharge on licenses and a $1 per bushel tax on harvested oysters help to fund this program. Maryland state officials point to studies indicating that three-dimensional reef shell piles are not effective means to restore bars. Instead, the Maryland strategy advocates spreading oyster shells in a thin layer (about 6 inches or 15.2 centimeters) to build bars on top of natural hills on the bay's bottom. It is believed that because sediment does not collect atop these hills, sediment suffocation is not an issue. Although Maryland watermen have long opposed private oyster leases, sentiment has changed somewhat in recent years.

Potomac River Oyster Bars. By law, all of the Potomac River oyster bars are public. Some sanctuaries have been designated, and some disease-free seed oysters and shells are being placed in some mid-river areas. Because the mouth of the Potomac River is very saline, it has a disease problem.

Maryland, the Potomac River Fisheries Commission, and Virginia are all using different strategies for restoring the oyster fishery of the Chesapeake Bay. Maryland sticks to its traditionally strong support for public bars. The Potomac River Fisheries Commission is not restoring reef habitats as in Virginia and Maryland. Virginia claims that the thin spreading of oyster shells, as in the Maryland program, is not effective. Scientists in Virginia point to historic evidence, from when early colonists arrived, of massive oyster reefs and plentiful oysters. Yet three-dimensional reef reconstruction is expensive and requires many tons of shells and heavy equipment.

The Oystering Process

Maryland oystering is done primarily from bay ports such as Annapolis, Baltimore, Cambridge, Crisfield, Deal Island, Rock Hall, Rock Point, Smith Island, Solomons, and Tilghman Island. The oyster season opens on October 1, when powerboat tongers are allowed to start oystering, and the hand-tonging season runs until April 15. Hand tonging is the method of taking oysters by scissoring two 20-foot-long (6.1 meters) wooden shafts with opposable steel baskets, or tongheads, at their lower end while at anchor. Hand tonging has changed little over the past 300 years. Most of the hand tongers are older men, as the younger watermen often opt for other methods. Most hand tonging is done in the rivers reserved for them. Patent tonging, which uses larger hydraulic tongheads, is a more economic and competitive method of oystering. One or two men, using foot pedals whereby the open tongs are dropped to the bottom, closed, and then wound in, operate patent tonger boats. The season for patent tongers also runs from October 1 through April 15.

On November 1, the aging yet graceful skipjacks are allowed to commence oyster dredging (or "drudging," as it is called on the Eastern Shore), which continues through March. In the 1890s there were as many as 2,000 skipjacks on the bay. The dwindling fleet of skipjacks was estimated at twenty-eight boats in 1980, but only thirteen working boats remained in 2010. As the last working commercial sailing craft in the United States, the skipjack has been placed on the National Register of Historic Places, quite a distinction for a seagoing vessel. This slowly disappearing boat, which has a crew of about six or seven, is expensive to operate and maintain. On Mondays and Tuesdays the crew are allowed to push their boat with a small motor-powered yawl, which helps to ensure that the crew gets a couple of days of oystering even when there is no wind. Skipjacks are restricted to certain oyster bars in the bay and in its tributaries. A Maryland law from 1865, still in force, requires that oyster dredging be done in motion under sail. Powerboats are not allowed to dredge in Maryland waters, with the exception of the small push boats carried by skipjacks.

Once landed at a port, oysters are loaded by conveyor belt onto a truck and taken to a processing plant to be shucked, cleaned, and canned by hand labor. Thomas Kensett of Baltimore was the first to can oysters in 1820. At one time, "buy boats" were plentiful on the bay. Buy boats would follow the skipjacks, buy their oysters, and then transport them to the shucking houses. In the off-season, the buy boats were used to haul grain, watermelons, and other freight. The few buy boats left on the bay follow the patent tongers to buy their catch. In the spring, buy boats catch, haul, and plant seed oysters. Another method of oystering is oyster diving. Although diving for oysters has been practiced in the Chesapeake Bay for the past thirty years, historians point out that Indians dove for oysters long before the first colonists arrived.

A big challenge for the oyster industry today is the lack of skilled shuckers. Although attempts have been made to develop mechanical shuckers, oysters are still shucked by hand. (An Eastern Shore tongue twister reads, "How many oysters could an oyster shucker shuck if there were enough shuckers to shuck oysters?") By 2000, with the tremendous decline in the bay oyster catch, oyster shucking as an occupation and way of life had faded to only a shadow of its past prominence.

In the 2007 to 2008 season, only 82,958 bushels of oysters were harvested in Maryland, with a dockside value of $2.6 million. The harvest can be disaggregated by gear type: hand tongs, 24,175 bushels (29%); divers, 11,745 bushels (14%); patent tongs, 15,997 bushels (19%); power dredge, 25,324 bushels (30%); and skipjack, 4,243 bushels (5%); not categorized, 1,564 bushels (3%).

Oyster Farming versus Oyster Hunting

In the first quarter of the twentieth century, it seemed that oyster farming, or aquaculture, would be the wave of the future. But by 2000, only 1,000 acres (405 hectares) of the mere 10,000 acres (4,047 hectares) of Maryland oyster bars under private lease were being actively cultivated. Strong opponents of aquaculture point to these figures as proof of the failure of oyster aquaculture to take hold in the Maryland waters of the Chesapeake Bay.

By the early nineteenth century, New England oyster grounds had been depleted. Taking this depletion as a warning, Maryland in 1885 legislated licensing, seasonal restrictions, and methods of oystering (e.g., prohibiting taking oysters from steamboats by dredging). Maryland had already passed leasing legislation in 1830 allowing citizens one acre of bottom for planting and growing shellfish, and in the late nineteenth century the oyster catch was bountiful. Most watermen who leased oyster grounds did not plant and raise oysters; they used the grounds to bed their oysters until the market price improved. Then, in 1875, the oyster catch suddenly dropped from fifteen million bushels to ten million. Alarmed, the Maryland General Assembly called for a study of the bay's oyster fishery.

From 1876 to 1877, Lieutenant Francis Winslow of the US Coast and Geodetic Survey did an intensive study of the oyster bars of the bay. Citing oyster depletion, Winslow recommended leasing large areas of oyster bars to private individuals to encourage oyster farming. A commission was set up to manage the oyster fishery, and one of its members was William K. Brooks, a prominent oyster biologist from Johns Hopkins University. Brooks took a strong stand in support of oyster aquaculture. He noted the severe opposition by the region's watermen, who seemed to believe that natural oyster bars were inexhaustible.

Virginia took the lead in Chesapeake Bay aquaculture in 1894 when it set aside 110,000 acres (44,516 hectares) for leasing. Those oyster bars accounted for most of the Virginia oyster harvest until the 1960s, when MSX attacked. Even today, only 10 percent of leased grounds are producing, but they account for 40 to 60 percent of the Virginia harvest (Leffler 2000). Maryland watermen continued to regard the leasing of oyster bars to private individuals as outrageous. Although they lobbied in Annapolis, the watermen were not able to prevent the passage of the 1906 Haman Oyster Act, which allowed a private planter to lease 30 acres (12.1 hectares) in tributaries, 100 acres (40.5 hectares) in Tangier Sound, and 500 acres (202.3 hectares) in open bay water. This act is still the basis for oyster farming in Maryland today, and yet most watermen remain unconvinced of its success. At the heart of the watermen's initial opposition was the fear of losing independence and becoming hired hands for large industry. Oystercatchers, after all, were men who sailed when they pleased and took what they wanted—theirs, they believed, were "ancient privileges" and "common law rights" (Leffler 2000). Watermen continued to lobby for amendments to weaken the Haman Oyster Act. State regulations favored the reseeding of public bars; they virtually prevented the sale of Maryland oyster seed to private leaseholders, who were forced to go to Virginia to buy seed. Until recently, private leaseholders were not allowed to harvest their bars with dredges; they had to use tongs. Oystermen were pitted against planters in what became a sort of range war on the Chesapeake. It was common for watermen to raid private bars; there were too few maritime police to cover the entire bay. Even today, poaching of planted oysters is still a problem. The threat of oyster poaching is a major impediment to more private leasing, as making such an investment is taking a big risk.

Despite the lengthy period during which oyster farming was legal in Maryland waters, state support was until recently heavily in favor of restoring public bars. Oyster farmers contend that farming makes them more independent. They can bed oysters from public bars (many farmers who lease private bars still work the public bars) until market prices are better, and in rough weather they can still work farmed plots because most are in tributaries. The many years of opposition to oyster farming hampered the development of other forms of infrastructure for the bay oyster fishery: hatcheries, seed bars, and sanctuary areas. By the late 1990s, efforts to plant seed oysters in the Maryland part of the bay had improved, but today there is still a severe shortage of seed oysters for both the public and private bars.

The Maryland Department of Natural Resources released encouraging news in March of 2011. The number of spat in Maryland waters was at its highest level since 1997, and more Marylanders planned to expand their oyster farming (aquaculture) businesses. Since 1939, the DNR has monitored the Maryland oyster population through annual sample surveys—one of the longest-running such surveys in the world. The survey tracks three important components of the oyster population: spatfall intensity, disease infection levels, and mortality rate. The encouraging news from the 2010 survey is that the increased spat count is widely distributed throughout the Chesapeake Bay and its tributaries. The heaviest counts are in the lower bay where salinity is higher, but moderate spatfall was also found in locations as far north as Pooles Island and the upper reaches of the Chester, Choptank, and Patuxent Rivers. Dermo infections were below the long-term average for the eighth consecutive year, and MSX seems to be in retreat once again.

Although the Maryland oyster fishery is much diminished from what it was in the late nineteenth century, the oyster has proven to be remarkably resilient. Even so, the successful restoration of the Chesapeake Bay oyster fishery will take more than relying on nature. Cooperation is needed by the state, farmers of private oyster bars, and watermen who favor public bar restoration.

7 The Minerals Industries

Maryland has a long and rich metallic mineral heritage. Beginning in the colonial period and continuing into the late 1800s, Maryland was a significant producer of metallic minerals, principally ores of iron, copper, and chrome. As new and more abundant sources of these ores were found elsewhere in the United States and around the world, Maryland's production gradually declined.

Captain John Smith reported the occurrence of iron ore in the clays along the Patapsco River estuary as early as 1608. During Maryland's days as a new colony, iron items such as tools, hinges, and nails were imported from England. Local iron production did not begin until 1681 (Edwards 1967). Early iron was made in small furnaces (called bloomeries) that produced spongy bloom masses of iron. This crude form of iron was then hammered into shape at a forge. Maryland had the requisites for producing iron: iron ore on the surface (often in bogs), hardwood to make charcoal needed for smelting, and easy water transportation. In 1719, the Maryland Assembly passed an "Act for the Encouragement of an Iron Manufacture within this Province" (Edwards 1967), and soon Maryland was producing enough iron to meet its own needs. Exports were sent to England, which was running short of hardwood trees for making charcoal.

Cecil County was the site of the Principio Furnace, Maryland's first furnace producing good-quality, slag-free iron. A second furnace was opened in 1723 on Gwynns Falls. Between 1719 and 1885, about fifty-five furnaces using local bog iron ores found in clay were operated in Maryland. As the United States industrialized and its railroads expanded, the demand for iron increased dramatically, especially from 1830 to 1855. The principal furnaces in Maryland included Curtis Creek Furnace near Glen Burnie in Anne Arundel County (established in 1759), Antietam Furnace near Antietam in Washington County (1765), Catoctin Furnace near Thurmont in Frederick County (1774), Friendsville Furnace near Friendsville in Garrett County (1828), Nassawango Furnace west of Snow Hill in Worcester County (1828), Georges Creek Coal and Iron Company Furnace at Lonaconing in Allegany County (1837), Ashland Furnace near Cockeysville in Baltimore County (1837), and Muirkirk Furnace in Prince George's County (1847); (Thompson 1977).

The last furnace to produce iron from local ores was the Muirkirk Furnace, which closed in 1916. Discoveries of vast deposits of rich iron ores in the Mesabi Range on the Superior Upland of Minnesota, along with cheaper and more efficient transportation, were the chief factors in the demise of the Maryland iron industry.

Early Maryland furnaces processed four different kinds of iron ore minerals:

1. *Hematite* was present in Allegany County on Evitts Mountain, Wills Mountain, and south of Tussey Mountain; this deposit was part of the famous Clinton iron ore of the Appalachian chain. Hematite was also mined with copper from a body of igneous rock between Sykesville and Finksburg in Carroll County.
2. *Limonite* was found throughout Maryland, most often in swamps or bogs; it came to be called "bog iron."
3. *Magnetite* was found on the east side of Catoctin Mountain near Thurmont, Frederick County, and in the schists of the western portions of Carroll, Harford, and Howard Counties.
4. *Siderite* occurred in the clays of the Arundel formation on the western edge of the Coastal Plain in Anne Arundel, Baltimore, Cecil, Harford, and Prince George's Counties.

Maryland was an important copper producer from before the Revolutionary War into the 1840s. Copper had been found in Maryland in 1722, but profitable extraction did not begin until 1750. By 1890, the rich copper ores of Lake Superior had driven all of the major Maryland copper mines out of business. Four areas produced copper: the Bare Hills District of Baltimore County, the Linganore District of eastern Frederick County and western Carroll County, South Mountain, and the Sykesville District of southeastern Carroll County. The Linganore District was the largest copper ore producer and was mined for the longest period of time. The last copper ores were mined in 1917 at New London mine in the Linganore District.

At one time all of the world's chromite (chrome ore) came from Maryland and Pennsylvania. Chromite resides in veins of serpentine in the eastern Piedmont running from southeast Pennsylvania to the Potomac River. In Maryland, the chromite-bearing districts were Jarrettsville-Dublin in Harford County, Montgomery County, Soldiers Delight and Bare Hills in Baltimore County, and State Line in Cecil County. Of these districts, Soldiers Delight was the most productive. In 1810, chromite was found at Bare Hills on the property of the Tyson family, and mining began in 1822. From 1827 to 1860, chrome mining was the monopoly of Baltimore industrialist Isaac Tyson Jr., who also controlled the Jarrettsville locality where the Reed mine became the second-largest chromite mine in the United States. The Live Pit mine in the State Line District lay astride the Mason-Dixon Line; the mine opening was in Pennsylvania but the operation occurred beneath Maryland. After the discovery of chromite ores in Turkey in 1848 and in California in 1860, Maryland's production and export declined. By the 1880s, most of the Maryland mines had closed, although some minor production continued until 1920 in the Soldiers Delight area.

Sparrows Point

Local metallic ores are no longer produced, but Maryland has a modern metallurgical industrial complex centering on the steel operations at Sparrows Point in Baltimore County. Operating at the same site from 1890 to 2013, Sparrows Point was founded after iron ore was discovered in Santiago, Cuba. This ore was imported into the United States to the new coastal mill at Sparrows Point, which was initially financed by Philadelphia businessmen. An extensive company town soon emerged, including a company store, churches, public schools, and about 500 houses.

In 1916, Charles Schwab, owner of Bethlehem Steel, bought Sparrows Point. During the Great Depression of the 1930s, Bethlehem Steel faced serious financial trouble, but it recovered during and after World War II. During the war, Sparrows Point turned out gun barrels, steel plate to make liberty ships in the nearby shipyards of Baltimore, steel for landing craft, barbed wire, grenades, and airplanes. After the war, Sparrows Point continued to prosper. Its location in the huge steel-consuming market of megalopolis was a key asset. Low-cost iron ore came by ship directly to the mill from new mines in Venezuela. Bethlehem Steel made huge profits, but it failed to invest in research and the new technologies being employed by European and Japanese steel makers.

In the 1950s, Sparrows Point was at its prime, the flagship of Bethlehem Steel. It was then the largest steel mill in the world. The mill consumed 1/500th of the total US output of electricity, burned 6 million tons (5,443,080 tonnes) of coal annually, and poured 22,000 tons (19,958 tonnes) of hot metal a day (15 tons per minute). During the post–World War II boom, Sparrows Point churned out steel bridge sections for the new Bay Bridge (opened in 1952); steel beer cans for National Beer (makers of National Bohemian, a true Baltimore favorite that is affectionately called "Natty Boh"); steel for big-finned American cars of the 1950s; and sheets of steel for stoves, refrigerators, and countless other products.

Sparrows Point epitomized the blue-collar economy of the Baltimore region. The area around the mill had its own microclimate. "Clouds of sulfuric acid rose from the pickling vats, catching the uninitiated by the throat. By day a shroud of gritty orange smothered the peninsula, while at night the heat and flames projected a quavering glow across the horizon that was visible for miles" (Reutter 2004). Baltimore was a well-defined urban heat island, largely because of its industries.

In 1960, Sparrows Point employed 30,000 workers. But by 2000, there were only 4,000 working at the mill. What happened? The 1960s brought increasing competition from companies making substitutes for steel. Reynolds Metals in Richmond, Virginia, produced low-cost aluminum cans, taking over the beer and soft drink market. The steel can was out, and Bethlehem Steel lost a major market. Then came steel imports from Japan at comparable quality but lower prices. Plastics and aluminum continued to cut into the steel market. Bethlehem Steel and other US steelmakers responded by raising prices and lobbying Congress for protection from imports. The 1982 recession hit Bethlehem Steel and the communities of Dundalk and East Baltimore hard. After $1.5 billion in losses, the company shut down many of the mill's units and laid off thousands of workers. Most of these workers never returned.

There was a brief upsurge in steel sales in the early to mid-1990s, and eventually Bethlehem Steel invested over $600 million to renovate the mill. The 110-year-old mill's revival centered on a new $300 million cold mill. The strong US economy of the 1990s fueled strong steel demand. But strong foreign competition and increasing steel imports muddied the waters. Steel imported into the United States jumped from about 25 million tons (22.7 million tonnes) in 1995 to about 42 million tons (38.1 million tonnes) in 1998. Steel manufacturers, including Bethlehem, continued to lobby the US Congress to maintain tariffs on steel imports. The opening of the new cold mill in May 2000 helped Bethlehem Steel cut costs at Sparrows Point, but one of the major tradeoffs was a reduction in the number of workers. The union agreed to 400 job cuts, mainly through attrition; the new mill required half the 800 workers of the old mill (Henry 2000). The future of the Sparrows Point steel mill remained uncertain. The new cold mill had state-of-the-art technology, but foreign competition drove down prices near the point where the mill was finding it difficult to remain competitive.

The pressures of steel imports, lack of new steel markets, and depressed prices led to severe financial losses. In 2001, Bethlehem Steel filed for Chapter 11 bankruptcy. At the time *Baltimore Sun* reporter Mark Reutter described what he found traveling down the spine of Patapsco Neck, past Dundalk to the mill on the fringe of Baltimore Harbor: "There you'll find a land of fantastic shapes and sizes—deltoid furnaces, long-armed cranes and broad-beamed buildings, some spilling out masses of railroad tracks, other eerily quiet and empty-handed, industrial hulks in search of work" (Reutter 2001). Bethlehem Steel, founded in 1857, dissolved in 2003, and the Sparrows Point mill was sold to International Steel Group, which merged with Mittal Steel in 2005.

The Mittal Sparrows Point steel mill became part of what is now the largest steel company in the world. Mittal Steel operates in sixteen countries on four continents and employs 165,000 people. The company is based in the Netherlands and has its US headquarters in Chicago. The founder of the company, Lakshmi Niwas Mittal, is known for his strategy of acquiring distressed steel mills and making them profitable by stringent management. Mittal now controls four of the five major steel mills in the Great Lakes region (only US Steel in Gary, Indiana, has escaped Mittal's control). He owns 40 percent of US production of flat-rolled steel used to make vehicles and appliances.

By early 2008, the Mittal Sparrows Point steel mill remained the only fully integrated steel mill in the United States with direct ocean access. Its large and efficient blast furnace was capable of making 3.9 million tons (3,538,000 tonnes) of raw steel annually. The main markets for the mill's rolled steel sheets, galvanized sheets, and tin mill products are the construction, automotive, container, and appliance industries. In late February 2007, the US Justice Department ordered Mittal Steel to sell Sparrows Point. Citing antitrust laws, the Justice Department determined that Mittal Steel had too much control over the US market for tin plate used to make cans for food, aerosol spray cans, and other products. The event that led to this decision was the merger of Mittal Steel with Luxembourg-based Arcelor SA to create the world's largest steelmaker. Mittal wanted to sell its plant in Weirton, West Virginia, but the Justice Department decided otherwise. In August 2007, Arcelor-Mittal agreed to sell the Sparrows Point mill to E2 Acquisition Corporation, a consortium of steel production companies headed by Esmark, based in Chicago. But this deal was not to be. E2 could not pull together the financing nor come to an agreement with the United Steelworkers. The deal fell through, and Sparrows Point was once again on the market.

In March 2008, Russian steelmaker OAO Severstal announced that it would buy the Sparrows Point mill and invest about $500 million over the next five years to modernize the plant, using Sparrows Point as a "swing plant." Swing plants like Sparrows Point step up or reduce production (and employment) to complement production at other plants owned by the parent company. But changing Sparrows Point into a swing plant resulted in a roller coaster ride for workers and the local economy as production steps up and slows down to meet demand, resulting in unpredictable employment for workers.

Severstal (which means "northern steel" in Russian) is staking out its place in the global steel industry. It also owns a mining division and has controlling interest in Grupo Lucchini, an Italian-French iron and steel company. Although only one-seventh the size of Arcelor-Mittal, Severstal is growing rapidly. Its principal owner is Alexei Mordashov, a Russian oligarch touted to be one of the twenty wealthiest people in the world. There are several reasons why Severstal sought to expand into the United States. Domestic suppliers cannot meet the demand for steel in the United States. The precipitous decline of the value of the dollar makes US purchases a bargain for many foreigners. As the costs of transporting imported steel into the United States rise, an in situ plant becomes even more cost effective. Under Severstal, however, the mill experienced repeated shutdowns and layoffs (Dance 2011). In March 2011, Severstal announced that it had reached a deal to sell the Sparrows Point steel mill to the Renco Group, a New York–based, family-owned investment company. But by January 2013, this new owner gave up on the steel mill, turning over the mill and its equipment to a liquidator for sale.

The Sparrows Point steel mill is part of the history of American industrialization. Its rise to success, decline, demise, and adjustments to globalization reflect the course of the nation's economy over the past 150 years.

Contemporary Mining

Commercially important mineral resources extracted in Maryland are those used for building materials and fuels. In 2007, the value of nonfuel minerals extracted in Maryland was $673 million. Maryland then ranked thirty-third among the fifty states in total nonfuel mineral production value, accounting for about 1 percent of the US total. By value, crushed stone was Maryland's leading nonfuel mineral, followed by Portland cement and construction sand and gravel (table 7.1). Maryland's mineral resources reflect its varied underlying bedrock and geology. Some of these resources were more important in the early economy of Maryland than they are today.

Coastal Plain

The Coastal Plain is a resource area for the basic minerals of which it is composed: sand, gravel, and clay. Sand and gravel are used in construction and accounted for about 16 percent of the value of all nonfuel mineral production in Maryland in 2008. Much of Maryland's extraction of these items comes from the Coastal Plain because it requires only sorting or sifting rather than more expensive crushing. Another important resource of the Coastal Plain is clay, which is used for making bricks, terra cotta items, and pottery (this latter type is called "balled clay"). Iron was once produced from local ores, especially in the 1700s to the early 1900s. Kaolin (white clay used in ceramics) was once extracted in large quantities in northeastern Maryland. Coastal Plain clay was once used extensively to manufacture firebricks that can withstand heat inside furnaces and kilns. The Calvert Formation was once the sole source in the United States for diatomaceous earth, which consists of hard siliceous shells or microscopic plants called diatoms. This material, used as an abrasive or in filtering, now comes mostly from California (Schmidt 1993).

Table 7.1. Nonfuel mineral production in Maryland

	2002		2003		2004		2005		2006		2007	
Mineral	*Quantity (metric thousand tons)*	*Value (thousands of dollars)*	*Quantity (metric thousand tons)*	*Value (thousands of dollars)*	*Quantity (metric thousand tons)*	*Value (thousands of dollars)*	*Quantity (metric thousand tons)*	*Value (thousands of dollars)*	*Quantity (metric thousand tons)*	*Value (thousands of dollars)*	*Quantity (metric thousand tons)*	*Value (thousands of dollars)*
Clays	268	550	269	550	262	571	317	686	286	851	173	NA
Crushed stone	22,300	141,000	26,200	165,000	29,900	185,000	33,500	277,000	32,000	317,000	31,100	282,000
Dimension	21	2,120	24	2,700	27	9,580	26	3,010	14	1,750	17	2,680
Gemstones	NA	1	NA	1	NA	1	NA	1	NA	1	NA	1
Portland cement	1,880	140,000	2,200	147,000	2,520	175,000	2,550	210,000	2,650	327,000	3,000	265,000
Sand and gravel	12,200	83,500	11,800	79,900	12,700	75,500	12,300	89,500	11,900	96,700	12,400	123,000
Other		33,500		31,700		35,400	NA	NA	NA	NA	NA	NA
Total	36,669	400,671	40,493	426,851	45,409	481,052	48,693	580,197	46,850	743,302	46,690	672,681

Source: US Geological Survey and the Maryland Department of Environment, Mineral, Oil, and Gas Division

NA, not available.

Piedmont and Blue Ridge

Much of these provinces consist of hard metamorphic rocks used in building stones. Granites—including Ellicott City, Guilford, Port Deposit, and Woodstock types—are still quarried, but to a lesser degree than in the past. A number of roofing slate quarries operated in this region in the past, but none do today. During the nineteenth century, Maryland was a supplier of building stone such as granite, marble, and sandstone. In the twentieth century, however, concrete and steel largely replaced stone as a common building material, and Maryland's production of cut stone nearly ceased (Weaver, Croffoth, and Edwards 1976).

The Piedmont has supplied building stone since the early history of Maryland. Defined as any massive, dense rock suitable for use in construction, building stones are chosen for durability, attractiveness, and economy. A dimension stone is quarried and cut in blocks to specific dimensions. A decorative stone can be quarried, cut, or carved. It is highly valued for its appearance and is used mostly on the interior of buildings or for monuments. *A Geologic Walking Tour of Building Stones of Downtown Baltimore, Maryland*, by Sherry McCann-Murray, details the various buildings around downtown Baltimore and explains the types of building stones used in their construction and where in Maryland the stones came from. Some of the building stones of the Piedmont are found in famous landmarks. Cockeysville marble was used in the Washington Monument in Baltimore (which predates the Washington Monument in Washington, DC), the national Capitol, as well as other famous public buildings in the eastern United States. The white marble steps for which Baltimore row houses are so renowned were usually made of Cockeysville marble. The western Piedmont's sedimentary sandstones have been used in many buildings. Usually called "Seneca Red Stone" or "brownstone," they are prominent in the Smithsonian Institution's castle in Washington, as well as in many of the brownstone houses of Washington and Baltimore.

The Texas quarry in Cockeysville, Baltimore County, is still one of the largest producers of crushed stone in the United States. Sedimentary limestone from the Monocacy Valley of Frederick County has been an important component in the manufacture of cement for many years. Limestone is also used as crushed stone and as agricultural lime. The Piedmont was once a source of metals including iron, copper, chromite, and small amounts of gold, silver, manganese, molybdenum, and titanium. The aluminum plant in Frederick County uses imported bauxite. In 2004, Maryland ranked ninth among the twelve states producing primary aluminum.

GOLD MINING

You might be surprised to learn that gold has been mined and prospected in Maryland for years. Gold lies in the metamorphic rock belt of the Piedmont, mainly in quartz veins. The concentration is sporadic, ranging from 0.1 to 5 parts per million (Reed and Reed 1969). The first record of gold being found in Maryland was from the farm of Samuel Ellicott near Brookville, Montgomery County, in 1849. During the Civil War, a union soldier stationed just outside Washington, DC, found gold while washing a skillet in a stream near Great Falls. By 1867, the first shaft of the Maryland Mine was sunk nearby (Kuff 1987). Although the Maryland Geological Survey lists forty-five sites stretching along the Piedmont from Harford County to Montgomery County where gold was mined, prospected, or found, most of Maryland's gold resides around Great Falls in Montgomery County. Gold was mined in this location until just prior to World War II, but production never exceeded 1,200 troy ounces in any given year.

No commercial gold mining is currently being done in Maryland, but some recreational gold prospecting still occurs. The Maryland Geological Survey recommends prospecting downstream from a quartz formation. A vein of gold cutting through the quartz formation, much like a human vein, might appear as grains, sheets, or in a wiry form. Small amounts of gold break off, mainly from weathering and flow downstream. Because gold is heavier than sand, it sinks to the bottom of the streambed. There is a surge of interest in gold panning every few years when the price of gold rises. Most of the gold found by panning is very fine grained, but it can range up to the size of coarse sand. Only rarely are nuggets found, but they do exist and have been found weighing up to 4 ounces. Some of the best gold areas near Great Falls are on nationally owned parklands; panning here requires permission of the state superintendent of parks. Panning on Maryland state parklands requires permission by law from the Maryland Forest and Park Service.

Valley and Ridge, and Allegheny Plateau

Nearly all of Maryland's coal resides in the geologically younger rock formations of the Allegheny Plateau (fig. 7.1). Some natural gas can also be found in the plateau and is actively being extracted. Each of the gas fields is in an anticline or, as near Accident, on a geologic dome. Gas can collect in permeable rock high in the anticline fold, similar to hot air rising. The gas is held by an impermeable rock overlayer (often shale). Limestone is found throughout the Valley and Ridge as well as the Allegheny Plateau. It is used mostly as crushed stone for building and for cement making. In Washington County, weathered clay is used to make bricks.

The most important mineral extracted in Maryland, by total value, is bituminous coal, which has been mined in Maryland since the late eighteenth century. All of the coal production in Maryland occurs in Allegany and Garrett Counties. Present estimates of coal, recoverable under present economic and technological conditions, are over 850 million tons (771 million tonnes).

Coal

Maryland is not well known nationally for its coal production, especially when compared to the prominence of nearby Pennsylvania and West Virginia. Many people from outside the state are surprised to find that coal is mined in Maryland at all; however, coal has long been an important industry for Allegany and Garrett Counties. As oil and gas prices skyrocketed in the late 1970s and again at the beginning of the twenty-first century, increasing attention was focused on domestic coal reserves.

History of the Maryland Coal Industry

The first actual discovery of coal in western Maryland is not known, but George Washington suggested in 1755 that the fuel of the future might lie beneath the soil of western Maryland (Bode 1978). Limited coal production began in the 1780s, when small amounts were mined to supply the frontier outpost at Fort Cumberland. A map published in 1782 indicated a coal mine near the mouth of Georges Creek, and by the early nineteenth century a small number of companies were tunneling into the coal seams. One of these early localities was about a mile north

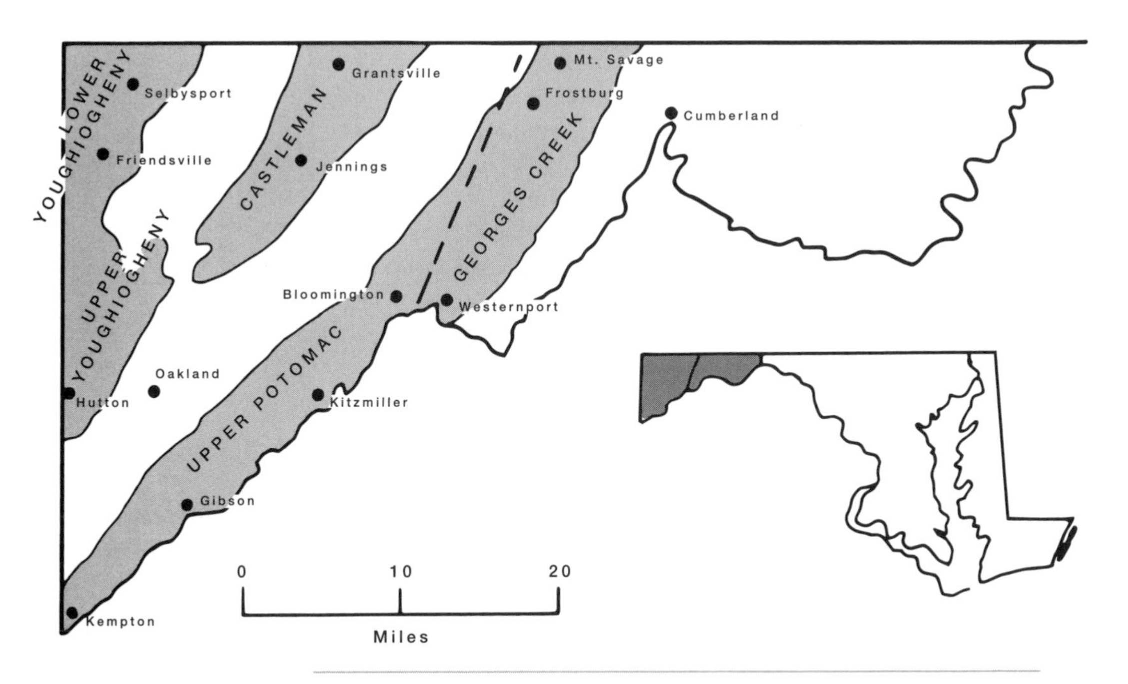

Figure 7.1 Maryland Coal Basins

of Frostburg in the valley of Jenning's Run. Between 1828 and 1838, the Maryland Assembly incorporated a dozen coal companies even though coal production remained small.

The remote location of coal in western Maryland relative to the markets to the east was at first a problem. Geographers often cite complementarity, transferability, and intervening opportunity as the three major principles of spatial interaction (i.e., the back-and-forth movement among places of services, goods, information, and other items). The supply of coal in western Maryland and the market demand in Baltimore, Washington, DC, and beyond certainly provided complementarity. Coal could flow east, and goods from the east were needed in western Maryland. Transferability was a major challenge in these early years, for it was not yet possible to easily move the coal in commercially profitable large amounts to the east before the canal and railroad were built. The last principle of spatial interaction, intervening opportunity, was also evident at this time. Until transferability could be established, other coal sources in Pennsylvania provided coal to Baltimore and Washington over the waterways and roads linking them. It would not be long before western Maryland itself was to become an intervening opportunity for that market. The construction of a new transportation infrastructure changed the relative location of western Maryland by affording its coal the opportunity to intervene geographically between other coal sources and the markets to the east.

The chief difficulty was getting the coal from remote western Maryland to customers. At first, coal traveled down the Potomac River in flatboats, with the first recorded shipment in 1820. Coal worth 6¢ to 8¢ a bushel in Cumberland brought 50¢ to 60¢ at the dock in Georgetown (Bode 1978). The situation for Baltimore was not as good. In 1828, construction of the Chesapeake and Ohio Canal began, but the canal was being built from Georgetown to Cumberland. In that same year, construction commenced on the Baltimore and Ohio Railroad, and by 1842, the railroad had reached Cumberland, 178 miles (286 kilometers) west of Baltimore. The Chesapeake and Ohio Canal was opened between Georgetown and Cumberland in 1850. Over 21 million tons (20,865,140 tonnes) of coal were shipped over the canal before it finally closed in 1923. Maryland coal production increased rapidly from 1,708 tons (1,549 tonnes) in 1842 to 5,532,628 tons (5,018,094 tonnes) in 1907. Given good transportation, coal was nothing less than black gold. During the last half of the nineteenth century, demand grew in the industrializing city of Baltimore. Coal was needed to keep the growing number of Baltimoreans warm and to produce steam power demanded by the growing number of factories.

The men who mined the coal had difficult, dirty, and dangerous lives. They burrowed into the hillsides, putting up timber braces as they went, since the bituminous coal was soft and apt to crumble at any time. Because they were paid by the weight of coal extracted, as little time as possible was spent on constructing braces, and as a result accidents were common. Miners were paid by the long ton (2,240 pounds). A miner digging two long tons a day in 1839 received 50¢ per ton (Garrett County Development Corporation 1978). Company towns were established at mining sites, such as the Georges Creek Coal and Iron Company at Lonaconing in Allegany County (Weaver, Croffoth, and Edwards 1976). At Lonaconing, company rules specified that every employee must work from sunrise to sunset on every day of the year except Sundays and Christmas Day. Employees were paid once a month, after deductions for debts at the company store, mills, and post office. Although the first mining settlements consisted of company-owned housing, the mining companies found that it was more profitable to sell houses to miners than to own and maintain them, and a high proportion of Maryland coal miners became homeowners. A Maryland state law in 1868 outlawed company stores in the Allegany coalfields, and many miners in Maryland managed to avoid the chronic indebtedness that ensnared many coal workers in Pennsylvania and West Virginia (Brugger 1988).

After peaking in 1907, coal production averaged 4.5 million tons (4,081,310 tonnes) per year until 1918. It declined after World War I and reached a low of 1,281,413 tons (1,162,242 tonnes) in 1938. The severe decline after 1929 reflected the economic downturn during the Great Depression, recurrent labor problems, and the replacement of much of the coal consumption by petroleum. A slight revival during World War II was followed by a major labor strike in 1948, which caused many mines to close. By 1954, fewer than 500,000 tons (453,500 tonnes) were being mined. From 1954 to 1981, the trend of coal production was upward, primarily because of the increased use of coal to generate electricity. From 1981 to 2005, total coal production in Maryland fluctuated between about 3 million and 5.8 million tons (2,721,540 and 5,261,644 tonnes). Total coal output reached a high point in Maryland in 2005, even exceeding the long-held record of 1907.

Deep Mining and Surface Mining

The mix of deep and surface mine production in Maryland has changed over the decades (fig. 7.2). Prior to World War II, deep underground mines accounted for nearly all Maryland coal production. The rising demand for coal during the war as well as a shortage of labor led coal companies to turn to surface mining. Surface mine production continued to increase, surpassing deep mine production in the mid-1950s. By the mid-1970s, surface mining reached its peak production, while deep coal mining had seriously declined.

Maryland's total coal production increased rapidly in the 1970s to meet demand during the energy crises. By 1975, about 97 percent of coal mined in Maryland was surface stripped. By the late 1970s, deep coal mine production also began to rise rapidly, largely accounted for

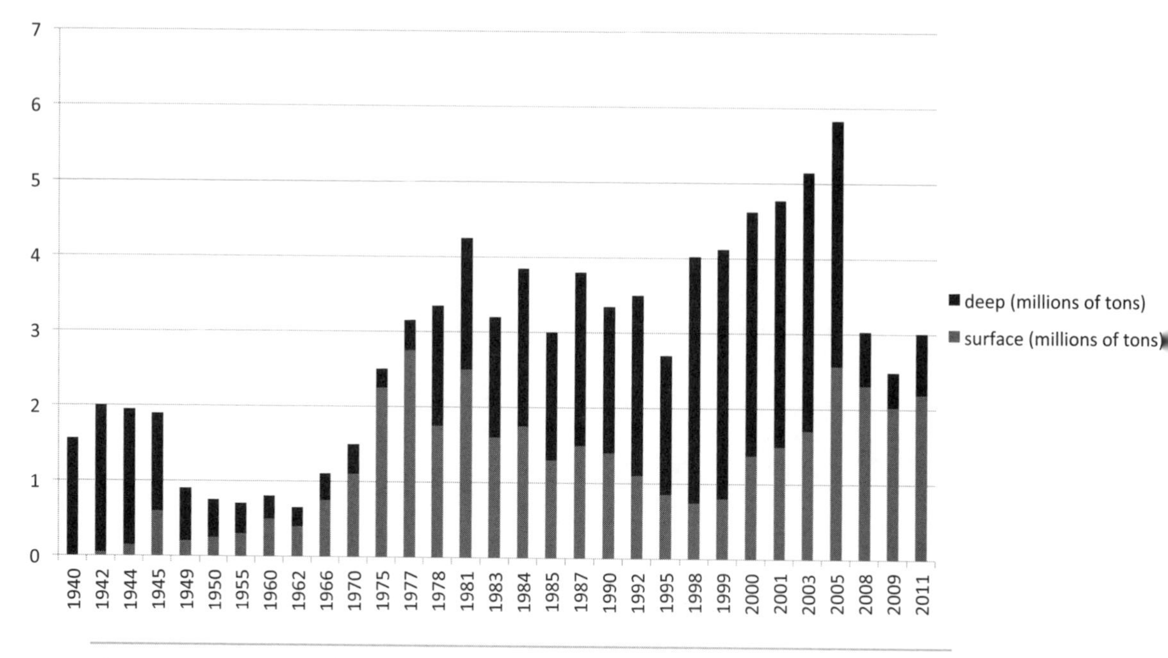

Figure 7.2 Maryland Surface and Deep Coal Production, 1940–2011

by the state's biggest underground mine, the Mettiki mine (Mapco Coal) in Garrett County. By 1983, deep coal production nearly equaled surface production, and by 1984 it surpassed it. Since 1983, surface coal production in Maryland has decreased while deep coal production has increased. All of Maryland's operating deep coal mines in 2005 were in Garrett County. Just three years later, that pattern of surface versus deep mining had changed once again. The total quantity of coal mined dropped to just over three million tons, but surface coal mining now out-produced deep coal mining. The number of people employed in Maryland coal mining reflects the history of Maryland's coal mining activity and changing technology (table 7.2). Employment peaked at about 5,000 in 1907; by 2009 there were a mere 364 people working in Maryland coal mines.

Coal Basins

Maryland's coal deposits reside in the westernmost portions of Garrett and Allegany Counties. All the coal being mined there is bituminous and is found along the eastern fringe of the great Appalachian coalfield extending from Pennsylvania to Alabama. The coal reserves of Garrett and Allegany Counties are found in five synclinal basins: Casselman, Georges Creek, upper Potomac, and the lower and upper Youghiogheny (fig. 7.1). The Georges Creek and Potomac Basins together account for 67.5 percent of Maryland's estimated recoverable coal reserves. In 2009, Maryland coal mines produced 2,491,189 tons (2,259,957 tonnes) of coal, all of which came from the Georges Creek and the upper Potomac Basins; none of the other three basins reported any production (table 7.3).

The layers of rock in these basins have been downfolded into elongated troughs (synclines) so that the beds on each limb dip toward the centerline of the basin (fig. 7.3). The basins, composed of Pennsylvanian and Permian age sediments, traverse the two counties in a northeast to southwest direction. Most of the rock strata in the coal basins are shale, siltstone, sandstone, or limestone; the coal seams themselves constitute only a small part of the total rock volume. The coal of western Maryland is typical of low volatile coals of the Appalachian Basin, and has a high British thermal unit (BTU) value and sulfur content.

The Casselman Basin

Situated in central Garrett County, the Casselman Basin is actually the southern end of the Somerset (or Berlin) basin of Pennsylvania. It is 18 miles (29 kilometers) long and 5 miles (8 kilometers) wide, running from Deep Creek Lake north to the state line. In contrast to the Georges Creek and upper Potomac Basins, the coal industry in Casselman Basin began to diversify earlier. Agriculture is more prevalent on the less rugged terrain and richer soils. The coal seams here are not especially

Table 7.2. Employees in Maryland coal mines

Type of mine	*Allegany County*	*Garrett County*	*Total*
Deep	0	65	65
Strip	154	145	299
Total	154	210	364

Source: Maryland Bureau of Mines, Annual Report 2009

Table 7.3. Maryland coal production, 2009

Basin	*Production (tons)*	*Percentage of state production*
Georges Creek	1,588,162	63.8
Potomac	903,027	36.2
Cassleman	0	0
Lower Youghiogheny	0	0
Upper Youghiogheny	0	0
Total	2,491,189	100

Source: Maryland Bureau of Mines, Annual Report, 2009

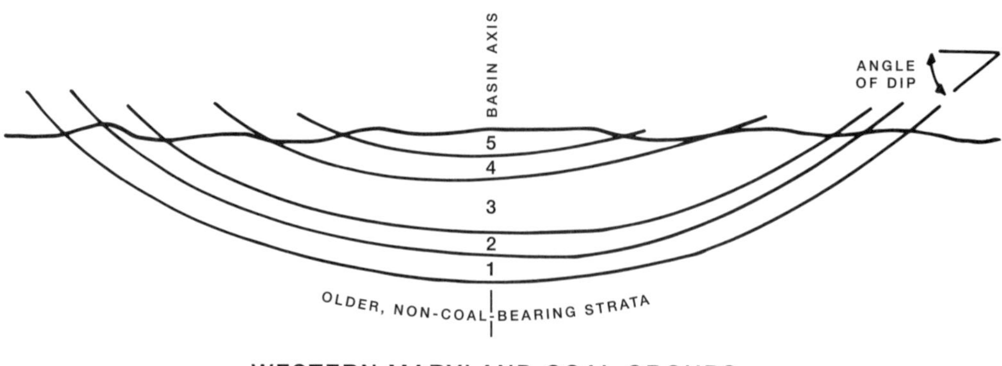

WESTERN MARYLAND COAL GROUPS

1- Pottsville (oldest)
2- Allegheny
3- Conemaugh
4- Monongahela
5- Dunkard (youngest)

Figure 7.3 Geologic Structure of a Coal Basin

thick, and a good portion of the land is neither suitable nor available for surface mining. Growing residential and commercial areas around Grantsville and Deep Creek Lake and those along Maryland Routes 40, 48, and 495 limit the areas for surface mining, as do the state forest lands on the flanks of Negro and Meadow Mountains. The Casselman Basin still ranks third in total land area and reserves among the five coal basins, but 1994 was the last year that any coal production was reported from this basin (Maryland Bureau of Mines 1994–2009).

The Georges Creek Basin

The Georges Creek Basin is located between Big Savage Mountain on the west and Dan's Mountain (Allegheny Front) on the east. This basin, which measures 5 miles (8.05 kilometers), wide stretches 21 miles (33.8 kilometers) from the Pennsylvania line (continuing into Pennsylvania as the Wellersbury Basin) south to the Savage River, which forms the boundary between the Georges Creek and the upper Potomac Basins. About three-fourths of the Georges Creek Basin lies within Allegany County, and it is where the most complete cross section of coal-bearing rock in Maryland is found. Most of the past mining activity in Maryland took place here, and numerous inactive deep mines dot the area. Over two-thirds of the coal mined in Maryland in the first half of the twentieth century came from the Georges Creek Basin. The presence of the high-quality Pittsburgh seam measuring 14 feet (4.27 meters) thick explained the high production. The Pittsburgh seam has been largely exhausted, accounting for only 1.5 percent of Maryland production.

Overall, the Georges Creek Basin accounts for 41.4 percent of Maryland's coal reserves and about 53.5 percent of state production.

The Upper Potomac Basin

The upper Potomac Basin is 30 miles (48.3 kilometers) long, averages 3 miles (4.8 kilometers) wide, and is transected by the north branch of the Potomac River between Maryland and West Virginia. The remaining coal in this basin is a thinner geologic section than in the Georges Creek Basin, but the upper Potomac Basin is the largest contiguous reserve of coal in the state. This basin, a continuation of the same synclinal structure as the Georges Creek Basin, lies between the Allegheny Front to the east and Backbone Mountain (a southern extension of Big Savage Mountain) to the west. The upper Potomac Basin contains 26.1 percent of Maryland's estimated recoverable coal reserves, yet it accounted for an impressive 46.5 percent of total state production in 2008. The only significant settlements in the upper Potomac Basin are four residential communities along the Potomac River: Bloomington, Gorman, Kitzmiller, and Shallmar. Most other development is found along Maryland Routes 38, 50, 135, and 560. Considerable amounts of land suitable for mining still exist in the basin, although some land is publicly owned (e.g., the Potomac State Forest and federal land around the Bloomington Dam on the Potomac River; Clark 1978). Surface mining on public lands is prohibited by law.

The Lower Youghiogheny Basin

Located in the extreme northwest corner of Garrett County, the lower Youghiogheny Basin extends north into Pennsylvania and west into West Virginia. Winding Ridge is the eastern boundary, and Sang Run separates the basin from the upper Youghiogheny Basin. In Maryland, the lower Youghiogheny Basin is roughly triangular in shape, 12 miles (19.3 kilometers) long and a maximum of 6 miles (9.7 kilometers) wide. Mining and development in this basin lagged behind other basins. The rugged area of the lower Youghiogheny Basin, with the Youghiogheny River flowing through it, is flanked by a deep gorge. The state's Wild and Scenic Rivers Program protects part of the gorge south of Friendsville from surface mining. The National Freeway (US Route 68) crosses the area, providing easy access for coal trucks going to Cumberland, Maryland, and to Morgantown, West Virginia. Although the lower Youghiogheny Basin contains 12.5 percent of Maryland's estimated recoverable reserves, significant coal production ceased in the late 1980s. Presently, there is no coal production in this basin.

The Upper Youghiogheny Basin

The upper Youghiogheny Basin is located immediately south of the lower Youghiogheny Basin. Sang Run forms the northern boundary, where a structural saddle separates the two synclinal basins. The basin extends west into West Virginia's Mt. Carmel Basin. To the east is a series of knobs (e.g., Roman Nose Mountain). The basin measures 10 miles (16.1 kilometers) long and 5 miles (8 kilometers) wide. There was minimal mining activity here in the late 1970s, and the Maryland Bureau of Mines has recorded no production since the late 1980s. The thin coal seams (the thinnest in Maryland) and restrictions on surface mining by the large areas in state forests do not point to a renewal of coal mining. Instead, this region has turned its attention away from its mining past and toward growth and commercial development. Nearby is the growing town of Oakland, the county seat of Garrett County, which is located on a main rail line of the Chessie System (also known as the Chesapeake and Ohio Railroad, which was combined with the Seaboard Coast Line in 1980 to form CSX Corporation).

Coal Mining Methods

Coal mining in Maryland began as underground deep mining. Surface mining, which began in the early 1940s, surpassed deep mining by the mid-1950s. By 1984, coal production from deep mining had surged once again and surpassed surface mining. Since 2008, surface mining has outpaced deep mining. The long decline of deep coal mining from the early 1940s to the mid-1970s was capped by the Mine Health and Safety Act of 1969, which resulted in the closing of almost all deep mines in Maryland. Growing demand, arising from the energy crises of the 1970s, refocused attention on deep mining, especially in the upper Potomac Basin. The largest deep mine—the Mettiki mine complex in western Garrett County, financed by Japanese interests—accounted for 86 percent of all deep coal mine production and 48 percent of total coal production in Maryland in 2005. By 2008, the Mettiki Coal Corporation in Maryland was not mining any coal from deep mines; its modest production of 124,565 tons (113,003 tonnes) came entirely from surface mining.

Surface mining is accomplished by successive parallel cuts in the earth. Maryland restricts the length of a cut and the amount of a site that may be exposed at one time. Surface miners backfill a completed cut as they dig another cut, rather than removing the whole top of a mountain at once to expose the coal. A surface mine resembles a gigantic, but orderly, construction site. Coal that is surface mined usually occurs too near the surface to be deep mined. The overburden, or soil and rock above a body of coal, is too shallow and too weak to provide a safe roof for underground mining (Edgar 1983). There are three general types of surface mining: contour, area, and open pit. In mountainous terrain where the slopes are greater than 15 degrees, contour surface mining is often used. As one area is cleared and mined, the previous area is reclaimed. In relatively flat areas with a slight dip in seaming, area surface mining is more common. Machinery scrapes away the overburden and sets it aside for later reclamation. When the coal has been removed, the area is filled in with the stored overburden, and the next strip is cleared for mining. Open pit surface mining occurs where thick coal seams are steeply inclined. This method, which combines the methods of contour and area mining, allows for nearly complete removal of the coal. Reclamation of open pit mines is often difficult because the overburden may not be adequate to fill in the depression from which so much coal has been extracted.

At one time, some of Maryland's coal was mined using augers in areas where the thick overburden made surface mining too costly. Because of the remaining thinner seams in Maryland today and the high percentage of coal left behind by auger mining (only 35% is recovered), this technique has disappeared from Maryland coal mining in the last decade.

In the United States, deep coal mining is used in Illinois, Kentucky, Maryland, Pennsylvania, Tennessee, and West Virginia. This type of mining reaches seams that are too deep for surface mining. One deep mining method is called room and pillar, in which a deep, vertical shaft is sunk down to the coal seam, and horizontal shafts are dug into the seam, leaving pillars of coal to give support to the mine. Machines undercut the coal face, and sometimes blasting is used. Various deep mining techniques include the more elaborate longwall method and the simpler shortwall method for smaller seams.

The ratio of overburden thickness to coal seam thickness largely determines the method of extracting coal. Surface or strip mining typically extracts coal seams lying less than 150 feet (45.7 meters) below the surface. Deeper seams require deep mining methods (Maryland Department of Natural Resources 1994). When conditions are right, surface mining has advantages over deep mining, including lower costs, fewer labor requirements, greater safety, and a higher percentage of coal recovery.

Abandoned Mine Reclamation

Maryland has nearly 9,500 acres (3,800 hectares) of land and 450 miles (724 kilometers) of waterways that were negatively affected by both deep and surface mining for many years before state and federal mining and reclamation regulatory programs began. Today, the Maryland Abandoned Mine Reclamation Program is responsible for the eventual reclamation of all the abandoned mining areas of Maryland that endanger the public safety and health as well as the environment, agriculture, historic sites, and recreation areas. The program is designed to address the problems of past and present coal mining through land reclamation, water cleanup, new facilities, and research.

A number of abandoned mines in western Maryland continue to be hazardous to human health and safety, not just locally but also beyond their immediate location in far western Maryland. The flow of both surface and groundwater links to watersheds beyond Garrett and Allegany Counties. Costs for reclamation and the impact on health and the economy affect the entire state. During the peak production years between 1900 and 1918—prior to state and federal environmental regulations—most deep coal mines in Maryland used gravity drainage to keep water from accumulating in mines. Acid, iron, sulfur, aluminum, and other toxic materials polluted much of the water draining from these mines, which eventually found its way into streams and groundwater. The most serious pollutant from old abandoned mines is acid drainage. Subsistence problems are also serious in parts of the Georges Creek Basin's Pittsburgh coal seam, which saw heavy deep mining in the early 1900s. As coal pillars in old deep mines collapse, the ground surface above caves in. Any buildings or other development over these sites can be severely damaged or destroyed.

Water that comes into contact with coal and overburden experiences chemical reactions that lead to greater acidity. Underground, this acid picks up iron, manganese, and aluminum in solution. When acidic mine drainage water comes into contact with the air, it deposits these metals on streambeds and banks. Acid mine drainage in moderate concentration is toxic to fish and aquatic plant life; higher concentrations can be fatal to all life in the waterway. Acid drainage also attacks bridge abutments, drainage culverts, and other structures that come into contact with the water, thus diminishing their lifespan.

Maryland surface mining began in the early 1940s and increased until the mid-1970s. At the same time, deep mining decreased until making a comeback in the early 1980s as the major form of coal mining in Maryland. Surface mining up to 1955 operated without any reclamation laws, and even the minimal reclamation requirements enacted in 1955 were not satisfactory. With increased surface mining activity, new problems emerged. Highwalls, pits with standing water, spoil piles, landslides, erosion, and acid drainage are prevalent remnants of the surface mines in the coal basins of western Maryland. The Maryland Strip Mining Law of 1955 was eventually amended in 1967 and again in 1969. More stringent amendments were put in place in 1972; others have been added almost annually since 1974. The decreasing acreage in abandoned mines and the increased acreage in improved mine lands since 1972 reflect the results of these efforts.

The environmental movement of the 1970s focused attention on the need to mitigate and prevent the negative impacts associated with coal mining, especially surface mining. New regulations emerged, requiring revegetation of land disturbed by strip mining, treatment of acid mine drainage, and sedimentation affecting streams (Thoreen 1998). These efforts culminated in the passing of the federal Surface Mining Control and Reclamation Act (SMCRA), which was signed by President Carter in 1977. Prior to the Maryland Strip Mining Law of 1955 and SMCRA in 1977, there was no governing legislation in place to prevent negative environmental effects from coal mining.

For almost 200 years, mine operators in Maryland had not been held accountable for the impact mining had on the environment. Cleaning up the damage done by prior operations has become the responsibility of the government and, by extension, the general society that must bear its costs. Even today, there are 450 abandoned mines in western Maryland that predate the SMCRA. It is no surprise that there is a direct correlation between locations of streams with the most severe acid pollution and the locations of abandoned mines.

In Maryland, the Department of the Environment takes the lead in addressing environmental issues associated with coal mining. Other notable efforts include the Chesapeake Bay Abandoned Mine Lands Conservation Initiative (developed by the National Fish and Wildlife Foundation) and the Western Maryland Coal Combustion By-Products/ Acid Mine Drainage Initiative (a partnership of the Maryland Department of the Environment Bureau of Mines, three local companies, a power plant, a consulting company, and the Maryland Department of Natural Resources; Petzrick 1999). The Geography Department and Lewis J. Ort Library at Frostburg University in Allegany County is an important contributor to the Maryland Coal Mine Mapping Project, started in 2003 to save and digitize hundreds of maps, photos, reports, and others archives of the mining industry. The project is a cooperative effort of Frostburg State University, the Maryland Bureau of Mines, and the Maryland Power Plant Research Program to collect, catalog, preserve, interpret, and provide access to the historical record of mining in Maryland, focusing on the mining of coal in western Maryland. It is producing maps for public safety, identifying subsistence prone areas, and facilitating ongoing environmental restoration efforts by the Maryland Bureau of Mines to mitigate the environmental impacts of prelaw historic coal mining (i.e., mining that occurred prior to the 1977 SMCRA) in Maryland's Appalachian Mountain region.

Coal Markets

Coal from western Maryland is shipped in all directions to a number of different markets. The two major markets are for *steam coal*, which is used to generate electricity, and for *metallurgical coal* (or met coal), which is used by steel companies to make coke. Before coal is loaded and shipped to market, a number of critical factors must be considered. Coal companies must look at market size and location, coal output, location of the mine, quality of the coal, total cost of extraction, and their ability to secure purchase orders. Electric power plants are the major consumers of Maryland coal, consuming about 75 percent of output. Other industries consume small amounts of coal, mainly cement plants and the West Virginia Company (Westvaco) paper mill at Luke, Maryland. Bethlehem Steel at Sparrows Point in eastern Baltimore County was a significant consumer until 1992, when its coke plant ceased operations.

Coal contributes about 25 percent of Maryland's energy needs; oil is the other major source of energy in the state. Maryland power plants get about 10 percent of their coal needs from coal mines within the state, with the rest coming mostly from West Virginia and Pennsylvania with small amounts from Kentucky and Virginia. Maryland's largest coal power plant, Baltimore Gas and Electric's Brandon Shores plant in Anne Arundel County, receives most of its coal from West Virginia, with a small amount coming from Kentucky. Maryland provides all of the coal needs for three West Virginia utility plants: Albright, Ft. Martin, and Mt. Storm.

8 The Shifting Geography of Manufacturing

After the service sector, manufacturing is the second-largest portion of the Maryland economy. In the manufacturing sector, raw materials and components are given what is known as *form utility* (i.e., they are made into a usable object such as sheet steel or a car part). A wide variety of manufacturing activity takes place in Maryland, including traditional industries such as food processing and heavy industries such as metal fabrication, shipyards, petroleum refineries, chemical plants, and truck assembly lines. Some of the newest and fastest-growing light industries include biotechnology, guided missile systems, and space vehicle propulsion systems. The hallmark of manufacturing in Maryland is diversity. No single type of manufacturing dominates. Eighteen of the twenty major manufacturing groups as defined by the US Census of Manufactures are well represented in the state.

Many people believe that the United States is becoming a postindustrial society and that the US landscape will soon be devoid of industrial plants. Those who hold this view maintain that the United States will become a largely service-oriented economy, with manufacturing being left to countries like China, where labor is cheaper and there are fewer regulations, including environmental ones. The reality today is quite different, however. US industrial output is rising. Contrary to popular belief, manufacturing's share of the US economy (measured by real gross domestic product, or GDP) has been stable since the 1940s, ranging from 16 to 19 percent of the GDP. In fact, the US manufacturing sector, if taken alone (and thus excluding the overall US economy), would have ranked as the world's seventh-largest economy in 2005. In 2010, Maryland had a gross state product (GSP) of about $300 billion, the thirteenth highest in the country. The counterpart of GDP at the national level, GSP is the gross state output (sales or receipts and other operating income, commodity taxes, and inventory change) minus intermediate inputs (consumption of goods and services purchased). Maryland derives a significantly smaller share of its GSP from manufacturing than does the nation.

The United States has lost manufacturing jobs since the 1990s, but at a much lower rate than Belgium, France, Japan, South Korea, Sweden, the United Kingdom, and other industrialized countries. Losses in US manufacturing employment have been offset by gains in productivity and overall output. Nationwide, losses of jobs in manufacturing have led many policymakers and citizens to diminish the importance of manufacturing in the economy. Although agriculture has experienced greater losses in employment in the United States and Maryland over the past century, manufacturing remains viewed as a critical sector of the economy and is supported by numerous federal and state programs.

Maryland mirrors the competitive advantages of the US economy: world-class research laboratories, research and development spending, protected intellectual property rights, close links between higher education and manufacturers, the immense scale of demand in its domestic economy, and functional capital markets. The manufacturing sector is an important part of Maryland's economic past, but it is also a vital part of its economic future. Some sectors of Maryland manufacturing have experienced a notable decline in employment (e.g., textiles and apparel as well as primary metals), reflecting Maryland's poor competitive position in low-tech manufacturing segments. Yet Maryland is increasingly competitive nationally and globally in a number of high-tech industries, especially in aerospace and biotechnology.

Since World War II, Maryland has transitioned from an older "smokestack" economy to a new economy characterized by technological processes that have replaced most assembly lines. Firms involved in the manufacture of defense electronics, fiber optics, and biological products are representative of this new economy. The vital importance of Maryland's industries to the nation has become even more evident in the post-9/11 world. With a disproportionate share of the nation's defense manufacturing capacity, Maryland is home to the headquarters of ATK's Advanced Propulsion, BAE Systems, Lockheed Martin, Northrop Grumman's Electronic Systems, Space Systems Group, and many other defense-related firms. In Maryland, as in the whole of the United States, global competition has squeezed out most labor-intensive, low-tech manufacturing industries. Today, chemicals, food processing, computers and electronics, and precision machinery lead Maryland manufacturing. The pattern of manufacturing exports has also changed. Today, transportation equipment (including aerospace), computers and electronics, chemicals, and machinery account for about two-thirds of Maryland's manufacturing exports. These manufacturing activities alone account for over half of all Maryland exports.

Although employment within the manufacturing sector in Maryland has steadily decreased in recent decades while the total labor force has grown, manufacturing still accounts for 6 percent of the state's employment. At the same time, manufacturing output in Maryland has steadily increased in recent years. The transition to the new economy has not come without pain for many Maryland workers. In 2005, Maryland lost 2,000 manufacturing jobs. It was the fifth consecutive year of jobs losses in manufacturing; job losses had been as high as 11,700 in 2002. Some major closures included General Motors Corporation's Broening Highway Assembly Plant near Baltimore in 2005 (a loss of 1,100 jobs) and the steady downsizing of the workforce of the steel mill at Sparrows Point. From 2000 to 2005, Maryland lost 33,300 manufacturing jobs—19.1 percent of its base. For the entire United States, from 2000 to 2005, over three million manufacturing jobs were lost, representing 17.7 percent of the base total of manufacturing jobs (Atkinson et al. 2012). But manufacturing does not exist in isolation. The tremendous growth in the service sector of the Maryland economy relates, in part, to the changing mix of manufacturing. As newer types of manufacturing become more prominent in Maryland, they in turn require various additional services. Economic geographers refer to this kind of regional growth as a *multiplier effect*.

Maryland has a competitive manufacturing business environment. About 8,000 manufacturers have discovered Maryland's locational advantages within the American Manufacturing Belt and the high-demand markets of megalopolis. In addition to its excellent relative location, Maryland has other factors necessary for a strong manufacturing sector, such as capital, labor supply, and transportation connectivity. The Washington–Baltimore region by itself is the fourth-largest consumer market in the United States. Three major airports serve the state, and the Port of Baltimore is the leading port in the United States for roll-on/roll-off cargo. Over 29,500 miles (47,474 kilometers) of public roads and six interstate rail lines link Maryland to other places. Manufacturers in Maryland have access to eighty-seven million people (a third of the US population) by overnight truck. The corporate tax is a flat competitive rate of 8.25 percent (lower than nearby Pennsylvania and Delaware, but higher than Virginia), and the state charges no sales tax on manufacturing equipment, no tax on gross receipts, and no personal property tax at the state level.

The Maryland Department of Business and Economic Development (2013) notes some of the many "Maryland Superlatives," including:

- Maryland ranks second among the states in the percentage of professional and technical workers (27.7%) in the workforce.
- Maryland ranks first in the nation in receiving research and development contract awards from the National Institutes of Health (NIH).
- Maryland has the second-lowest poverty rate in the nation, with 10.3 percent of the population living below the poverty level compared with 15.9 percent for the United States.
- Maryland ranks second in federal obligations for research and development ($15.9 billion). On a per-capita basis, Maryland ranks first among the states in federal research and development obligations.

- Maryland has the highest median family income among the fifty states at $74,122, which is 38% above the national median.
- According to Quality Counts, *Education Week*'s annual assessment of key indicators of student success, Maryland's K–12 public school system ranked first in 2013 for the sixth year in a row.
- Maryland ranks first in the nation in the rate of high school completions at 95 percent; the national average is 86 percent.
- Maryland ranks fourth in the United States in the percentage of residents with a bachelor's degree or higher; 36.9 percent of the population aged twenty-five and over hold such degrees.
- Maryland has the highest concentration of employed doctoral scientists and engineers. The state ranks first in employed PhD scientists and engineers per 100,000 employed workers. Maryland also ranks first in mathematical sciences (71), first in biological sciences (398), first in health (63), and third in physical sciences (197) per 100,000 employed workers.
- The Baltimore metro region ranks fifth and Washington, DC, which includes parts of Maryland, ranks second in a *Forbes* ranking of the best cities for technology jobs, based on growth of science and technology-related employment.
- Maryland is one of only nine states with a "triple triple" general obligation bond rating from the three major bond rating houses. Maryland has the longest-running Triple-A rating with a "stable" outlook from Standard & Poor's that dates to 1961.

Maryland's close proximity to the nation's capital—which offers direct access to embassies, federal agencies, national and international associations—also makes the state special. Its proximity to major federal institutions is an important factor in deciding on a location for many businesses. For example, the Northrop Grumman Corporation headquarters and plant in Linthicum employs 7,650 people producing electronic sensors systems, surveillance radar, fire control systems, electronic countermeasure equipment, electro-optical space-borne sensors, sonar, and air traffic control systems. Its location, which is practically next door to the National Security Agency (NSA), one of its major customers, is important to its operations. Northrop Grumman also has a plant in Annapolis that houses its Ocean Systems Division, which produces sensors and sonar systems. The closeness of the US Naval Academy was a major factor in Northrop Grumman's decision to locate the Ocean Systems Division in Annapolis.

Maryland Manufacturing

The Maryland economy began as a one-crop economy in the seventeenth century with tobacco farming that depended on slaves and indentured servants to provide the labor. As settlement moved west into central and western Maryland, agriculture diversified. By 1820, manufacturing was rivaling agriculture for economic dominance, and in the late nineteenth century, shipbuilding, steel making, and commerce had transformed Baltimore into a major US city. The Maryland economy has traditionally been rooted in manufacturing, trade, and transport. Today, manufacturing has shifted toward high-tech and knowledge-based industries. While manufacturing output continues to rise, its share of the GSP has fallen slightly (from 9% in 1997 to about 8.5% in 2010). Manufacturing's regional distribution is also in transition. The Baltimore region core is still prominent, but other parts of the state are seeing growth in manufacturing, too.

The Baltimore Area

Baltimore—including nearby areas such as Anne Arundel, Baltimore, Harford, and Howard, Counties—has traditionally had the largest concentration of manufacturing in the state, with nearly 2,900 companies employing nearly 100,000 workers (46% of Maryland manufacturing employment). Baltimore City alone has 720 manufacturing establishments employing over 26,000 workers. Thirty years ago, heavy industries linked to basic iron and steel, fabricated metals, and metal-using industries dominated with their durable goods (durable goods are items with a life span of three or more years; e.g., automobiles, furniture, household appliances, heavy machinery, etc.). While the heavy durable goods industries are still prominent, the rise of other types of goods over the past thirty years has changed the manufacturing scene of the Baltimore area. Today there is a growing prominence of printing and publishing, biosciences, defense electronics, and light tools.

Baltimore versus Washington, DC

Fifty years ago, the manufacturing patterns of Baltimore and Washington, DC, stood in sharp contrast. There were 200,000 manufacturing employees in the Baltimore metropolitan area but only 37,000 in the Washington area. Metals, engineering, and food products dominated in Baltimore, while Washington was stronger in printing and publishing as well as in light industries. Today, that stark contrast has softened as the Baltimore area has attracted many new high-tech and information-age firms while its heavier industries have declined. As the number of employees in Baltimore has declined severely since 1960, the number in Washington has grown (fig. 8.1). The numbers of employees in manufacturing in both metropolitan Baltimore and metropolitan Washington are now closer, but Baltimore still maintains a lead.

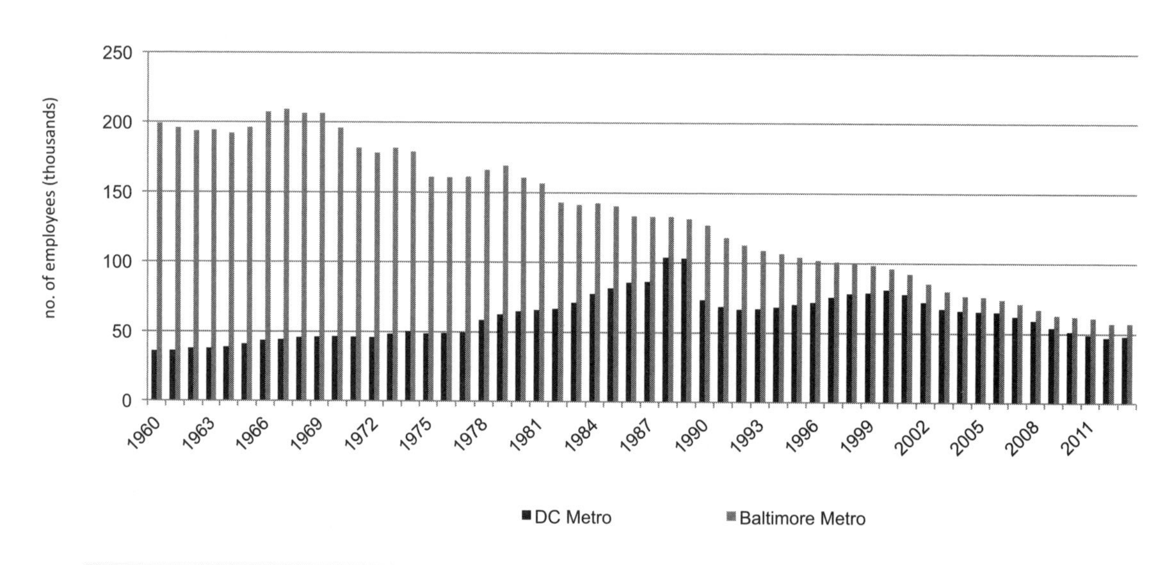

Figure 8.1 Number of Manufacturing Employees in the Baltimore and Washington, DC, Metro Areas, 1960–2013

Montgomery and Prince George's Counties

Two Maryland counties situated near Washington, DC, have emerged as industrial growth areas. Montgomery County now ranks first among the state's counties in population, number of manufacturing establishments (only Baltimore City, which is not a county, has more), and number of manufacturing employees (table 8.1). The largest manufacturing employer in Montgomery County is Lockheed Martin (1,000 employees), whose headquarters in Bethesda produces aeronautics, space systems, information technology, and communications products. Lockheed Martin employs 140,000 worldwide and had sales of nearly $32 billion in 2010; it is a major contractor with the US Department of Defense. Additional prominent employers produce radio and television communications systems (e.g., COMSAT, which is headquartered in Bethesda), guided missile and space systems (e.g., Orbital Sciences in Germantown), and other high-tech firms. Phillips Publishing, also located in Montgomery County, represents one the fastest-growing segments of Maryland industry. Historically, industries such as machinery, chemicals, food products, and miscellaneous goods dominated in Prince George's County. Today, Prince George's has a more diversified manufacturing sector; one example is the NASA Goddard Space Flight Center in Greenbelt, a major attraction for many related private commercial establishments both in Prince George's and surrounding counties. Beretta USA Corporation, which manufactures small firearms, is headquartered in Accokeek, where it employs 325 people and had $100 million in sales in 2010.

Table 8.1. Maryland manufacturing establishments, 2008

County	*Establishments*		*Employees*		*Population*	
	Number	*Rank*	*Number*	*Rank*	*Number*	*Rank*
Baltimore City	1,321	1	26,035	3	632,681	4
Montgomery	674	2	38,235	1	852,174	1
Prince George's	664	3	18,849	4	781,781	2
Baltimore	592	4	35,401	2	723,914	3
Anne Arundel	555	5	18,243	5	480,483	5
Frederick	264	6	9,679	7	190,869	8
Washington	221	7	8,462	8	127,791	10
Carroll	219	8	5,944	11	152,468	9
Harford	209	9	6,516	10	217,908	7
Howard	197	10	13,313	6	243,112	6
Wicomico	132	11	5,205	12	79,560	14
Allegany	93	12	4,514	14	71,162	16
Talbot	86	13	1,909	15	33,550	19
Charles	85	14	1,861	17	120,946	11
Cecil	85	15	4,630	13	84,238	13
Worcester	68	16	885	23	43,672	17
Calvert	63	17	1,093	22	73,748	15
Garrett	62	18	1,887	16	29,389	22
St. Mary's	61	19	1,810	18	88,758	12
Queen Anne's	58	20	1,525	19	40,668	18
Caroline	55	21	1,507	20	29,708	21
Dorchester	51	22	6,614	9	29,709	20
Kent	38	23	1,494	21	19,089	24
Somerset	29	24	471	24	24,236	23
Total	5,882		216,082		5,171,614	

Source: Harris Maryland Manufacturers Directory, Harris Infosource, 2010 and 2011

Salisbury and Cambridge

On the Eastern Shore, the cities of Salisbury and Cambridge are major centers of manufacturing. The top employer in Salisbury is Perdue Farms, whose two plants and headquarters in Salisbury employ 1,640. Salisbury, too, has begun to diversify its manufacturing activities. Its manufacturing sector has experienced a rapid rise in the number of high-tech businesses in electronics, pharmaceuticals, and precision metal products. The enhancing roles of Salisbury University and the University of Maryland Eastern Shore have been critical to attracting these businesses. Several other manufacturing plants of significant size are located in Salisbury, including K & L Microwave, a Dover, Delaware, company producing electronic communications filters for both military and commercial use that employs 306. A fairly new addition with 800 employees is Harvard Customs Manufacturing, a Boston-based electronics firm producing complex electronic interconnect assemblies used by the military and commercial customers. There is also Cadista Pharmaceuticals; Lorch Microwave, which produces electronic radio filter components; Pepsi Cola Bottling Company; MaTech precision machinery; Tishcon, which manufactures vitamins; and several publishing and printing companies represented by the *Daily Times* and the Standard Register Company.

Cambridge has far less industry than Salisbury, but its manufacturing sector is still diversified. Cambridge International produces wire springs, metal conveyor belts, and architectural mesh. Along with its subsidiary, Maryland Wire Belt, it employs 425. Unlike Salisbury, smaller companies—especially those in food products and processing and boat industry-related firms—dominate manufacturing in Cambridge. These companies represent the new look of manufacturing on the Eastern Shore.

Western Maryland

Manufacturing in western Maryland is centered on Hagerstown in Washington County (159 manufacturing plants in 2011) and Cumberland in Allegany County (fifty-eight manufacturing plants in 2011). The largest industrial employer in Hagerstown is Volvo Powertrain Company (a division of Volvo North America), which employs 1,100 (down from 4,400 in 1980, when it was Mack Truck). In addition, LE High Phoenix, a printing and book manufacturing firm, employs 460. Western Maryland also has a wide variety of other industries, including food processing (e.g., Good Humor-Breyers Ice Cream, which employs 525) and industrial machinery. The largest manufacturing employer in Cumberland is Hunter Douglas Northeast Company, which produces windows and blinds and employs 600. Other major Cumberland manufacturing plants include Superfos (plastic packaging), with 200 employees, and Yoder and Sons (metal fabrication), with 100 employees. Additional significant employers produce blankets and pillows, erosion control products, and publishing products.

Measures of Manufacturing Activity

Researchers use a number of standard variables to determine the level of manufacturing activity in a region, including earnings, employment, number of firms, and value added. The shifting geographical patterns and trends of these variables reflect the changing manufacturing landscape of Maryland.

Earnings

In 2009, total wages and salaries for Maryland workers in all jobs totaled $125.7 billion. Of this amount, workers in manufacturing jobs earned $7.1 billion (5.6%). Over the long term, the total earnings in Maryland from manufacturing jobs have been trending downward; in the late 1970s, manufacturing jobs accounted for 21 percent of all earnings. In 1978, the highest weekly and hourly earnings in manufacturing were in the primary metals industry, followed by transportation equipment. Workers producing textile, apparel, and leather products earned the lowest wages. In 1999, the highest weekly and hourly wages were still in the heavily unionized primary metals industry, followed by transportation equipment and then industrial machinery. The lowest wages are now paid in the industries of lumber and wood as well as fabricated metals. Even so, Maryland's average weekly and hourly earnings in manufacturing in 2009 were higher than the US average.

Employment

Major changes in the mix of manufactures in Maryland have occurred in the past thirty years (table 8.2). The biggest losers as measured by declines in the number employed have been the primary metals industry (–91%), tobacco products (–85%), apparel and textile products (–64%), paper and allied products (–61%), and transportation equipment (–64%). No other single manufacturing plant better typifies these declining industries than the steel mill at Sparrows Point, which employed 19,000 in 1980 and only 2,500 by 2011.

The biggest increase in the number of employees has been in the production of measuring-analyzing-controlling instruments for medical, photographic, navigation, timekeeping, and other sectors (240%). Another significant gain in employment (in terms of percentage) since 1976 has been in furniture products (34%). As the industrial mix in Maryland has changed, it has maintained its high degree of diversity. No one industry dominates the economy of the entire state, although an industry might dominate in a region of the state. In its regional setting in the eastern United States, Maryland is among the smaller industrial states (table 8.3). When grouped with states to the north (in what the US Census Bureau calls the "Mideast"), Maryland exceeds only Delaware in manufacturing employees. When compared to states to the south (the "Southeast"), Maryland exceeds only West Virginia.

Number of Manufacturing Establishments

In 2011, there were 5,882 manufacturing establishments in Maryland. The US Census Bureau defines an establishment as a single plant site or factory. It is not necessarily identical to the business unit, company, or firm, which may consist of one or more establishments at more than

Table 8.2. Changes in manufacturing employment in Maryland

Industry	Number employed 1976	2011	Percentage change
All	160,800	104,199	-35.2
Apparel	10,500	3,757	-64.0
Electric and electronic equipment	18,200	18,351	1.0
Fabricated metals	8,800	8,221	-6.6
Food and kindred products	15,900	14,628	-8.0
Furniture and fixtures	2,800	3,762	34.4
Industrial machinery	11,100	6,215	-4.4
Lumber and wood	2,200	1,766	-19.7
Measurement instruments	2,700	9,090	240.0
Paper and allied products	7,700	3,042	-60.5
Petroleum and coal products	1,000	843	-20.0
Primary metals	28,700	2,500	-91.3
Printing and publishing	11,800	12,284	4.1
Rubber and plastics	5,700	3,551	-37.7
Stone, clay, and glass	6,500	4,873	-30.0
Tobacco, textiles, and leather products	1,000	153	-85.0
Transport equipment	14,400	5,233	-63.7

Source: US Census Bureau, Annual Survey of Manufactures, Geographic Areas, 1976 and 2011

one location. For example, Perdue Farms has two plants in Salisbury, and each is considered a separate establishment in the US Census Bureau's official data. The shifting spatial pattern of manufacturing establishments continues to diffuse as the number in Baltimore City fluctuated over the years, while those in the counties of the Washington–Baltimore urban corridor increased steadily. In 1950, Baltimore City had 1,738 manufacturing establishments, Baltimore County had 130, and Montgomery County had only 79. By 1978, the number of establishments in Baltimore City had dropped dramatically to 841, while those in Baltimore County increased to 264 and Montgomery County to 203. In 2010, Baltimore County had 592, Baltimore City 1,321, and Montgomery County 674.

These changing numbers reflect the growth of highway transport-oriented industrial parks, such as Hunt Valley in central Baltimore County. Much of this industrial decentralization has taken place in northern Anne Arundel County, eastern Howard County close to Interstate 95, and in eastern and central Baltimore County close to Interstates 83, 95, and 695. The heavy growth of manufacturing establishments in Montgomery and Prince George's Counties is in close proximity to Interstates 95 and 495 as well as the rapidly growing corridor along Interstate 270, stretching northwest from Washington to Frederick. There is a noticeable concentration of manufacturing firms in the Germantown area of Montgomery County, as the urban corridor continues to dominate manufacturing in Maryland. The moderately high numbers of manufacturing establishments in Washington, Frederick, Carroll, and Wicomico Counties reflect concentrations in Hagerstown, Frederick, Westminster, and Salisbury, respectively. Although the spatial pattern of manufacturing in Maryland is decentralizing from Baltimore City, it is still one of regional concentration, occurring at newly rising industrial centers within the urban corridor.

Value Added

To determine the value added by manufacturing, one subtracts the cost of materials (raw materials, parts, supplies, fuel, goods purchased for resale, power, and purchased services) from the value of outgoing

Table 8.3. Number of employees and value added by manufacturing in Maryland and selected eastern states, 2011

Region and State	*Number of employees (thousands)*	*Rank*	*Value added (millions)*	*Rank*
United States	10,650		$2,295,219	
	Mideast			
Pennsylvania	530	1	$100,737	1
New York	419	2	$77,618	2
New Jersey	235	3	$47,285	3
Maryland	**104**	**4**	**$23,203**	**4**
Delaware	27	5	$7,421	5
	Southeast			
North Carolina	388	1	$102,559	1
Georgia	313	2	$65,055	3
Tennessee	277	3	$132,135	2
South Carolina	258	4	$81,630	4
Virginia	222	5	$54,557	6
Alabama	213	6	$43,374	7
Kentucky	202	7	$38,181	5
West Virginia	50	8	$10,140	8

Source: US Census Bureau, Annual Survey of Manufactures, 2011

shipments. Value added in Maryland increased by 48 percent between 1990 and 2011 (table 8.4). The tremendous growth in value added in computers and electronic equipment (643%) is particularly prominent. Although the strength of growth in furniture and fixtures was strong (200%), it is still a modest segment of Maryland's manufacturing economy in absolute dollar figures. Another notable high-growth sector is chemicals (135%), which is also high in absolute dollar figures. The largest segment of the chemical industry represented in these numbers is pharmaceuticals.

While valued added for Maryland's manufacturing sector has increased overall, it has not been accompanied by an increase in number of workers employed; in fact, manufacturing employment has decreased. The development of more modern mechanization, which has replaced many workers while at the same time increasing productivity, explains much of the decline in jobs.

Compared to states to the north and south, the total value added in Maryland was lower than all except Delaware and West Virginia. Figure 8.2 shows the geographic pattern of value added by Maryland counties. As might be expected, the highest value added for manufacturing corresponds with the highest employment and number of manufacturing establishments. The concentration stretches along the

Table 8.4. Value added for selected Maryland manufacturing industries, 1990–2011

Industry	*1990 (millions)*	*1996 (millions)*	*2004 (millions)*	*2009 (millions)*	*2011 (millions)*	*1990–2011 percentage change*
All	$15,723.7	$17,454.6	$19,131.0	$20,848.2	$23,203.0	47.6
Chemicals	$2,011.7	$2,360.8	$3,347.0	$3,801.6	$4,734.3	135.3
Electronic equipment	$714.8	$1,142.4	$3,527.0	$5,295.3	$5,308.6	642.7
Fabricated metals	$632.8	$744.7	$1,343.0	$1,249.2	$1,031.8	63.1
Food and kindred products	$2,315.9	$2,729.7	$2,759.0	$3,070.7	$3,386.3	46.2
Furniture and fixtures	$157.2	$204.3	$362.0	$397.9	$471.5	199.9
Industrial machinery	$932.6	$1,429.7	$1,026.0	$916.0	$1,338.4	43.5
Lumber and wood	$174.5	$164.4	$380.0	$201.1	$192.2	10.1
Nonmetallic mineral products	$406.0	$513.1	$703.0	$602.0	$606.7	49.4
Paper and allied products	$577.0	$730.3	$461.0	$448.9	$438.9	-23.9
Printing and publishing	$1,818.3	$2,175.6	$1,082.0	$2,067.0	$1,093.0	-39.9
Rubber and plastics	$445.1	$787.0	$802.0	$767.3	$851.8	91.4
Transport equipment	$1,516.4	$844.2	$859.0	$523.6	$891.8	-41.2

Source: US Department of Commerce, Bureau of Census, Annual Survey of Manufactures, various years

Figure 8.2 Value Added by Manufacturing, 2010
Source: Data from US Bureau of the Census, Economic Census, Manufacturing, Geographical Area Series, 2012

urban corridor from northeast to southwest, including Anne Arundel, Baltimore, Harford, Howard, Montgomery, and Prince George's Counties as well as the large value added for the industries of Baltimore City. The figures are much smaller for western Maryland, southern Maryland, and the Eastern Shore, where secondary areas of importance are centered on the urban centers of Cambridge, Cumberland, Frederick, Hagerstown, and Salisbury.

Measuring Industrial Location

Economic geographers have developed a number of indices to measure industrial location. One of these, the location quotient, measures the degree to which a specific region has more or less than its share of any particular industry (Alexander and Gibson 1979). The location quotient is a ratio of ratios. To illustrate, in 2011 the ratio of the number of US employees in primary metals industries to the total number of employees in manufacturing was 351,863 to 10,649,378, or 3.3 percent. The ratio of these two variables for the state of Maryland was 2,500 to 104,199, or 2.4 percent.

The location quotient for the primary metals industry of Maryland is determined by dividing the state ratio (2.4) by the national ratio (3.3). A location quotient of 1.0 means that Maryland has exactly its share—that is, manufacturing strength—of the nation's employment in primary metals manufacturing. A quotient greater than 1.0 means that Maryland has more than its share; a quotient less than 1.0 indicates that Maryland has less than its share. For primary metal manufacturing in the year 2011, Maryland's location quotient was 2.4/3.3, or 0.726. This value indicates that Maryland had less than its share in primary metals manufacturing, reflecting the decline of this industry in Maryland. In 1980, the Maryland location quotient for primary metals was 2.08, indicating a significantly greater share of the nation's employment in primary metals at this earlier time. The location quotients for a number of durable and nondurable goods manufacturing industries appear in table 8.5. Maryland has more than its share of the nation's employment in electronics; stone, clay, and glass; measurement instruments; printing and publishing; furniture and food products; apparel; leather; and petroleum and coal products. In the remaining industries shown in table 8.5, Maryland has less than its share of the nation's employment. The Maryland share is particularly low in textiles, wood products, rubber and plastics, and transport equipment.

These results generally reflect industries in which Maryland has regional comparative advantages (those above 1.0) and those for which Maryland is weaker (below 1.0). These location quotients give only a general picture of the relative strength of various categories of manufacturing within Maryland, as it is still possible to have a low quotient but an individual manufacturing plant that is nationally competitive. The quotients used here are based on employment and not on value added, productivity, or profits. Also, production plants in any category could be a subsidiary of a much larger firm operating nationally or internationally. The Maryland plant could be producing components for other plants in the firm. Although smaller, the local plant could be an important cog in the entire manufacturing process that stretches beyond the borders of the state.

Manufacturing in Maryland accounts for a smaller share of total employment than the manufacturing sector at the national level. The state has also experienced faster-than-average rates of decrease in manufacturing jobs over the past ten years. But that is not the end of the story. The ongoing shift in manufacturing in Maryland includes superstars in the new high-tech, knowledge economy worldwide. The rise of high-tech electronics industries in Maryland over the past twenty years is clearly reflected in its location quotient of 2.296 for this industry, which is significantly above the national rate.

High Technology: The Knowledge-Based Sector

The story of manufacturing in Maryland to this point has mostly been one of traditional industries forming the historical base of manufacturing, but with strong shifts taking place on the Maryland manufacturing landscape. Maryland today is experiencing the rapid rise of high-tech and knowledge-based industries. This sort of transforma-

Table 8.5. Location quotients of selected Maryland industries, 2011

	Maryland employment	*US employment*	*Location quotient*
	A	B	
All manufacturing	**104,199**	**10,649,378**	
Durable goods	C	D	
Primary metals	2,500	351,863	0.726151266
Electronics	18,351	816,813	2.296136941
Transport	5,233	1,235,431	0.432905329
Industrial machinery	6,215	964,668	0.658451744
Fabricated metals	8,221	1,285,707	0.653496598
Stone, clay, and glass	6,128	455,147	1.37602956
Measurement instruments	9,090	389,226	2.386836836
Wood products	1,766	324,870	0.555573833
Furniture	3,762	319,869	1.202007923
Nondurable goods			
Food products	14,628	1,358,996	1.100088095
Printing and publishing	12,284	456,897	2.747781485
Apparel	4,842	456,500	1.084038471
Rubber and plastics	3,551	674,690	0.537906869
Paper	3,042	346,439	0.897414588
Leather	634	61,700	1.05018252
Textiles	824	503,600	0.167225376
Petroleum and coal production	971	98,249	1.010070722

Source: Data from US Census Bureau, Annual Survey of Manufactures, Geographic Areas, 2011

Note: The location quotient is determined by (C/A)÷(D/B).

tion does not happen just anywhere. A place must have the necessary factors to support the rise to prominence in the "new economy."

The Knowledge Economy

In 2002, C. D. Mote Jr., then the president of the University of Maryland at College Park, presented a paper to the Maryland Economic Development Commission that read in part:

> At the beginning of this millennium, the most powerful person in the world was Genghis Khan, whose empire extended from Hungary across Asia to Korea and from Siberia to Tibet. At the beginning of

EXPORTS OF MANUFACTURED GOODS

Basic economic activities provide goods and services not only to the people, businesses, and institutions within Maryland but also to those beyond Maryland's borders. This external trade brings income into Maryland, creates new jobs, and generates revenue. Maryland exports goods and services to other states and to other countries, too. In 2012, Maryland exported $21.7 billion worth of goods to foreign countries, 90 percent of them manufactured goods valued at $19.5 billion.

Transportation equipment, followed closely by computers, electronics, and chemicals, led Maryland's manufactured exports in 2012. That year, Maryland exported to 202 foreign countries and dependencies. Canada is the leading recipient of manufactured exports from Maryland (19%), followed by Egypt (11%), Japan (7%), Belgium (5%), the United Kingdom (5%), and Mexico (4%). The remaining 4 percent of Maryland manufactured exports go to various other countries of the world. Manufacturing exports account for about 2.0 percent of the total number of jobs in Maryland's private sector. The US Bureau of the Census estimates that 13.2 percent of all manufacturing jobs in Maryland depend on exports.

Foreign trade and export links have other multiplier effects for the Maryland economy. While it is true that Maryland has lost manufacturing jobs when companies outsource jobs to other countries, foreign companies have also insourced jobs into Maryland. In 2010, foreign-controlled companies employed over 99,000 workers in Maryland. Of these jobs, about 14,600, or 14.7 percent, were in manufacturing. Major sources of foreign direct investment in Maryland in 2010 were Canada, France, the Netherlands, and the United Kingdom. By 2010, foreign-controlled companies accounted for 9.9 percent of total manufacturing jobs in Maryland (International Trade Administration 2014). These figures reflect the importance of international trade and the impacts of globalism on the Maryland economy.

> the 19th century, the most powerful person in the world was Queen Victoria, who ruled an empire that encompassed 20% of the land and 25% of the people on earth. At the end of this century (20th), many would say that the most powerful person in the world is Bill Gates. Bill Gates' power does not derive from his armies, from his land, from his rule over people, or from controlling vast natural resources. Bill Gates' empire sits on your desk.

In the present day, knowledge and information are produced much in the same way cars and steel were produced at the beginning of the twentieth century. Those who produce knowledge best will become the magnates of the twenty-first century. The transformation from manufacturing to the information age of the Maryland and US economies today is comparable to the industrial revolution in the nineteenth century, which transformed agrarian economies into manufacturing ones (albeit still based on natural resources). But the current transformation—from a manufacturing to a knowledge economy—has never before happened in world history. For the first time, something that can be neither directly seen nor touched has become more valuable than physical resources such as land, gold, and oil. Just as electricity powered the manufacturing economy, the research university powers the knowledge economy. Research universities supply educated workers for the information market. They are also sources of new ideas, new technologies, creativity, and entrepreneurial drive. It is unlikely that the knowledge economy would even exist if not for research universities. Stanford University and the University of California, Berkeley, feed the most imitated powerhouse of the knowledge economy, Silicon Valley. Similarly, the University of North Carolina, North Carolina State, and Duke University surround the Research Triangle in North Carolina. Massachusetts's knowledge-based economic powerhouse along Route 128 is linked to Harvard and the Massachusetts Institute of Technology.

Maryland's prominence in the knowledge economy should be no surprise. Johns Hopkins University and the University of Maryland (both its Baltimore and College Park campuses) are strong research universities. Maryland has the potential to become one of the great centers of the knowledge economy. In the region a strong and growing number of knowledge-based industries reside, especially information technology, communications, and biotechnology. This region also has the advantage of a number of federal government laboratories. In the knowledge economy, repositories of information such as NASA, NSA, and NIH are as valuable as oil in the manufacturing economy.

Maryland's knowledge economy is especially strong in biotechnology. The United States has an aging population with increasing demands on science and technology to prolong good health and to improve quality of life. The biotechnology industry needs universities even more than the information and communications industries. Biotech development is an expensive and long-term process that requires teams of exceptionally well-educated and trained scientists with state-of-the-art laboratories. It can take years or decades before a product is ready to market. The financial risks are great and the payback can be long delayed. Biotechnology companies noticeably cluster around research universities. They need that research commitment and long time horizon that only research universities can support. Biosciences go far beyond biology to include biochemistry, biophysics, bioengineering, neurosciences, mathematics, computer science, and other fields. Only a university (or a group of universities clustered in a region) with broad strengths in these areas can support research in biotechnology.

Maryland has a "research backbone." At one end are Montgomery and Prince George's Counties and the I-270 corridor to Frederick. The University of Maryland at College Park anchors that end. At the other end, up I-95, is the Baltimore region, anchored by Johns Hopkins University and its various branches as well as the University of Maryland at Baltimore. The University of Maryland at College Park and Johns Hopkins University complement and balance each other's strengths. Johns Hopkins is nationally prominent in biomedicine, and the University of Maryland at College Park leads in information technology. In between are powerful government laboratories, knowledge economy businesses, and other universities.

The rise of the knowledge economy has presented challenges for Maryland's government and universities. Both have come to realize that they must work together and in partnership with businesses. One example of such cooperation is the Continuous Innovation Initiative, part of the A. James Clark School of Engineering at the University of Maryland. This program brings together businesses and academia. Highly competitive businesses integrate processes to continually generate new ideas in order to deliver new products, enter new markets, increase productivity, add customers, and forge new partnerships. The Continuous Innovation Initiative is based on the idea that companies must continually innovate to stay ahead. In a globalized economy, many Maryland companies cannot compete on cost alone. The university is well positioned to spread new knowledge about innovation—its goal is to share its information with every business it can. Continuous Innovation Initiative guides select Maryland companies in establishing innovation-driven processes for sustained growth.

High-Tech Industries

By 2010, there were 14,856 high-tech firms in Maryland, a significant increase (21%) from 12,360 in 2001 (table 8.6). Nationally, there are 181 sectors that are considered high-tech industries according to the North American Industrial Classification System (NAICS). Of the 181 high-tech sectors that exist in the United States, 150 can be found in Maryland. The classifications Computer Systems Design and Related Services as well as Management, Scientific, and Technical Consulting Services have dominated growth in the state since 2001.

Montgomery County is home to the most high-tech firms by far. In 2010, it had 4,499, or 30.1 percent of the state total. Baltimore County was second with 1,668 firms (11.1%), and Howard County was third with 1,654 firms (11.1%). There are several major geographical concentrations of high-tech firms in Maryland. The Gaithersburg-Germantown-Rockville area in Montgomery County along the I-270 corridor from the Capital Beltway (I-495) to Frederick County is a heavy concentration. Also in Montgomery County are significant concentrations inside the Capital Beltway around Bethesda and Silver Spring. Montgomery County borders Washington, DC, and is about 30 miles (48.3 kilometers) from Baltimore City. In Montgomery County alone, the nearly 4,500 high-tech firms are strong in biotechnology, information

Table 8.6. High-tech industry establishment in Maryland, 2001–2010

	Number of high-tech firms		*Percentage of high-tech firms*		*Increase*	
County	*2001*	*2010*	*2001*	*2010*	*(no.)*	*(%)*
Allegany	61	69	0.50	0.40	8	0.13
Anne Arundel	1,241	1,568	10.00	10.40	327	0.26
Baltimore City	877	988	7.10	7.10	111	0.13
Baltimore County	1,495	1,668	12.10	11.10	173	0.12
Calvert	119	154	1.00	1.00	35	0.29
Caroline	19	22	0.20	0.10	3	0.16
Carroll	293	336	2.40	2.10	43	0.15
Cecil	89	95	0.70	0.60	6	0.07
Charles	156	213	1.30	1.40	57	0.37
Dorchester	24	29	0.20	0.20	5	0.21
Frederick	486	665	3.90	4.40	179	0.37
Garrett	42	46	0.30	0.30	4	0.10
Harford	347	481	2.80	3.20	134	0.39
Howard	1,181	1,654	9.60	11.10	473	0.40
Kent	31	32	0.30	0.20	1	0.03
Montgomery	3,980	4,499	32.20	30.10	519	0.13
Prince George's	1,205	1,432	9.70	9.50	227	0.19
Queen Anne's	86	124	0.70	0.80	38	0.44
Somerset*	19	17	0.20	0.10	-2	-0.11
St. Mary's	183	295	1.50	2.00	112	0.61
Talbot	88	98	0.70	0.70	10	0.11
Washington	149	190	1.20	1.30	41	0.28
Wicomico	120	158	1.00	1.10	38	0.32
Worcester	61	86	0.50	0.60	25	0.41
Not geocoded	8	37	0.10	0.20	29	3.63
Total	12,360	14,956	100.00	100.00	2,596	0.21

Source: Tables prepared by the Maryland Department of Planning, March 2009 and 2011. Data from US Census Bureau, Zip Code Business Patterns 2006, March 2009

*Showed a decrease over the reporting period.

technology, telecommunications, and aerospace engineering. Montgomery County is the third-largest biotechnology center in the entire United States. It is also the location of nineteen major federal agencies, including the US Food and Drug Administration and NIH.

Large concentrations of high-tech firms also exist in Anne Arundel County (Annapolis and around Baltimore-Washington International Thurgood Marshall Airport), Baltimore County (Lutherville-Timonium area along I-83 and Owings Mills along I-795), and Howard County (Columbia area). In Prince George's County, there is a concentration of high-tech firms along Route 450 in Lanham and in Beltsville near the Capital Beltway. Outside of the Washington-Baltimore corridor, there are secondary concentrations in Frederick near Fort Detrick; in Lexington Park, St. Mary's County, near the Patuxent River Naval Air Station; and in Washington County around Hagerstown.

The Maryland Department of Planning's list of high-tech industry establishments in the state shows that the largest number was concentrated in Computer Systems Design and Related Services (29.4%), followed by Management, Scientific, and Technical Consulting Services (24%); Architectural, Engineering, and Related Services (16%); Management of Companies and Enterprises (7.4%); and Wired Telecommunications Carriers (4.8%). The majority of these businesses are part of the services sector of the economy, not in secondary manufacturing, although the Maryland Department of Planning refers to the grouping as an "industry." The location of these businesses is important because high-tech businesses usually cluster, with services linked to manufacturers, universities, government, and to each other.

There are over 500 bioscience firms alone in Maryland, representing 8 percent of the US industry. In 2012, Maryland ranked fourth nationally in the number of biotechnology firms. Prominent biotech firms in Maryland include Astra Zeneca/MedImmune, Lonza, Martek, Nabi, Otsuka, Quiagen, Shire, and Teva. Several segments of Maryland's high-tech industry—namely information technology and telecommunication—benefit from the geographical proximity of major federal agencies, including the Federal Communications Commission, NSA, and the NASA Goddard Space Flight Center. The Maryland aerospace industry, with more than forty companies employing more than 13,000 employees, also benefits from the dense network of federal agencies and allied high-tech businesses in the region.

The Aerospace Industry

Maryland boasts an impressive array of space industry assets. The NASA Goddard Space Flight Center has been in Maryland for over fifty years and manages NASA's observation, astronomy, and space physics missions. The Hubble Telescope, the first major optical telescope placed in space, was built in Maryland and was carried into orbit by the Space Shuttle Discovery in 1990. The Space Telescope Science Institute

(STScI), located at Johns Hopkins University in Baltimore, manages the Hubble Space Telescope. A consortium of thirty-seven US and seven international organizations operates STScI, employing over 400 workers. STScI is working on the replacement James Webb Space Telescope to be launched in 2016.

The space industry, born in the 1950s, is still young. Nationally, this industry employed about 266,000 in 2009. In February 2010, President Obama announced plans to cancel NASA's Constellation Program, the most recent human spaceflight initiative. This decision has caused tremendous uncertainty in the industry, especially in those states heavily invested in manned space flight such as California, Florida, and Texas. Fortunately for Maryland, its aerospace industry is focused on other aspects of space exploration than manned flight—namely science and research as well as defense, security, and intelligence gathering. Overall, in 2009, Maryland's space industry directly generated over 10,000 jobs and about $1 billion in wages in aerospace (table 8.7). ATK's Propulsion and Controls operation in Elkton, Cecil County, which manufactured solid fuel rocket motors, was one exception, laying off about 200 workers in 2010.

The Applied Physics Laboratory at Johns Hopkins University has designed, built, and launched sixty-four spacecraft and more than 150 instruments since 1959; it also built and launched the New Horizons Mission to Pluto. Maryland universities performing cutting-edge aero-

Table 8.7. Maryland employment in the space industry, 2009

Occupation	*Number employed*	*Mean salary*
Aerospace and operations technicians	250	$60,980
Aerospace engineers	2,690	$115,310
Astronomers	270	$125,520
Atmospheric and space scientists	400	$109,530
Avionics technicians	340	$54,070
Chemical engineers	660	$98,750
Materials engineers	560	$108,840
Materials scientists	210	$92,670
Mechanical engineers	5,090	$89,410
Postsecondary atmospheric, earth, marine, and space science teachers	70	$100,780
Total	10,540	$100,103

Source: US Bureau of Labor Statistics, Occupational Employment Statistics, May 2009

space research include Bowie State University, Morgan State University, and the University of Maryland campuses in Baltimore County, College Park, and the Eastern Shore. Other notable space-related entities in the state include the US Naval Research Laboratory's Center for Space Technology, which has two facilities in southern Maryland, and the US Naval Academy's outstanding aerospace engineering program. Maryland's private sector assets in the aerospace industry include ATK, Hughes Communications, Lockheed Martin, Northrop Grumman, Orbital Sciences, Raytheon, and others. Each year NASA contracts about $1.4 billion of business with Maryland companies.

Maryland has entered the twenty-first century with considerable social and economic assets. It has the highest concentration of astronomers in the United States (eleven times the national average) and ranks second in the concentration of physicists (three and a half times the national average). Its overall income and education levels are high, and the poverty rate is the second-lowest in the nation (with 10.3% of the population living in poverty, compared to the national rate of 13.3%). Maryland likewise has high percentages of professional and technical workers and residents who hold a bachelor's, master's, or doctoral degree; a high quality of life; and other achievements. But while Maryland is superlative it these areas, its wealth and all it can command is not distributed evenly both by economic class nor geographical region.

PART III

Human Footprints on the Marylandscape

9 The People of Maryland

Although a small state, Maryland is home to a large, heterogeneous population. Marylanders enjoy higher incomes, better job prospects, and greater diversity than residents of most other states. But these benefits have not come without their share of challenges.

Population

In 2011, 5,828,289 people called Maryland home. As a whole, Marylanders are prosperous, earning the highest median household income (above $60,000 per year) in the nation (following Maryland, in descending order, were New Jersey, Connecticut, Alaska, Massachusetts, Hawaii, and New Hampshire). About 10.3 percent of Marylanders live below the poverty level, ranking forty-ninth among the states (only New Hampshire has a lower rate of poverty). When these figures are coupled with its strong diversified economy, Maryland comes across as a prosperous state with much to offer its people (US Census Bureau 2010). The population geography of Maryland helps to explain the social and economic landscape. People, wealth, poverty, amenities, facilities, and opportunities are not evenly spread across Maryland. The story begins with some general aspects of the population: numbers, distribution, and growth patterns, followed by a consideration of racial and ethnic diversity and then a look at Maryland's foreign-born population.

First, some details of Maryland's population from the US Census of Population
Statewide:

- Between 2000 and 2011, Maryland's population grew by 531,803 (9.7%), ranking Maryland fifteenth among the states by population size.
- The 9.7 percent growth over these years was identical to the national rate, but it was still the lowest Maryland growth rate since the 1970s.
- All of Maryland's population growth is among minority populations; the non-Hispanic white population has declined.

- The largest absolute increase since 2000 was in the Hispanic population (242,716, or 106.5%), followed by non-Hispanic African Americans (209,494, or 14.3%), and multiracial persons (42,894, or 51.7%).
- The non-Hispanic white population has dropped by 128,589 (–3.9%) since 2000; their share of Maryland's population dropped from 62.1 percent in 2000 to 54.7 percent in 2011. But it is still the single largest group in the population, followed by non-Hispanic African Americans (29.0%), Hispanics (8.2%), non-Hispanic Asians (5.5%), and multiracial non-Hispanics (2.2%).

Breaking down the data by jurisdiction, population changes from 2000 to 2011 include:

- The four inner suburban counties have had largest absolute increases in population since 2000: Anne Arundel (52,733, or 10.7%), Baltimore County (54,343, or 7.2%), Montgomery (112,316, or 12.8%), and Prince George's (68,122, or 8.5%).
- The four second-tier counties also experienced significant growth: Charles (27,904, or 23.0%), Frederick (40,182, or 20.4%), Harford (26,692, or 12.1%), and Howard (43,552, or 17.4%).
- Baltimore City declined in population, but its loss of 29,593 people (–4.6%) was below the average loss of the 1990s (–84,860, or –11.5%) and the smallest decline since the 1950s. Still, Baltimore City did gain in population of Hispanics (14,899) and non-Hispanic Asians (4,573).
- In general, the second- and third-tier counties in Maryland grew more than cities and inner counties, where escalating housing prices and real estate taxes have spurred growth and migration into the outer counties.
- Counties that grew most were Charles (23%) and St. Mary's (24.3%) in southern Maryland; Caroline (10.8%), Cecil (17.6%), and Queen Anne's (18.6%) on the upper Eastern Shore; Wicomico (16.8%) on the lower Eastern Shore; and Washington County (12.2%) in western Maryland.
- There was historic growth for some counties. St. Mary's and Washington Counties grew more than in any other ten-year census period since the first US Census in 1790, with gains of 18,940 and 15,507 residents, respectively.

Data on the jurisdictional population changes by race and Hispanic origin reveal the following:

- Most of the non-Hispanic white population loss was in the four inner suburban counties, plus Howard and Charles. While the non-Hispanic white population declined statewide, it grew in Calvert, Carroll, Frederick, and Washington Counties as well as on the Eastern Shore.

- Two counties moved to "minority-majority" status in 2010, meaning that a majority of the population is made up of people from minority groups: Charles (51.6%) and Montgomery (50.7%). Montgomery County has the largest population of any jurisdiction in the state.
- Hispanic growth took place statewide, but it was largest in six jurisdictions: Prince George's (71,915), Montgomery (64,794), Anne Arundel (20,000), Baltimore County (19,961), Baltimore City (14,899), and Frederick (12,471).
- All of Howard County's population change occurred because of minority growth, but unlike other counties the largest component was non-Hispanic Asians (22,124), who now make up 14.3 percent of the county's population—the highest in Maryland. The overall minority share in Howard County's population is now 40.8 percent. It is highly likely that it will become a minority-majority county by the 2020 census or soon after.

Maryland's changing demographic structure clearly reflects the recent national demographic trends of slower population growth, steadily rising immigration rates, increasing number of nonfamily households, an older age profile, and an accelerated rate of suburbanization and nonmetropolitan growth. The population core of Maryland extends from Harford County southwest to Montgomery County. It includes the Baltimore and Washington regions, which together are home to 82 percent of Maryland residents. Western Maryland and the Eastern Shore have much smaller populations. This geographic pattern has many ramifications for the state, particularly when it comes to statewide elections, political representation, and the allocation of resources. Population size data for each county are given in table 9.1.

Population Density

The regions with the highest population densities are Baltimore and Washington, DC. Baltimore City has over 7,648 people per square mile, and the next-highest densities are in Montgomery and Prince George's Counties (the first- and second-largest counties by population, respectively). The tremendous growth in population in these two counties in recent decades marks a shift of the center of population from the Baltimore region to the Washington region, which has important significance in Maryland politics. Not only is the Washington–Baltimore urban corridor dominant within Maryland, but also within that corridor a shift is taking place toward the southwest. The low densities on the Eastern Shore and in western Maryland reflect their relatively small populations and large areas. The moderate—yet rising—densities in southern Maryland reflect its growing population.

Within the entire nation, Maryland is larger in area than only eight other states, yet its population is larger than thirty-one other states. With a relatively large population for its small land base, Maryland's

Table 9.1. Population, density, and land area changes 2000–2011

Region	*Land area (square miles)*	*Population in 2011*	*Population density (per square mile)*	*Population change, 2000–2011 (%)*
Maryland	9,774	5,828,289	596.3	9.7
Baltimore region	2,237	680,756	304.3	6.5
Anne Arundel County	416	544,403	1,308.7	10.7
Baltimore City	81	619,493	7,648.1	-4.6
Baltimore County	599	809,949	1,352.2	7.2
Carroll County	449	167,288	372.6	10.5
Harford County	440	246,489	560.2	12.1
Howard County	252	293,142	1,163.3	17.4
Washington region	1,644	2,097,772	1,276.0	11.8
Frederick County	663	236,745	357.1	20.4
Montgomery County	496	989,794	1,995.6	12.8
Prince George's County	485	871,233	1,796.4	8.5
Southern Maryland	1,037	345,870	333.5	22.3
Calvert County	215	89,256	415.1	18.8
Charles County	461	149,130	323.5	23
St. Mary's County	361	107,484	297.7	24.3
Western Maryland	1,531	252,946	165.2	6.9
Allegany County	425	74,692	175.7	-0.2
Garrett County	648	30,051	46.4	0.7
Washington County	458	148,203	323.6	12.2
Upper Eastern Shore	1,588	241,262	151.9	14.8
Caroline County	320	32,985	103.1	10.8
Cecil County	348	101,694	292.2	17.6
Kent County	279	20,204	72.4	4.9
Queen Anne's County	372	48,354	130.0	18.6
Talbot County	269	38,025	141.4	12.2
Lower Eastern Shore	1,735	209,683	120.9	12.1
Dorchester County	558	32,640	58.5	6.7
Somerset County	327	26,339	80.5	6.6
Wicomico County	377	99,190	263.1	16.8
Worcester County	473	51,514	108.9	10

Source: US Census of Population, 2000, 2010, and 2011(estimate)

population density is exceeded only by the four northeastern states of Connecticut, Massachusetts, New Jersey, and Rhode Island.

Population Growth

Maryland's population grew by 9.7 percent between 2000 and 2011 (fig. 9.1). The Baltimore region grew by a modest 6.5 percent overall. Baltimore County grew by only 7.2 percent, but Howard County grew by a whopping 17.4 percent, Harford by 12.1 percent, and Carroll by 10.5 percent. Baltimore City lost 4.6 percent of its population, and Allegany County lost 0.2 percent of its population. Keep in mind that these are percentages of growth, not absolute numbers, and that growth rates should be compared to actual population numbers. Still, growth rates identify those areas to which Marylanders and incoming new residents are moving in increasing numbers.

Growth rates in the Washington region (11.8%) were led by Frederick County (20.4%). The accelerated growth and development around Frederick and along its direct rail and road links to Washington, DC, are important parts of the changing demographic map of Maryland. There are a growing number of businesses, especially of a high-tech

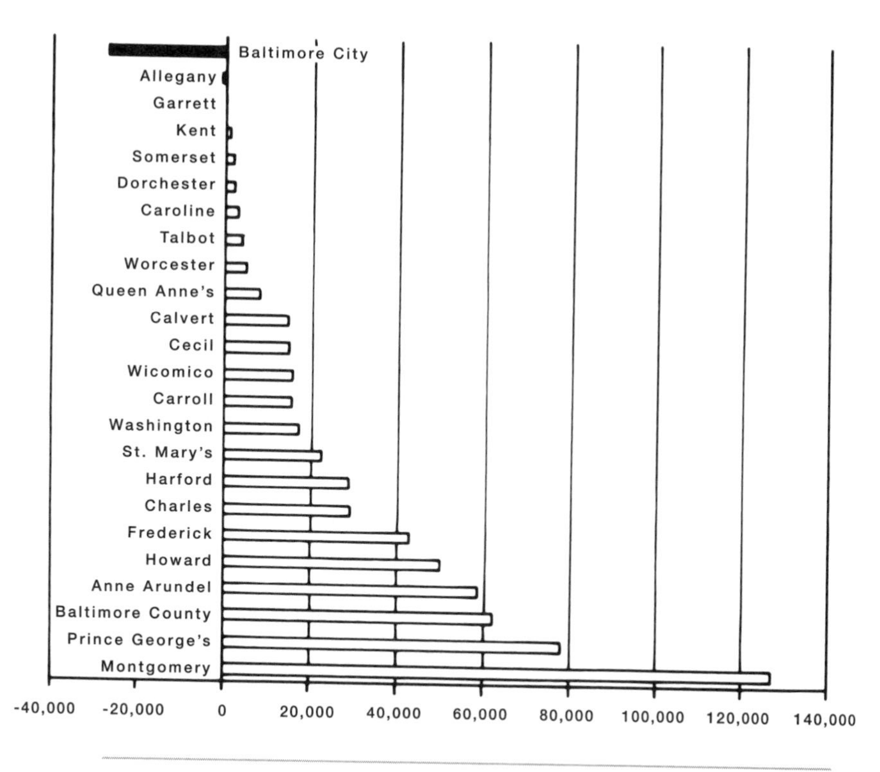

Figure 9.1 Population Change, 2000–2012

nature, along the Frederick–Washington corridor along I-270. Southern Maryland, which as a region had the highest overall growth rate in the state, continues to grow at a rapid rate (22.3%). St. Mary's County added 24.3 percent to its population during this period. Charles grew by 23.0 percent and Calvert by 18.8 percent. Southern Maryland is experiencing growth both southeastward from the Washington metropolitan area as well as southward from Anne Arundel County.

On the Eastern Shore, the counties that stand out for growth are Queen Anne's at 18.6 percent and Cecil at 17.6 percent. Queen Anne's County is attracting more people from the Baltimore metropolitan area, which includes Annapolis on the other side of the Bay Bridge. The US Census Bureau now considers Queen Anne's County to be part of the Baltimore metropolitan area. In Wicomico County, the population grew at a rapid pace of 16.8 percent, especially in and near Salisbury. This region continues to attract new businesses as well as retirees and purchasers of second homes.

Recent data on population changes in Maryland counties reported by the US Census Bureau show both the absolute and relative (percentage) changes between 2000 and 2010. In some instances, a county can rank higher in percentage change but lower in absolute numbers. A county with a large population might have a smaller percentage change, but a smaller percentage of a larger population base can result in a larger absolute increase in population. For example, Queen Anne's County had a 17.8 percent increase between 2000 and 2010, while Prince George's County increased by only 7.7 percent. Yet, in absolute numbers, Queen Anne's increased by only 7,235 people while Prince George's increased by 61,905.

Maryland's population grew steadily from the first US Census of Population in 1790 (320,000 people) to 1920 (1,450,000 people). The 1920 census was the first to report that more people in the United States were living in urban than rural areas, but 50 percent of Maryland residents were already living in urban areas by 1900. In the years following World War II, Maryland experienced rapid growth. Its population increased from 1,821,000 in 1940 to well over 5.7 million by 2010.

Population Age

The structure of the Maryland population reveals some important patterns. For Maryland's white and nonwhite populations, the increasing percentage of females relative to males in the older age cohorts reflects the generally longer life span for women. The white population also has an older median age. The nonwhite population is younger, but the white population lives longer, having a greater percentage of its people in older age cohorts. The nonwhite population is more broadly based in the younger age cohorts.

Table 9.2 shows several selected vital statistics for Marylanders. The median age for all Marylanders is thirty-six; for whites, however, the median age is over thirty-eight. The younger black and Hispanic populations have median ages of 32.2 (blacks) and 27.2 (Hispanics). Not only is the white population older and longer lived, but it also has a low infant mortality rate (i.e., the number of infant deaths in the first year of life per 1,000 births). For the entire Maryland population, the infant mortality rate is 7.3, while the white infant mortality rate is a much lower 4.7. The Hispanic infant mortality rate of 4.6 is also significantly lower than the overall rate and lower than the white rate; the infant mortality rate for Maryland's black population is a high 12.7. These differences reflect many factors, including prenatal care, access to health-care facilities, health insurance, environment, diet, and education level.

By 2010, the life expectancies of whites, blacks, and Hispanics were converging, but the life expectancy for blacks was still lower than for the other two groups. Birth and death rates reveal the natural growth of these population groups, which is determined by subtracting the death rate from the birth rate. For the white population, the birth rate exceeds the death rate by 5.0, meaning that the white population will continue to grow at a moderate rate. (Keep in mind that natural growth rates do not include net migration figures; out-migration of whites explains the drop in numbers and percentage of this population from 2000 to 2010.) Maryland's black population has the highest

Table 9.2. Vital statistics of selected population groups

	Total (%)	*White (%)*	*Black (%)*	*Hispanic (%)*
	Median age			
Both sexes	36.0	38.5	32.2	27.2
Male	34.9	37.5	30.5	26.5
Female	37.0	39.5	33.7	28.0
	Life expectancy			
Both sexes	77.0	78.5	72.5	80.4
Male	74.3	76.0	68.7	77.1
Female	79.7	80.8	76.0	83.7
	Birth/death rates			
Infant mortality	7.3	4.7	12.7	4.6
Birth	13.4	12.5	14.6	27.2
Death	7.9	7.5	9.6	5.9

Source: Maryland Department of Health and Mental Hygiene, 2010
Note: Hispanics are an ethnic and not a racial minority. Hispanics can be of any race. In Maryland, over 90 percent of Hispanics are white.

death rate at 9.6 per population size of 1,000, but its high birth rate of 14.6 means that this population will also continue to grow moderately. The Hispanic population has a death rate of only 5.9 per population size of 1,000, but also a very high birth rate of 27.2. The differential of 21.3 points to a high natural growth rate among Hispanics.

Median age also varies geographically across Maryland (fig. 9.2). There is a thirteen-year difference between the oldest and the youngest counties by median age. Worcester (48.1 years) and Talbot (47.4 years) are the oldest counties, exceeding the median age for Maryland as a whole (38.0 years). Talbot County has a population that is aging in place. There is also significant in-migration of relatively wealthy retirees who often buy high-end waterfront housing. Worcester County is also aging in place with significant in-migration. Unlike Talbot County in-migration, those coming to Worcester County are mainly middle-income retirees relocating to retirement communities near the Ocean City area.

Prince George's County has the youngest median age of Maryland counties at 34.9 years, which is a full 13.2 years younger than Worcester County. Prince George's County is experiencing a large population gain by natural increase, plus a large increase in the number of young

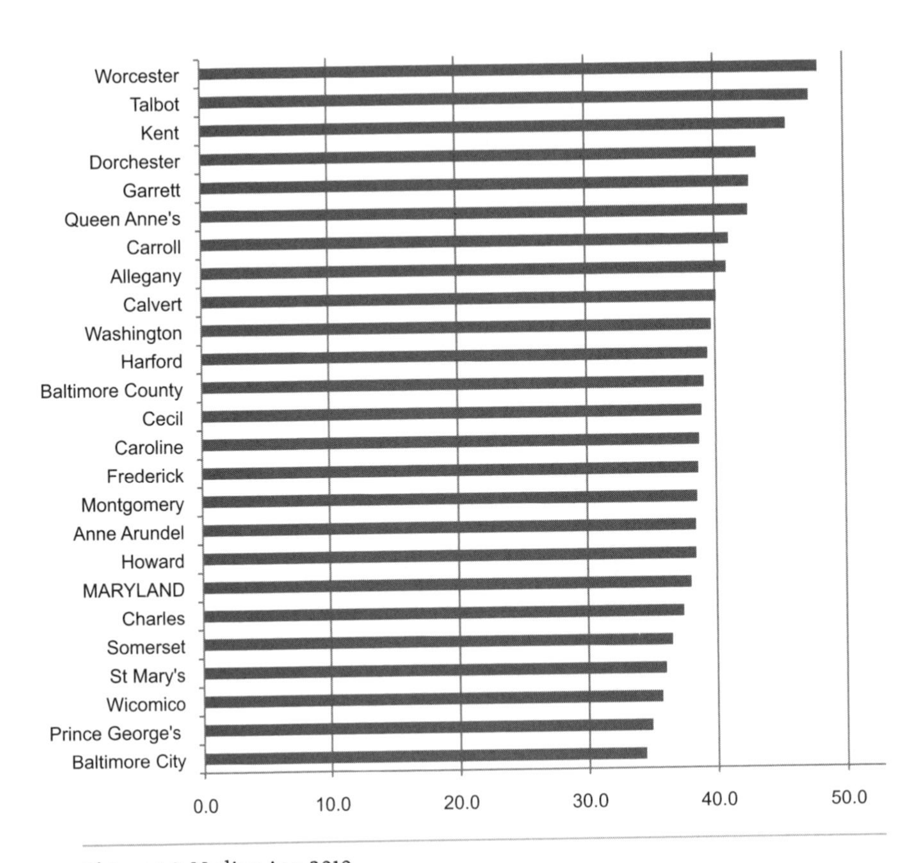

Figure 9.2 Median Age, 2010

(mainly in their twenties) foreign-born immigrants. The median age in Baltimore City (34.4 years) is lower than in all of the counties. Charles (36.5 years), St. Mary's (36 years), and Wicomico (35.7 years) Counties also have relatively young populations. Somerset and Wicomico Counties have attracted younger populations as the greater Salisbury area continues to grow. These two counties contrast sharply with the older population of its immediate eastern neighbor, Worcester County. The population in southern Maryland is younger, mainly owing to natural increase plus in-migration of young families, many moving from Prince George's to Charles, Montgomery, and St. Mary's Counties. In the late 1990s, St. Mary's experienced an influx of young residents relocating to the Patuxent River Naval Air Station, mainly from New Jersey, Pennsylvania, and Virginia.

The percentage of the population aged sixty-five and older ranges from nearly 24 percent to under 10 percent, which roughly corresponds to the median age pattern (e.g., Talbot County has 23.7 percent aged sixty-five and older; Worcester 23.2 percent). The younger median age in Prince George's County mirrors the low 9.4 percentage of people aged sixty-five and older. Other counties with about 10 percent are Charles (9.5%), Howard (10.1%), and St. Mary's (10.3%). All of these areas are growing because of natural increase plus significant in-migration of younger families, mainly from the older suburban areas of Baltimore, Montgomery, and Prince George's Counties. These growing areas are the new suburbs, but by 2020 they will likely have a larger percentage of elderly residents. There are higher percentages of elderly west of Frederick County in western Maryland, representing a population aging in place, especially in Allegany County, but with some in-migration of retirees to Garrett County and to a lesser degree to Washington County.

Growing Diversity

Already a diverse state, Maryland is becoming even more so. The net out-migration of whites (non-Hispanic) and in-migration of African Americans, Hispanics, and other minorities—in addition to steady foreign immigration—are heightening the state's diversity. Foreign immigration is a significant factor in the population growth of Maryland, especially for certain counties. Although the minority population is growing in most jurisdictions, the geographical pattern of racial and ethnic composition is not uniform. For some counties, the growth from 2000 to 2010 of Asians and Hispanics matches or exceeds growth for the entire previous decade.

Maryland's Minority Population

In this chapter I define a minority as a person belonging to any race that is not non-Hispanic white. This includes African Americans, Asians and Pacific Islanders, American Indians, multiracial individuals, and

Hispanics. Hispanics are an ethnic—not a racial—minority. Hispanics can be of any race, but in Maryland over 90 percent of Hispanics are white (Goldstein 2006). Marylanders are 58.18 percent white and 29.45 percent African American; these two racial groups constitute nearly 88 percent of the population.

Maryland's population of 5,828,289 people is far from homogeneous. Yet the geographic distribution by race reveals a high degree of homogeneity in some counties and high diversity in others (table 9.3). Statewide, the US Census Bureau classifies 97 percent of Marylanders as being of one race, and the remaining 3 percent are of two or more races. By any measure of diversity, Garrett County (97.8% white) is the least diverse of all Maryland counties. But it is not only the jurisdictions with high percentages of white population that lack significant diversity. Several jurisdictions have high percentages of blacks or African Americans, including Baltimore City (64%) and Prince George's (64%), Somerset (42%), Charles (41%), and Baltimore (26%) Counties. Although Maryland is considered diverse and on the verge of becoming a minority-majority state, at local geographical scales there are clearly highly homogeneous concentrations of racial groups.

Some counties across the northern border of the state have low percentages (under 20%) of minorities in their population: Allegany, Carroll, Cecil, Garrett, and Washington. Garrett County has the smallest relative minority population with only 2.2 percent. The county has a historic pattern of white European settlement, and there continues to be little in-migration of minorities or foreign-born immigrants. Carroll County has the second-lowest share of minorities with 8.8 percent. Unlike Garrett County, Carroll County is experiencing more rapid population growth and strong in-migration. Although whites—mainly those from Baltimore, Howard, and Montgomery Counties—dominate its in-migration, the minority segment of Carroll County's population is growing, too (it was 4.9% in 2000).

At the highest end of the spectrum are Baltimore City and Charles, Dorchester, Montgomery, Prince George's, and Somerset Counties, all of which have minority populations of over 47 percent. There is a noticeable pattern of counties with high percentages of minorities ringing Washington, DC. Prince George's County has the highest minority percentage in the state at 75.7 percent, which is mostly due to in-migration from Washington as well as recent foreign immigration, especially of Asians and Hispanics. Significant out-migration of white families has further concentrated the minority population in the county. Prince George's minority population is primarily African American, including professionals from around the country who have come to Prince George's County to work in Washington, at the many federal agencies in Maryland, and for private high-tech firms.

Baltimore City has a large African American population, especially since the immediate post-Civil War period, and the accelerated white

Table 9.3. Maryland's population by race

County	*Total*	*White*	*Black or African American*	*American Indian or Alaska Native*	*Asian*	*Native Hawaiian or other Pacific Islander*	*Other race*	*Two or more races*
Allegany	75,087	66,981	6,028	107	568	31	186	1,186
Anne Arundel	537,656	405,456	83,484	1,665	18,352	484	12,642	15,573
Baltimore City	620,961	183,830	395,781	2,270	14,548	274	11,303	12,955
Baltimore County	805,029	520,185	209,738	2,625	40,077	319	12,801	19,284
Calvert	88,737	72,231	11,930	329	1,260	44	578	2,365
Caroline	33,066	26,396	4,585	123	188	53	1,011	710
Carroll	167,134	155,282	5,332	328	2,418	56	1,199	2,519
Cecil	101,108	90,189	6,284	294	1,097	48	1,019	2,177
Charles	146,551	73,677	60,031	957	4,366	103	1,963	5,454
Dorchester	32,618	22,065	9,042	112	301	9	463	626
Frederick	233,285	190,306	20,148	730	8,946	107	6,684	6,464
Garrett	30,097	29,440	301	43	76	0	39	198
Harford	244,826	198,763	31,058	614	5,826	199	2,318	6,048
Howard	287,085	178,523	50,188	866	41,221	123	5,709	10,455
Kent	20,197	16,169	3,056	42	165	6	393	366
Montgomery	971,777	558,358	167,315	3,639	135,451	522	67,847	38,645
Prince George's	863,420	166,059	556,620	4,258	35,172	541	73,441	27,329
Queen Anne's	47,798	42,397	3,298	149	469	12	651	822
St. Mary's	105,151	82,636	15,030	424	2,596	73	1,054	3,338
Somerset	26,470	14,170	11,192	85	184	7	371	461
Talbot	37,782	30,746	4,829	65	472	22	1,030	618
Washington	147,430	125,447	14,133	314	2,056	66	1,626	3,788
Wicomico	98,733	67,784	23,873	236	2,471	47	1,872	2,450
Worcester	51,454	42,194	7,022	145	573	11	632	877
Total	**5,773,552**	**3,359,284**	**1,700,298**	**20,420**	**318,853**	**3,157**	**206,832**	**164,708**
Percentage	**100**	**58.19**	**29.46**	**0.35**	**5.52**	**0.05**	**3.58**	**2.85**

Source: Data from US Census Bureau, Census Demographic Profiles, 2010

out-migration since 1950 has increased the percentage of Baltimore City's African American population even more. In recent decades, however, the growth of the minority share of the population has slowed, mainly because of out-migration of African Americans. Although both whites and African Americans have migrated out of Baltimore City, the white out-migration has occurred on a much larger scale. On the Eastern Shore, blacks make up Somerset County's largest minority population. Other counties with high minority percentages (30–45%) include Baltimore, Dorchester, Howard, and Wicomico. In absolute numbers, the minority population of Maryland is concentrated in the area from Baltimore City and County, south to the Maryland counties surrounding Washington, DC.

Maryland has one of the largest percentages of racial minorities among all the states. It is exceeded only by four states, all with minority-majority populations: California, Hawaii, New Mexico, and Texas. In addition to these four states, Washington, DC, also has a minority-majority population (African American), as does Baltimore City. In August 2006, the US Census Bureau's annual American Community Survey estimated that the "white only" population of Maryland had fallen below 60 percent. The population of Maryland that is "white only" stood at 64 percent in 2005 and 58 percent in 2010.

In nearly every state, population diversity is increasing. The minority-majority dialogue in Maryland has long been about the majority white population and the minority African American population. It may soon be the case that Maryland will be described more in terms of white, black, brown, and others. Given current trends, it may not be long before Maryland joins the list of states in which there is no majority group. In the spring of 2007, the US Census Bureau reported that Maryland's minority population was 51 percent of the total population under five years of age, with a decrease of the white population overall.

Ancestry and Foreign-Born Residents

The US Census of Population estimates the ancestry of state populations. The largest group in Maryland by far includes those claiming German ancestry, followed by, in order, the Irish, English, American, Italian, Polish, and Sub-Saharan African.

Foreign-born residents constitute another segment of the Maryland population that cuts across race and various socioeconomic groups. The term *foreign born* includes people living in Maryland who were not US citizens at birth. The foreign born (or immigrants) coming to Maryland include naturalized citizens, lawful permanent immigrants, refugees and asylum seekers, legal nonimmigrants (e.g., those coming to the United States on student, work, or other temporary visas), and those residing in Maryland without legal authorization. In contrast, *natives* include those living in the United States who are US citizens by virtue of being born (1) in one of the fifty states or Washington, DC; (2) in a US insular area (e.g., Puerto Rico or Guam); or (3) in another country to at least one parent who is a US citizen.

Among the fifty states, Maryland ranks twelfth in terms of the number of residents who were foreign born. During the 1990s, Maryland added nearly 205,000 foreign-born residents to its population and another 213,000 in the 2000s. It is estimated that in 2010 there were over 740,000 foreign-born residents in the state, representing about 13 percent of the total population (Migration Policy Institute 2010). This percentage has grown steadily from only 3 percent in 1960.The largest share of Maryland's foreign-born population comes from Latin America (37.6%), followed by Asia (32.8%), Africa (16.1%), and Europe (12.2%). Only 1 percent comes from places in North America (Bermuda, Canada, Greenland, and the French island possessions off the coast of northeast Canada, Miquelon, and St. Pierre). The overall number of foreign born in Maryland has increased steadily since 1990 (table 9.4). During this time, the percentage of foreign-born residents from Europe has decreased, and there have been significant increases in the shares coming from Latin American and Africa. The numbers of Asian foreign born have not increased significantly by percentage, but they still represent the second-largest group, holding steady between 32 and 37 percent.

In 1990, the top three countries of origin for Maryland's foreign born were Germany, India, and South Korea. Ten years later, the top three countries of origin for Maryland's foreign born were El Salva-

Table 9.4. Maryland's foreign born by birthplace

	1990		*2000*		*2010*	
	Number	*Percentage*	*Number*	*Percentage*	*Number*	*Percentage*
Total population of Maryland	4,781,468	100	5,296,486	100	5,669,478	100
Native born	4,667,974	93.4	4,778,141	90.2	4,969,078	87.2
Foreign born	313,494	6.6	518,345	9.8	730,400	12.8
Maryland foreign born	313,494	100	518,315	100	730,400	100
Born in Europe	74,318	24.0	86,870	16.8	89,185	12.2
Born in Asia	120,840	38.4	181,504	35.0	239,471	32.8
Born in Africa	23,173	7.4	62,688	12.1	117,315	16.1
Born in Oceania	1,198	0.4	1,957	0.4	2,085	0.3
Born in Latin America	86,678	27.5	176,026	34	274,679	37.6
Born in Northern America	7,287	2.3	9,300	1.7	7,665	1.0

Source: Migration Policy Institute, www.migrationinformation.org

dor (11.3%), India (6.4%), and China (excluding Taiwan; 4.7%). Over the years, the pattern of source countries of foreign born coming to Maryland, and the United States overall, has changed. In 2010, about 46 percent of Maryland's foreign-born residents were US citizens.

Also important to consider is where in Maryland the foreign born have settled. By far, the top county in attracting new foreign-born residents during the decade 2000 to 2010 has been Montgomery County (35.9%), followed by Prince George's County (24.0%). Together, these two counties accounted for nearly 60 percent of Maryland's arriving foreign residents during the past ten years. The urban corridor of Maryland along I-95 from Harford to Montgomery Counties and along I-270 to Frederick from Washington, DC, contains the bulk of Maryland's foreign-born population. Far western Maryland (Allegany, Garrett, and Washington Counties) have small foreign-born populations. The largest group of foreign born in that region is in Washington County, attracted by the growing Hagerstown Area. The Eastern Shore also attracts a small number of foreign-born immigrants. Cecil County has a small, but growing, foreign-born population. Its location adjacent to I-95 and close proximity to Wilmington, Delaware, and Philadelphia make Cecil County increasingly attractive for these new Marylanders. On the lower Eastern Shore, Wicomico County is attracting an increasing number of foreign born thanks to the growth of Salisbury and its surrounding area, enhanced by Salisbury University and growing business and industrial activities. Southern Maryland is also experiencing an influx of foreign-born residents. Calvert, Charles, and St. Mary's Counties are all close to Montgomery and Prince George's Counties as well as to Washington, DC, and Annapolis. A noticeable spread effect has occurred in recent years as the foreign born from these areas have moved into southern Maryland.

Immigrants have historically been indispensable in making Maryland one of the most prosperous of states in the country. Over the years, their countries of origin have changed, but the challenges of coming to a new land and integrating into society remain. Foreign-born immigrants coming to Maryland make critical contributions to the economy and enrich the state's culture—in fact, foreign-born immigrants account for the majority of the growth in Maryland's labor force in the last decade. These immigrants are bringing needed skills to the new high-tech information economy in Maryland, and they are more likely than native Marylanders to have a college degree (43% to 36%). Maryland's foreign born include 27 percent of the state's scientists, 21 percent of its health-care workers, and 19 percent of its computer specialists. They make up about a third of Maryland's blue-collar maintenance workers and about a quarter of construction, agricultural, food, and health-care workers (Capps and Fortuny 2008).

HEALTH CARE

In the aggregate, Maryland has an impressive array of health-care facilities and personnel both within the state and in nearby Washington, DC. The geographical patterns of health care reveal a strong hierarchical system, ranging from fewer facilities in several counties to highly sophisticated, specialized facilities in Baltimore. Most of the state's general and specialized hospitals are located in the Washington–Baltimore metropolitan region. The limited number of hospital beds in western and southern Maryland and on the Eastern Shore reflects the lower population densities in those areas.

Baltimore, at the top of the Maryland health-care hierarchy, is a recognized worldwide leader in health care. Two large teaching hospitals, Johns Hopkins University and the University of Maryland, in addition to over twenty-five general and nine specialty hospitals provide extraordinarily high-quality care to those who can access it.

> Baltimore is a medical town that has produced surgeons like Halsted and Blalock, internists like Osler, ophthalmologists like Wilmer, and pediatricians like Taussig. Biologists from Baltimore figured heavily in the conquest of polio. Bufferin comes from Baltimore and the increasingly popular "belly button" sterilization procedure for women was developed there as well. Medicine, like steel, is one of Baltimore's principal exports and its practice is viewed locally with a reverence approaching awe. (Franklin and Doelp 1980)

Baltimore has declined in recent years as a major exporter of steel, but it has prevailed in its eminence as a medical center. To appreciate more fully the entire range of medicine and its history in Maryland, take a look beyond the prominent institutions in Baltimore:

> When we look at the wider picture, we find a medical setting interlaced with class, race, gender, religious, and regional differences. Over the past century, Maryland medicine has maintained a high profile in the world, a profile chiefly associated with the University of Maryland Medical Centers and the Johns Hopkins Medical Institutions. These dominate the state's largest city. Renowned as they are, however, they have never been able to serve all of the state's population. Throughout this relatively small state, important differences have always been starkly defined: east and west, north and south, rural and urban; Baltimore itself is like a number of separate towns or villages, concentrated but never fully integrated. Maryland's particular conglomeration of differences gives it a distinctive position in the region and in the history of medicine. (Sewell 1999)

The geographical distribution of physicians per population of 1,000 and by the broad specialty categories of primary care, medical specialties, surgical specialties, and all others should not be too surprising. The US Health Resources and Services Administration (HRSA), which sets benchmarks for all types of physicians nationally and by states and regions, finds that Maryland is among the states with the highest physician-to-population ratio, about 27 percent above the national average.

The Baltimore metro region, with a total physician-to-population ratio of 2.85, exceeds the HRSA national benchmark of 1.93 by 48 percent. This region also exceeds the HRSA benchmarks for primary care, medical specialties, surgical specialties, and all others. Southern Maryland has a physician-to-population ratio below the HRSA benchmark for all categories of physicians. A closer analysis of the situation in southern Maryland shows that physicians in that region generally work more hours per week than the state average, and that many residents of southern Maryland travel to the adjacent National Capital region (Prince George's and

Maryland health-care supply by type of physician and region, 2009–10

Region	*Total*	*Primary care*	*Medical specialties*	*Surgical specialties*	*All other*
Maryland physicians per population of 1,000 (excluding residents and federal employees)					
Baltimore metro	2.85	0.86	0.48	0.61	0.9
Eastern Shore	1.86	0.62	0.27	0.39	0.57
Washington metro	2.25	0.72	0.41	0.48	0.64
Western Maryland	2.17	0.73	0.39	0.42	0.63
Southern Maryland	1.34	0.53	0.25	0.26	0.3
Total	2.44	0.77	0.42	0.52	0.74
HRSA baseline	1.93	0.69	0.27	0.43	0.53
Difference from the HRSA baseline (%)					
Baltimore metro	48	24	76	41	70
Eastern Shore	-4	-10	0	-11	8
Washington metro	17	4	49	11	21
Western Maryland	12	5	41	-4	19
Southern Maryland	-31	-24	-8	-40	-43
Total	27	11	54	19	39

Source: Analysis of Maryland 2009/2010 license renewal database, calculations from HRSA 2008, and population counts from US Bureau of the Census
Note: Baltimore metro includes Baltimore City and Anne Arundel, Baltimore, Carroll, Harford, and Howard Counties; Eastern Shore includes Caroline, Cecil, Dorchester, Kent, Queen Anne's, Somerset, Talbot, Wicomico, and Worcester Counties; Washington metro includes Montgomery and Prince George's Counties; western Maryland includes Allegany, Frederick, Garrett, and Washington Counties; and southern Maryland includes Calvert, Charles, and St. Mary's Counties.

Montgomery Counties and Washington, DC) for health care. The National Capital region exceeds the HRSA benchmarks for all types of physicians, yet only marginally (4%) for primary care.

The Eastern Shore ranks below the total HRSA benchmark by –4 percent. This region is more specifically below the benchmarks for primary care physicians (–10%) and surgical specialties (–11%). Overall, western Maryland is above the HRSA benchmark, but falls just below it for surgical specialties (–4%). These factors, underscored by the values in the table above, confirm that Maryland is a state with a substantial physician resource base. Although there is no severe shortage of physicians state wide, in some regions there is a low enough ratio to be of mild concern. At the national level, the American Medical Association reported in 2009 that Maryland had the second-best ratio of nonfederal physicians to population in the country; only Massachusetts had a better ratio (American Medical Association 2009). Maryland's ratio was 364 physicians per population of 100,000, while Massachusetts had 407. For the entire United States, the median was 240 per population of 100,000 (Association of American Medical Colleges 2011).

A word of caution concerning physician-to-population ratios. The seemingly favorable ratio in large population centers such as in Baltimore City, Baltimore County, and Montgomery County can be somewhat misleading. Many of the physicians are on the staffs of various hospitals and may be serving in research and teaching capacities rather than in the provision of direct care. The numbers of physicians in the above table are nonfederal (i.e., they do not include physicians employed by the federal government), a factor that is especially important for counties such as Montgomery that have many federal medical facilities. Also excluded from these figures are hospital residents. The goal of this brief regional analysis of physician-to-population ratios is to get an idea of the number of physicians directly treating patients. Although research physicians and residents are important assets for Maryland, they are filtered out in this type of analysis.

Another factor especially important in Baltimore City, but also of significance in other counties, is access to medical care. Simply dividing the population by the number of physicians gives a ratio. But this ratio can be misleading because it assumes an equal spread of people among the physicians in a county, which is simply not the case. Some physicians have larger practices than others. Many lower-income people who lack health insurance coverage may not visit physicians as often as middle- and upper-income people who have insurance. In 2012, nearly 14 percent of Maryland residents did not have health insurance coverage, ranking it twenty-ninth nationally for the percentage of citizens without insurance. Nationally, the average of citizens without health insurance is 15.6 percent. Texas has the highest percentage without health insurance (24.6%), while Vermont has the lowest (9.5%).

Although Baltimore City may have one of the most favorable physician-to-population ratios in Maryland, nearly 25 percent of the city's residents are classified as living in poverty, a higher poverty rate than all counties in the state except Somerset (26%). It is even more revealing to look at the Baltimore poverty numbers by race: while nearly 13 percent of whites live in poverty, the rate jumps to nearly 28 percent for African Americans. In short, health care is not equally accessible to many of the citizens of Maryland. The physician-to-population ratio is, at best, only a crude estimate of availability of physicians and is not a direct measure of access to physicians and health care. Access to health care is complex. It results from the interplay of socioeconomic factors such as income, race, education, and geographical location.

Income

Income data describe Maryland as one of the more affluent states. In 2012, the overall per-capita personal income in the United States was $43,735. Maryland's per-capita personal income averaged $53,816, ranking it second highest among the states (US Bureau of Economic Analysis 2012). The only state ranking higher than Maryland was Connecticut ($54,397).

Although the concepts of wealth and standard of living can be estimated many ways, they cannot be measured directly. Both are conceptual and mean different things to different people. One way to estimate wealth and standard of living is to use a surrogate variable, the most common of which is income. There are numerous income data sets collected by the US Census Bureau that include total personal income for a state, per-capita personal income, median household income, and median family income. According to the US Census Bureau, a "family" consists of two or more people (one of whom is the householder) who are related by birth, marriage, or adoption and who reside in the same housing unit. A "household" consists of all people who occupy a housing unit regardless of relationship. A household may consist of a person living alone or multiple unrelated individuals or related family members living together (US Census Bureau 2013). By median household income, Maryland ($70,976) ranked first in the nation in 2011, followed closely by New Jersey ($68,342) and Connecticut ($67,304). Overall, the US median family income in 2011 was $50,221. However income is measured, Marylanders are on average among the most prosperous people in the country.

Next we must take these impressive statewide income figures and determine how they are distributed geographically around the state. With 10.3 percent of Maryland residents earning incomes that fall below the poverty level (the second lowest among the states), it is obvious that wealth and affluence are not evenly spread. The *median* household income for the entire state in 2011 was $70,976 (fig. 9.3), meaning there was an equal number of households earning more than the median, and an equal number earning less. The *mean* family income for Maryland in 2011 was much higher, $90,250. Because of the way the mean is calculated (by adding up the total household incomes in Maryland and dividing by the number of households), mean income figures are usually higher than the median. The mean is readily skewed to a higher value even if there are a small number of households earning extremely high incomes. The median does a better job of describing income distribution.

The pattern of median household income reflects the prosperity of the Baltimore region (excluding Baltimore City), the Washington, DC, region extending west to Frederick County, and southern Maryland. On the upper Eastern Shore, Cecil and Queen Anne's Counties have respectably high incomes. The extremely low median household incomes in far western Maryland and the lower Eastern Shore concur with many of the economic patterns in the state. Howard County has the highest median family income in Maryland ($101,417), while Somerset County has the lowest ($35,621)—a range of $65,796 between the upper and lower ends of the distribution.

Examining per-capita income, which is found by dividing the total personal income in a county by the number of people in that county, gives us another perspective on income. It treats individual people rather than household units. Although the patterns for personal and median household incomes have a high degree of similarity, there are

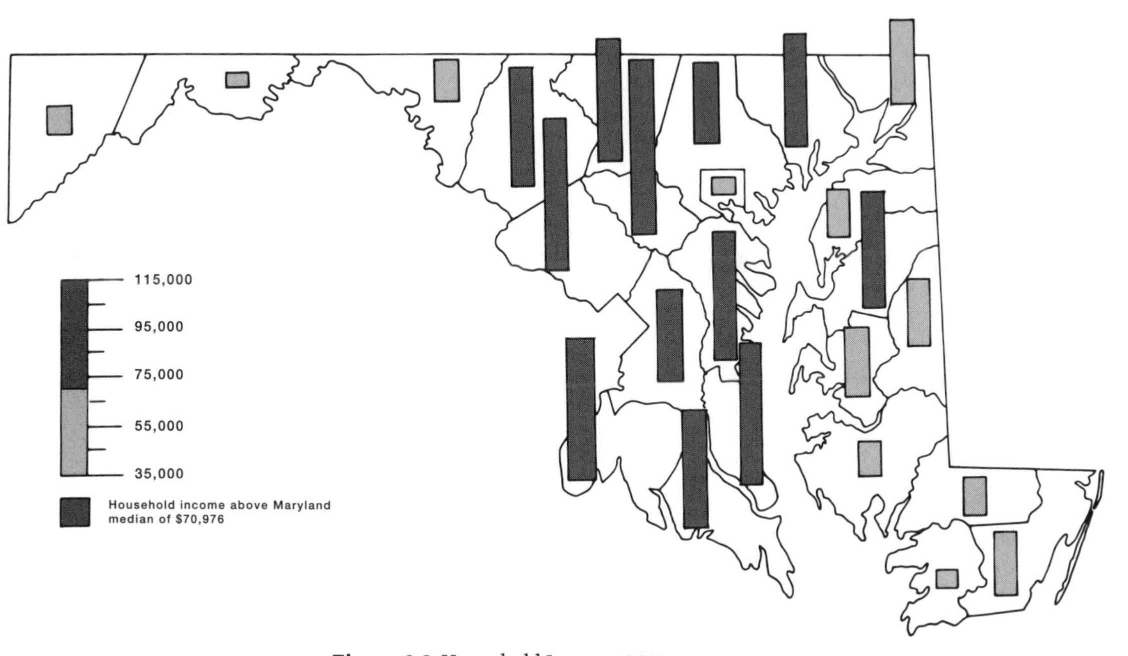

Figure 9.3 Household Income, 2011
Source: Data from US Census Bureau, American Community Survey, December 2012

some differences. Montgomery County has the highest per-capita income in the state ($46,122), and Somerset County ($17,378) has the lowest. Maryland counties boasting high per-capita incomes include Anne Arundel, Calvert, Charles, Frederick, Howard, Montgomery, Queen Anne's, and Talbot. Talbot County on the Eastern Shore is an interesting case. Although the median household income for Talbot is a modest $59,633, the per-capita income is $39,294 (recall that the first income amount is for the entire household; the second income amount is the mean average for each person in the county). We know that the higher incomes of some wealthy residents pull up the per-capita income for the entire population of the county. In addition to higher-income retirees who continue to move to Talbot County, a number of established wealthy families bought land and built estates here in the 1920s and during the Great Depression (Gibbons 1977). Wealthy residents in the other high-income counties also increase the per-capita income.

Contrasting Poverty

Geographical income data aggregated at the state and county levels mask numerous pockets of extreme poverty in Maryland. In 2010, there were over 80,000 Maryland families living below the poverty level. This translates into nearly half a million people, ranking Maryland twenty-eighth among the states in the absolute number of people living below the poverty level. On a percentage basis, 10.3 percent of individuals (or 6.1% of families) in Maryland live below the poverty level. By the percentage of its people who live below the poverty level, Maryland ranks forty-ninth among the states. Only one state has a smaller percentage of its population below the poverty level—New Hampshire (7.5%). For the entire United States, 13.3 percent of all individuals live below the poverty level.

Poverty level is a designation that the federal government uses as a means of identifying areas and persons with income levels so low that economic hardship is present. Factors such as family size, sex of family head, number of children, and type of residence all help determine the various levels of poverty. For a family of four, for example, the 2011 federal poverty threshold was set at $22,250 in the forty-eight states and Washington, DC, although it is higher in Alaska and Hawaii. Although the national poverty level is cyclical, it usually ranges from 13 to 17 percent. Across the nation, poverty rates dropped during the generally prosperous 1990s, but since that time they have crept back up. Like the rest of the nation, Maryland's highest poverty rates are in its urban centers and rural areas. Although Maryland has one of the lowest poverty rates in the nation, such stark income disparity in a state with so much affluence is troubling. One critical factor in reducing poverty is adult education, but Maryland ranks in the bottom third of states in spending on adult education. More attention needs to be paid to job training, but Maryland spends little beyond federal funds intended for this purpose.

Pockets of poverty exist throughout the state. In the southeast corner of Maryland in Worcester County, Stockton has a 45 percent poverty rate. Just to the west in Somerset County, Princess Anne has a 40.7 percent poverty rate. In western Maryland, Frostburg in Allegany County has a 30.5 percent rate. The rate of nearly 20 percent for College Park includes a number of students, including many resident graduate students who temporarily have low incomes. Baltimore City has the largest absolute number of people below the poverty level, with 20.9 percent.

The poverty rates for the counties of Maryland plus Baltimore City are shown in figure 9.4. The geographical pattern should not be surprising, given the above discussions on income. The highest poverty levels are found in Baltimore City, the Eastern Shore, and western Maryland. Levels below the state rate exist in the urban corridor from Harford to Montgomery Counties, with low rates extending into southern Maryland and out to Frederick County. On the Eastern Shore, there are two counties with lower poverty rates: Queen Anne's and Talbot. Queen Anne's County is part of urban Maryland, being tied directly by the Chesapeake Bay bridges. Heavy daily commuting occurs over the bridges connecting residents of Queen Anne's County to the urban

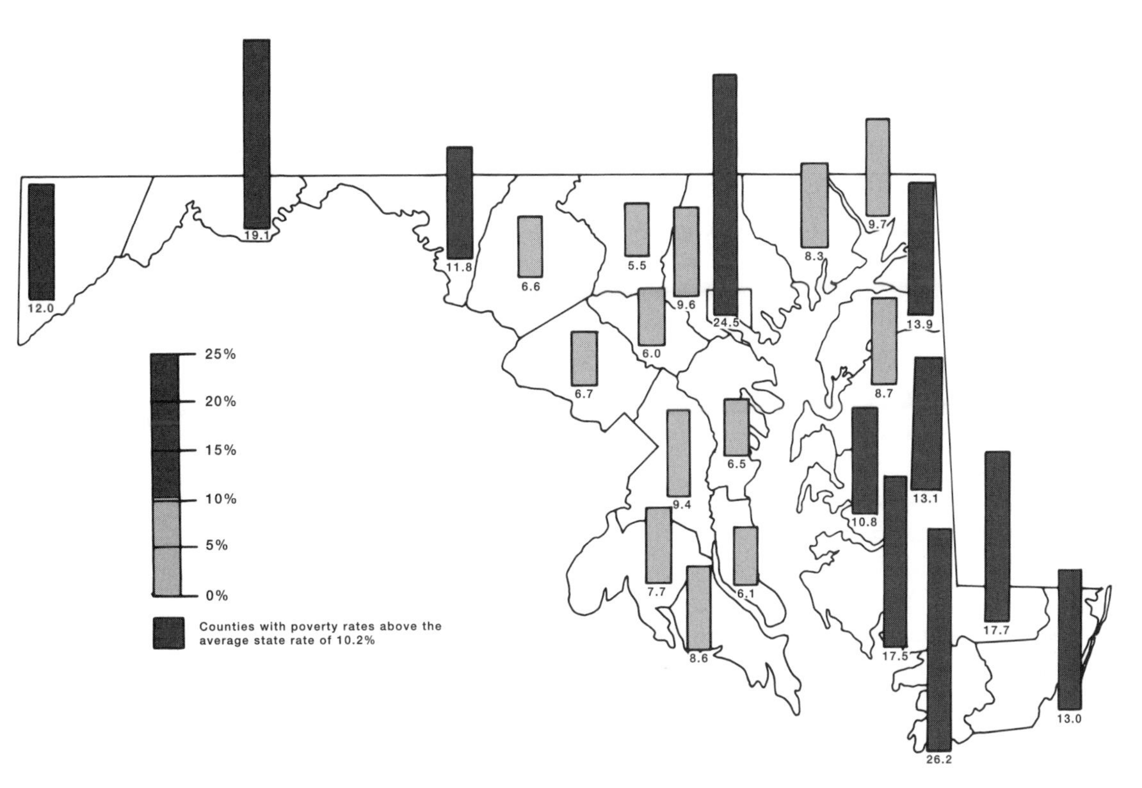

Figure 9.4 Individual Poverty Rates, 2011

Source: Data from US Census Bureau, Small Area Income and Poverty Estimates, July 2013

region on the Western Shore. Talbot County has a higher-income, older population with a lower poverty rate.

Residents of Maryland under the age of eighteen living in poverty reveal some interesting patterns. Maryland has the second-lowest poverty rate in the nation among this age group (10.9%), but even so there is some geographic variation. Baltimore City has a high of 30.9 percent of those under eighteen living in poverty. High rates of poverty for this age group are also found in Dorchester County (22%), Garret County (20%), and Somerset County (26.6%). In affluent states such as Maryland, youth from low-income families lag behind their more affluent counterparts, even more than in less affluent states. Because housing costs in Maryland are so high, a higher proportion of one's income goes toward paying rent. While the Maryland economy has been growing, wages have been stagnant except for those on the upper-income end of the scale. Maryland is one of the wealthiest states in the nation. It is also an expensive place to live, and it is getting more expensive every year. The cost of living for a family of three in Maryland grew by 31 percent between 2001 and 2007. To make ends meet, this family of three must earn more than twice as much as the federal poverty level. Using the same federal poverty level across the nation does not make sense because the cost of living varies significantly from region to region.

Income Inequalities

For a state that ranks so high in income, the absolute number of people living in poverty is a symptom of diverging incomes and wealth. As in the United States as a whole, a smaller percentage at the top of the income scale accounts for an ever-greater percentage of the state's income, resulting in greater income inequality. Complete income equality is not realistic nor is it desirable, but heightening levels of inequality can eventually lead to a more dualistic society with strong centrifugal (divisive) forces.

Statisticians usually measure income inequality by using the Gini coefficient (GC), which is calculated by looking at what percentage of income a certain percentage of the population earns. If every household of a geographic unit had the exact same income, the GC would be zero and there would be perfect equality of income; this is obviously not the case, nor is it ever likely to be. If only one household made all the income, then the GC would be 1.0, and inequality would be at its maximum. Determining income inequalities is critical for the measurement of social well-being. Very high GCs (i.e., where inequalities increase to a high level) can endanger the social cohesion of the population, with some people feeling as though they have been left behind. In 2009, an estimate of income distribution for the entire state of Maryland found that:

- the poorest 20 percent of Maryland households had only 3.6 percent of the state's total income;
- the middle 60 percent of households had 48.4 percent of total income; and
- the wealthiest 20 percent had 48 percent of the income.

Income inequality continues to grow in Maryland. In the last two decades, the broader middle class has seen its share of total income in Maryland decline. The wealthier households have made significant gains, while the poor continue to lose its already small share (Goldstein 2002). Maryland is following a national pattern of increasing income inequality. Various factors explain this troubling trend. One factor is the increasing number of one-person households, a number that has increased dramatically in Maryland over the past two decades, today accounting for more than 25 percent of all households in the state. Single female-headed households have also grown in number to over 8 percent. Both one-person and female-headed households typically have lower incomes. As discussed above, nearly all of Maryland's recent population growth has been in its minority population.

On average, minority population households have lower incomes than non-Hispanic white households, with the major exception being Asian households.

Since 1990, investment income in Maryland (from dividends, interest, and rent) has grown more rapidly than earnings from work. It should be no surprise that the overwhelming majority of investment income goes to the wealthiest households. Also, manufacturing jobs have decreased dramatically in Maryland, while service and retail jobs have increased. Manufacturing jobs usually offer good wages to blue-collar, semiskilled workers who often have only high school diplomas or less. Many new service-sector jobs pay low wages. Higher-wage service jobs usually require a college degree or some kind of specialized training (Goldstein 2002). In sum, there are many factors that can explain the growing income disparities in Maryland. The failure of the federal minimum wage to keep up with inflation erodes the wages of the unskilled. The decline of manufacturing and the number of jobs represented by unions, as well as the influx of foreign immigrants who often fill low-wage jobs, have also eroded middle- and lower-income wages.

The level of income inequality is not uniform across Maryland. An analysis of income inequality in the state by Mark Goldstein at the Maryland Department of Planning revealed that, with the exception of Baltimore City, all nine counties with the greatest degree of inequality (exceeding the state average) are either on the Eastern Shore or in western Maryland. These same nine counties also have household incomes below the state average (Goldstein used the GC in his study). The study showed that Talbot County had the greatest amount of income inequality; its wealthiest 20 percent of households earned nearly 55 percent of all income in the county. The bottom 20 percent of households earned just over 3 percent, and the middle 60 percent about 42 percent. Although Talbot County has a sizeable low-income population, it also has a group of high-income households. In contrast, Charles County in the growing southeast suburbs of Washington, DC, has the most equal (lowest GC) income distribution in Maryland. The top 20 percent of households earned nearly 40 percent of total income in the county; the bottom 20 percent had about 6 percent, and the middle 60 percent about 54 percent.

Goldstein found that the GCs for Maryland counties tell us that the poorer and more rural counties are more heterogeneous (i.e., have a greater mix of poor and wealthy households). The more affluent suburbs are more homogeneous and typically have a higher education level and two good incomes. For example, the share of families with two or more workers is highest in Howard (63.2%), Carroll (62.5%), Calvert (61.4%), and Frederick (61.3%) Counties. All of these counties have lower income inequality (lower GCs) than the state average. The lowest percentages of families with two or more workers are found in the jurisdictions with income inequalities greater than the state average, such as Baltimore City (29.8%), Somerset (42.8%), Allegany (46.0%), Dorchester (46.0%), Worcester (46.0%), and Kent (46.1%).

Goldstein also found that counties experiencing the largest growth in inequalities (largest percentage growth in GC) are poorer or rural, including Dorchester, Somerset, and Talbot Counties. Income inequality in some of the dynamic metropolitan suburbs has grown significantly, however, including Baltimore, Howard, Montgomery, and Prince George's Counties. It is thought that the rising inequalities in these suburbs are related to the changing racial and ethnic compositions of their populations. With the exception of Asians, the average minority households in these counties have lower incomes than their non-Hispanic white neighbors. Still, although inequalities might be increasing in these suburban counties, they still have among the lowest absolute levels of inequality.

Although Maryland is experiencing a growing divergence of incomes as indicated by rising income inequalities, there are some counties where inequalities are decreasing, especially those with relatively high levels of income inequality: Allegany, Caroline, Cecil, and Kent.

Education

In 1694, the Maryland General Assembly provided for a free school, the first in what is now the United States. The school, which opened as King William's School in Annapolis in 1696, is today St. John's College. In 1864, the schools of Maryland were placed under the supervision of the State Board of Education, and in 1900 a state superintendent of schools was appointed (Vokes 1968). Until the twentieth century, school populations in Maryland were almost exclusively white.

Pre-Kindergarten to Twelfth Grade

In 2010, Maryland had 848,412 children enrolled in its public schools (pre-kindergarten through the twelfth grade). An additional 188,733 children attended private schools in the state. School enrollments reflect the populations and diversities of Maryland's counties. For public schools statewide, 45.5 percent of students are white, 37.9 percent black, 10 percent Hispanic, and the remainder from other various groups. As recently as 2006, 54 percent of Maryland's public school students were white. The black student population has increased slightly from 36.8 to 37.9 percent, but the most significant growth by percentage has been in Hispanic students (4.4% to 10.0%) and Asians (4.2% to 6.1%) from 2006 to 2009. There are, however, stark contrasts in the mix across the state (fig. 9.5).

Representative of the new outer suburbs of metropolitan central Maryland, Carroll County's student population is heavily white (90.8%), with a small percentage of black and other minority students. On the Eastern Shore, Queen Anne's County has a dominant white

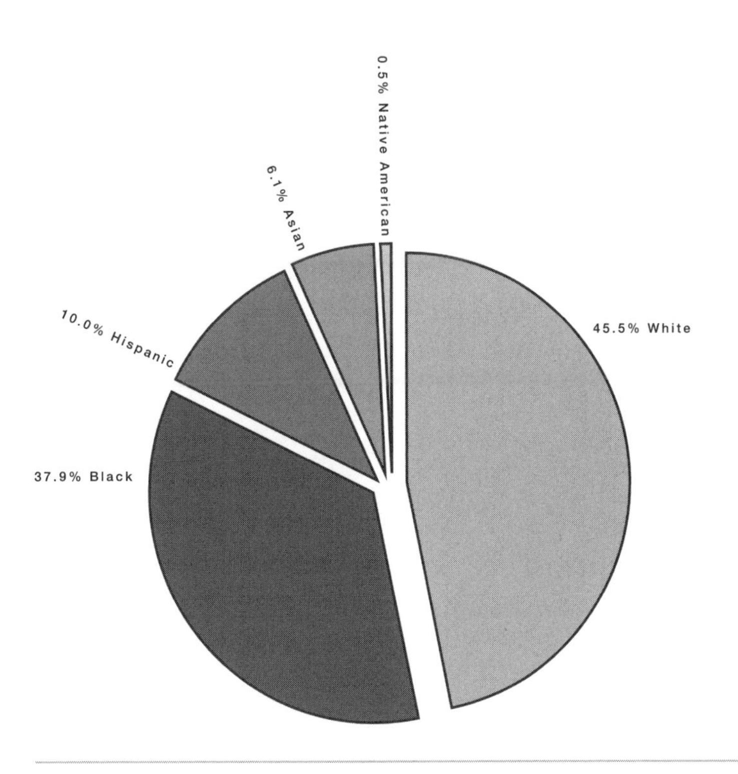

Figure 9.5 Maryland Public School Population by Race, 2010

school population (86.8%), but with a significant minority black school population of 8 percent. The overwhelmingly white general population of far western Maryland is reflected in the school population of Garrett County, which is nearly all white (98.6%), having the least diversified school population of all Maryland's counties.

The high school graduation rate also varies significantly across Maryland. In 2009 the statewide rate was 83.9 percent, which translates into 48,310 high school graduates, about 79 percent of whom had plans to go on to college (full or part time). The highest graduation rates in Maryland (over 90%) are in Carroll, Frederick, Howard, and Montgomery Counties. Together, these counties form a contiguous block in central Maryland. The lowest graduation rates (less than 80%) are in Baltimore City, Dorchester, Somerset, and Wicomico Counties. The remaining counties had high school graduation rates between 80 and 90 percent. The relationships among income, poverty, racial mix of school population, and other socioeconomic factors described above are obvious. Jurisdictions with lower incomes, higher poverty rates, and higher percentages of nonwhite students generally have the lowest graduation rates. The Washington, DC, and Baltimore City school populations contrast most starkly with the rest of Maryland. Both are overwhelmingly black, but there are differences between them. Baltimore City has over twice the percentage of white students as Washington, but a significantly lower Hispanic student population.

The Maryland Department of Education reports how Maryland ranks among the states and some of its education superlatives:

- Ranked thirty-second in per-capita state government expenditures for all education.
- Ranked eighteenth in per-capita expenditures of state and local governments for all education.
- Ranked sixth in average salaries of public school teachers.
- Ranked thirtieth in pupils enrolled per teacher in public school grades K–12.
- *Education Week*, the nation's leading education newspaper, ranked Maryland public schools first in the nation for four consecutive years.
- In 2010, College Board ranked Maryland first for the second year in a row for the percentage of public school students scoring 3 or higher on at least one advanced placement (AP) exam. (Scores of 3 through 5 are considered mastery of college-level work.) Maryland has also increased its AP participation of minority students, especially Hispanic and African Americans.
- In August 2010, Maryland received $250 million from the competitive federal Race to the Top fund for public education.
- *Newsweek* magazine's "Top High School List" for 2010 reported that Maryland has the highest percentage of rigorous high schools in the nation. Over half of Maryland's public high schools were on the list (98 of 185), covering all Maryland regions (the Eastern Shore, southern and western Maryland, metropolitan Baltimore, and metropolitan Washington, DC).

Educational Achievement and Higher Education

In 2010, of all 3,141 counties and county equivalents (e.g., Louisiana has "parishes," and Alaska has "boroughs") in the United States by percentage of people twenty-five years and older who have earned at least a bachelors degree, Montgomery County, Maryland, was first in the nation (57.4%), followed by nearby Fairfax County, Virginia (56.3%). Howard County was fourth in the nation (54.6%), Baltimore County sixty-second (35.6%), Anne Arundel County 105th (30.4%), and Baltimore City 174th (24.2%). At the state level, Maryland has the second-highest percentage of people aged twenty-five and older with a bachelor's degree (35.1%), exceeded only by Massachusetts (37.0%). Neighboring Washington, DC, exceeded all of the states with 45.9 percent. The rate for the entire United States is 27.2 percent. Marylanders are in impressive company when it comes to their educational attainment.

The geographical pattern of educational attainment is not even, however; the more highly educated are concentrated in a few places in large numbers. Only Anne Arundel, Howard, and Montgomery Counties exceed the statewide average of 35.1 percent. Howard (57.2%) and Montgomery (56.1%) greatly exceed the state average, while Anne Arundel County (35.3%) exceeded it by only 0.2 percent. The lowest percentages of people with bachelor's degrees are in western Maryland and the Eastern Shore, and Somerset County comes in last with only 14.4 percent. As with wealth, the concentration of more highly educated people is in central Maryland. Many of these people, who hold bachelor's and advanced professional degrees, have migrated to Maryland, especially its urban corridor, attracted by its many research institutions, high-tech companies, and other economic opportunities.

Educational attainment also varies by racial group and sex. Table 9.5 summarizes the degrees awarded in 2009 to 2010 for people twenty-five years and older in all of Maryland. Most of Maryland's recipients of college degrees are whites, followed in a distant second by African Americans. Women degree earners outnumber white men in each category. African Americans average about 20 percent of Maryland's associate's, bachelor's, and master's degrees earned, but only 8.3 percent for doctoral degrees. The much smaller population of Asian Marylanders accounts for nearly 11 percent of doctoral degrees.

The trends in the number of men and women in higher education in Maryland are also interesting. Between 1994 and 2010, the number of women enrolled in institutions of higher education in Maryland increased by over 20 percent. During the same period, the number of men increased by only 12 percent. By 2010, approximately 60 percent of all students in higher education in Maryland were women, reflecting a nationwide trend. Sociologists, demographers, and others are currently analyzing this trend to determine why young men are dropping out of education after high school in larger numbers.

The Maryland Higher Education Commission in 2011 reported that Maryland's public and independent colleges and universities had a total enrollment of 355,299 students. The largest four-year institution is the University of Maryland at College Park, followed by Towson University and the University of Maryland at Baltimore County. Anchoring the University System of Maryland on the Eastern Shore are Salisbury University and the University of Maryland Eastern Shore. Frostburg University in Allegany County is the major university system campus serving western Maryland. Among the independents, Johns Hopkins University in Baltimore is the largest. The second largest, Loyola University Maryland, is also in Baltimore. The greater Baltimore region with its complex of many colleges, universities, and other institutions is the dominant higher educational center in Maryland. In a special category with about 4,300 midshipmen is the US Na-

Table 9.5. Higher education degrees by race and gender, 2009–10

	Certificate		*Associate*		*Bachelors*		*Masters*		*Doctorate*		*Total*	
	Number	*Percentage*	*Number*	*Percentage*	*Number*	*Percentage*	*Number*	*Percentage*	*Number*	*Percentage*	*Number*	*Percentage*
White	2,230	46.2	7,197	59.6	16,024	58.7	8,161	52.2	1,415	53	**35,027**	56
African American	1,915	39.7	2,715	22.5	5,264	19.3	2,797	17.9	220	8.3	**12,911**	20.6
Hispanic	206	4.3	556	4.6	1,004	3.7	443	2.8	73	2.7	**2,282**	3.7
Asian	163	3.4	588	4.9	2,329	8.5	1,081	6.9	284	11	**4,445**	7.1
Native American	19	0.4	39	0.3	86	0.3	40	0.3	10	0.4	**194**	0.3
Foreign	62	1.3	361	3	594	2.2	1,459	9.3	456	17	**2,932**	4.7
Other	227	4.7	619	5.1	2,011	7.3	1,663	10.6	206	7.7	**4,726**	7.6
Total	**4,822**	**100**	**12,075**	**100**	**27,312**	**100**	**15,644**	**100**	**2,664**	**100**	**62,517**	**100**
Men	1,736	36	4,579	37.9	11,397	41.7	6,372	40.7	1,237	46	**23,585**	37.7
Women	3,086	64	7,496	62.1	15,915	58.3	9,272	59.3	1,427	54	**34,110**	54.6
Not designated											**4,822**	7.7

Source: Maryland Higher Education Commission. *Data Book 2011*, Baltimore: MHEC: 19

val Academy in Annapolis, which was founded in 1845 by Secretary of the Navy George Bancroft and was situated on 10 acres (4.05 hectares) of land at old Fort Severn, Annapolis. Since then, over 60,000 young men and women have graduated from the academy to serve in the US Navy and Marine Corps.

Overall, nearly 62 percent of Marylanders have at least some college or an associate's degree compared to 55.6 percent for the nation. Maryland's sixteen community colleges enroll about 54,000 full-time and 86,000 part-time students, and about 93 percent of these students are Maryland residents. Allegany College has the lowest percentage of Maryland residents (45%), reflecting its service area that extends into Pennsylvania and West Virginia. Maryland's four-year public colleges and universities enroll over 83,000 full-time and 31,000 part-time undergraduate students. The largest institution in this group is the University of Maryland at College Park (26,542), followed by University of Maryland University College (24,284) and Towson University (17,148). Maryland's independent colleges and universities enroll 28,035 full-time and 3,928 part-time undergraduate students. Of this number, about 53 percent are Maryland residents. Some independent colleges enroll higher percentages of Maryland residents (e.g., Baltimore International College, 88%; Notre Dame of Maryland University, 91%, and Sojourner-Douglas College, 95%). At the other end of the spectrum are independent Maryland institutions that enroll large percentages of out-of-state residents; for example, Johns Hopkins University (17.9%) and St. John's College (12.3%). Some of the specialized independents have small percentages of students who are Maryland residents (e.g., the Harry Lundeberg School of Seamanship, 4%, and the National Labor College, 7%). Overall, Maryland has a robust number of public and private institutions of higher education within close proximity to most Marylanders. As expected, the greatest cluster of institutions of higher education is in the urban corridor. Still, there are colleges and universities available to far western and Eastern Shore Marylanders.

10 Moving around Maryland

An excellent modally integrated transport network serves Maryland's people, institutions, and businesses. Historically, this transportation network developed concomitant with its settlement and growth.

The Port of Baltimore

The Port of Baltimore has long been the heart of the state's economy, with nearly everyone living in Maryland being affected in some way by its activities. Each year, more than 4,000 ships call at the port carrying diverse cargoes, including teak from Burma (Myanmar) and India to be used in cabinet making, hemp from the Philippines for the brush industries of Frederick County, and automobiles and trucks for the entire US market. The Port of Baltimore is a vast industrial complex that encompasses 45 miles (72 kilometers) of shoreline and 3,403 acres (1,377 hectares) of waterfront. Baltimore survives in a highly competitive market, including other East Coast ports and some Canadian ports (Maryland Port Administration 2012).

The export of many local and national products through the port puts Maryland on the interface for the international exchange of goods. What happens across the globe can easily have a significant impact in Maryland, and the ups and downs of the local economy reflect the integration of the Port of Baltimore and Maryland with the global economy. By most standards, the port is Maryland's largest single economic asset—an asset largely based on its geographic location.

Before there was a Port of Baltimore, there was a natural feature here called a harbor. The inner and outer harbors of Baltimore with their geographical situation relative to the Eastern Seaboard and potential links to the Midwest afforded a great opportunity for development and growth. But the natural feature of the harbor does not alone explain Baltimore's growth. It took vision and leadership to create the cultural component of this milieu that has become the port. A harbor is a natural feature, but a port is the composite of cultural features built around the harbor, including marine terminals, warehouses, shipping docks, dry docks, container shipping facilities, and myriad related activities.

HISTORY OF THE PORT OF BALTIMORE

The Port of Baltimore began operations in 1706, predating the establishment of the city by twenty-three years. Early competition from Annapolis, Oxford, and the tobacco ports of southern Maryland was keen, but Baltimore's latent geographical locational advantages eventually prevailed. Because of the configuration of the coast, Baltimore is located farther inland, putting it closer to the Midwest than other East Coast ports. As the interior of the country was developed and railroads built connecting Baltimore to the Midwest, the port's geographical advantages led to rapid growth.

In the 1700s, the port's inner harbor was a mud flat that could accommodate only small sailing vessels. At that time the only settlement was on the waterfront at Fells Point. In 1758, Doctor John Stevenson gave a lift to the small port when he began to ship grain from the port to customers up north and overseas. In the following years a port infrastructure developed—piers and warehouses, shipbuilders, financiers, merchants, and brokers. Fells Point's geography was critical to its success as a seaport. This narrow, hook-shaped peninsula about a mile east of the center of Baltimore jutted into the northwest branch of the Patapsco River, forming an excellent natural harbor.

During the Revolutionary War, the port remained open despite a British blockade, some historians believe because of the British taste for Maryland tobacco, which was exported through the Port of Baltimore (Maryland Department of Transportation 1978). During the War of 1812, Baltimore was not as fortunate. The British attack on the city culminated in September 1814 with the bombardment of Fort McHenry, which to this day sits at the entrance to the harbor. Another lasting outcome of the British attack is the national anthem of the United States, "The Star-Spangled Banner," written by Francis Scott Key shortly after the battle. Set to the tune of "Anacreon in Heaven," a popular drinking song of that period, the song (initially called "The Defence of Fort M'Henry") immediately became popular throughout the young nation. The failed British attack by both land and sea provided a sorely needed lift to the country, coming only days after the fall of Washington to the British, who had burned the Capitol, Library of Congress, White House, and other buildings. "The Star-Spangled Banner" became the official national anthem by an act of Congress and the signature of President Herbert Hoover on March 3, 1931.

After the War of 1812, the Port of Baltimore rebounded and developed a lucrative trade with South America and China. By the 1830s, the Baltimore shipyards at Fells Point were turning out the famous Baltimore clippers, the fastest sailing vessels then afloat. Even today, Baltimore retains some commercial shipbuilding and ship repair activities. Baltimore became a railroad port in the 1860s, with direct rail line connections to the Midwest. Baltimore soon benefited from the discovery of oil in western Pennsylvania. Refineries were built at the port, through which American oil was shipped worldwide. During this period, Baltimore became the world's largest refiner of copper as well as the country's leading fertilizer and chemical producer, based on the import of guano (bird droppings) from coastal Peru.

Although trade slowed down during the Civil War, it quickly revived soon thereafter. The port's main ship channel was deepened, mainly through the efforts of Major William P. Craighill of the US Army Corps of Engineers. By the late 1880s, the railroads had built huge grain elevators at the port; Baltimore is still a major grain shipping port. During this time the port also became a major shipper of coal and a fueling depot for steam-powered ships. These factors combined to make Baltimore the world's sixth-largest port by 1876.

When the Panama Canal opened in 1914, the Port of Baltimore once again benefited. For many years it handled more cargo westward bound through the canal than any other East Coast port. During World War I, the port grew even more. By the onset of World War II, Baltimore was ready to support the war effort in a major way. Between 1941 and 1945, Baltimore shipyards produced 608 ships. The first "liberty ship," the SS *Patrick Henry*, was built in Baltimore. After the war, the state became more involved in the administration of the port, and in 1956 the state established the Maryland Port Authority (now the Maryland Port Administration).

Much of the lifeblood of contemporary Maryland's economy pumps through the Port of Baltimore. By 2000, it was annually generating about $1.8 billion in revenue for the Maryland economy and employing nearly 126,700 Marylanders in maritime-related jobs, including truck drivers, longshoremen, tugboat operators, railroad workers, freight forwarders, bay pilots, warehouse workers, office workers, employees of the Maryland Port Administration, and many others.

General Export and Import Factors

The US Maritime Administration measures the amount (by short tons and value) of foreign ocean-borne cargo (both imports and exports) moving through US ports. Their figures do not account for the nearly 37 percent of all cargo moving through the Port of Baltimore that is domestic (i.e., going to and coming from another US port). About 71 percent of all cargo moving through the port is bulk cargo such as coal, ores, and petroleum.

Coal accounts for 48 percent of all the bulk cargo (Maryland Port Administration 2009). Since the mid-1990s, coal exports have declined owing to the availability of cheaper coal on the global market from foreign suppliers. From the mid-1990s into the twenty-first century, there was an overall increase in imported cargo, mainly in the form of automobiles, forest products (especially wood pulp and roll paper), and aluminum. In 2011, Baltimore surpassed New York as the leading US port for the import and export of automobiles and trucks. To underscore Maryland's interdependence with the global economy, the decline in exports in the late 1990s and again after the 2008 recession was related to less demand from the slowing economies of Japan and other important foreign markets. General trends in imports and exports between 1996 and 2005 can be seen in figure 10.1. Exports by tonnage nearly doubled between 2000 and 2008, and then decreased in 2009 (table 10.1). By value, exports nearly tripled between 2000 and 2008, decreasing in 2009, and then increasing again—by more than double—by 2012. From 2009 to 2012, exports climbed steadily to a new high volume. Imports through the Port of Baltimore by tonnage grew significantly from 2000 to 2005, decreased from 2005 to 2009, and then grew again by 2012. At the same time, imports by value nearly doubled from 2000 to 2008, decreasing in 2009 (table 10.1). From 2009 to 2012, imports also climbed steadily to a new high.

Exports

By most major standards, Baltimore is one of the leading ports in the United States. The numbers in table 10.1 show that Baltimore's trade declined in 2009, but this was the case for the entire nation, and Baltimore maintained its rankings among all US ports. These figures once again underscore the impact on Baltimore of the national and global economic recession. Among major US ports, Baltimore ranks eleventh in tonnage and ninth in dollar value of total waterborne trade, including imports and exports (table 10.2). Baltimore's neighbor and major competitor down the bay—Norfolk, Virginia—ranks ahead Baltimore in tons of total waterborne trade (fifth) as well as in dollar value (sixth). As recently as 1999, Norfolk ranked behind Baltimore in both categories.

Export trade is carried on with many countries (table 10.3). In 2012, China was Baltimore's major export customer by tonnage, followed by the Netherlands. Other major trading export customers included Brazil, France, Germany, India, Japan, South Korea, Ukraine, and the United Kingdom. Table 10.4 shows Baltimore's major trade lanes. Europe remains as the major trade lane for the port for both imports and exports (by value), followed by Asia. When measured by tonnage of imports plus exports, Asia moves ahead of Europe.

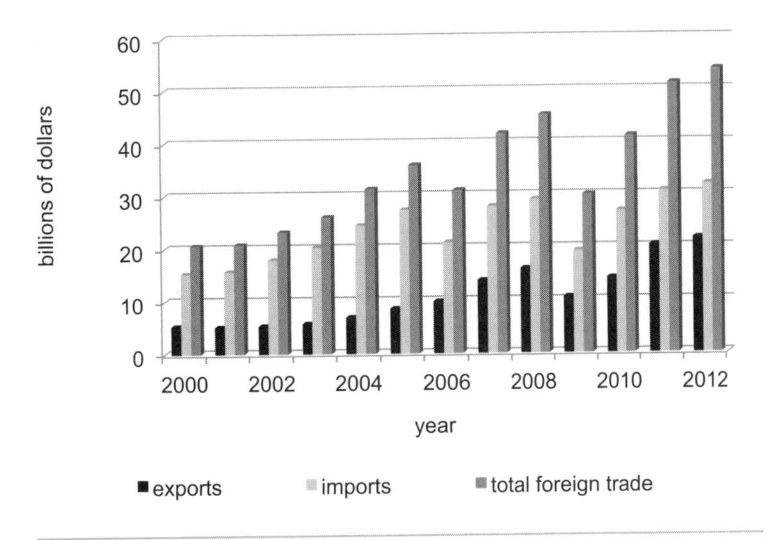

Figure 10.1 Port of Baltimore Trade, 2000–2012

Table 10.1. Port of Baltimore: foreign ocean-borne commerce, 2000–2012

Year	*Exports value*	*Short tons*	*Imports value*	*Short tons*	*Total foreign trade*	*Short tons*
2000	$5,298,649,000	8,733,526	$15,307,388,000	17,425,151	$20,606,037,000	26,158,677
2001	$5,130,868,000	6,937,119	$15,689,062,000	18,763,913	$20,819,930,000	25,701,032
2002	$5,251,082,000	5,710,957	$17,934,745,000	17,928,377	$23,185,827,000	23,639,334
2003	$5,685,838,000	4,937,263	$20,270,412,000	19,801,838	$25,956,250,000	24,739,101
2004	$6,900,998,348	6,915,622	$24,410,928,007	24,900,416	$31,311,926,355	31,816,038
2005	$8,612,301,282	7,420,411	$27,255,822,295	25,005,332	$35,868,123,577	32,425,743
2006	$9,625,675,074	8,365,495	$27,072,934,230	22,254,975	$36,398,609,304	30,620,470
2007	$13,979,157,856	11,291,633	$27,911,789,942	19,490,995	$41,890,947,798	30,782,628
2008	$16,183,900,998	15,052,545	$29,151,323,038	17,965,267	$45,335,224,036	33,017,812
2009	$10,836,545,113	10,216,971	$19,376,346,831	12,146,045	$30,212,891,944	22,363,016
2010	$14,267,216,521	17,323,675	$27,029,172,560	15,244,276	$41,296,389,081	32,567,952
2011	$20,485,408,678	23,637,396	$30,753,358,644	13,989,106	$51,238,767,322	37,626,502
2012	$21,773,030,546	23,757,853	$32,077,990,394	12,929,929	$53,851,020,940	36,687,782

Source: Maryland Port Administration, Foreign Commerce Statistical Report, 2012

Table 10.2. Top US port districts: imports plus exports, 2012

Rank	*District*	*Total tons*	*Rank*	*District*	*Value (millions)*
1	Houston, TX	162,525,907	1	Los Angeles, CA	$283,553
2	New Orleans, LA	107,244,642	2	New York/New Jersey ports	$205,202
3	New York/New Jersey ports	82,399,783	3	Houston, TX	$176,661
4	Los Angeles, CA	73,355,250	4	Long Beach, CA	$100,974
5	Norfolk, VA	70,209,622	5	Georgia ports	$92,990
6	Gramercy, LA	69,489,839	6	Norfolk, VA	$69,815
7	Long Beach, CA	49,215,484	7	South Carolina ports	$63,645
8	Morgan City, LA	48,542,498	8	New Orleans, LA	$59,171
9	Port Arthur, TX	45,796,812	9	**Baltimore, MD**	**$53,851**
10	Corpus Christi, TX	44,666,390	10	Tacoma, WA	$45,983
11	**Baltimore, MD**	**36,687,782**	11	Oakland, CA	$44,832
12	Georgia ports	36,285,828	12	Seattle, WA	$38,422
13	Texas City, TX	33,430,181	13	Morgan City, LA	$35,585
14	Lake Charles, LA	31,455,846	14	Port Arthur, TX	$28,867
15	Mobile, AL	29,545,767	15	Corpus Christi, TX	$28,504
	US Total	**1,424,757,788***		**US Total**	**$1,782,247***

Source: Maryland Port Administration, Foreign Commerce Statistical Report, 2012
*US total tons and value figures are larger than the sum of their columns because only the top 15 port districts are shown in the table.

Imports

In 2012, Baltimore ranked nineteenth in the United States by import tonnage but jumped up to tenth by value. Canada is the leading source of imports by tonnage and sends a wide range of items through the Port of Baltimore. Raw materials such as salt, sulfur, stone, lime and cement, ores, mineral fuels, and fertilizer dominate Canadian imports. Brazil is another major source of raw material imports, including ores, sugar, wood pulp, paper products, iron and steel, and rare metals. Imports from China through the Port of Baltimore have grown rapidly in recent years.

Brazil, China, France, Germany, Japan, Russia, South Africa, and the United Kingdom dominate imports by value. Vehicles dominate imports from Germany, but machinery, paper products, and iron and steel are also significant. The United Kingdom's imports are also dominated by vehicles, but they include a wide range of goods, including iron and steel, pharmaceuticals, rubber articles, alcoholic beverages, and footwear. Likewise, vehicles dominate Japan's list, but their imports also include significant amounts of machinery and mineral fuels. Machinery leads Swedish imports, followed by rare metals, vehicles, and iron and steel (Maryland Port Administration 2012).

Table 10.3. Port of Baltimore's top ten trading partners, 2012

Exports	*Short tons*	*Imports*	*Short tons*
China	6,535,168	Canada	1,997,559
Netherlands	3,224,703	Brazil	1,052,546
Japan	2,292,007	China	918,662
South Korea	2,173,656	United Kingdom	805,653
India	1,477,825	Chile	767,632
Brazil	1,369,553	Japan	687,833
France	876,891	Suriname	542,497
Germany	587,186	Germany	539,174
Ukraine	508,627	Russia	402,020
United Kingdom	501,186	Finland	334,273
Exports	*Value (thousands)*	*Imports*	*Value (thousands)*
Australia	$1,674,735	Germany	$6,087,000
Saudi Arabia	$1,602,314	Japan	$2,846,637
China	$1,311,124	United Kingdom	$2,718,798
Germany	$1,282,851	China	$2,631,248
United Arab Emirates	$1,093,339	Russia	$1,731,241
Russia	$1,039,031	Brazil	$1,620,491
Chile	$993,719	South Africa	$1,263,035
Belgium	$988,563	Sweden	$1,067,580
United Kingdom	$987,756	Belgium	$774,330
Brazil	$738,145	Netherlands	$707,445

Source: Maryland Port Administration, Foreign Commerce Statistical Report, 2012

Table 10.4. Port of Baltimore trade lanes

Trade lane	*Export tons*	*Export value (thousands)*	*Import tons*	*Import value (thousands)*
Africa	558,946	$1,595,672	458,021	$1,488,152
Asia	13,299,480	$7,098,917	2,695,081	$8,653,529
Australia	151,889	$1,779,940	73,478	$358,648
Europe	7,159,597	$8,157,476	3,274,306	$16,836,909
North America	358,367	$250,997	3,208,356	$1,533,130
South America	2,149,575	$2,890,029	3,220,064	$3,207,621

Source: Maryland Port Administration, Foreign Commerce Statistical Report, 2012

The entire list of imports and exports through the Port of Baltimore is extensive. The items mentioned here highlight only some of the major economic activities related to the port: vehicles, iron and steel making and trading, sugar refining, paper and publishing, and rare metals used in high-tech industries and in aerospace, petroleum, and agriculture. Each year the Maryland Port Administration publishes its Foreign Commerce Statistical Report, which lists in detail all of the imports and exports by type and country. As a major port serving the wider region, cargo moving through Baltimore goes to and comes from many states in the mid-Atlantic and Midwest, not just from Maryland-based companies and individuals. In addition to foreign commerce, the Port of Baltimore handles domestic cargo transported within the Chesapeake Bay or to nearby East Coast ports. Most of this domestic cargo is raw materials (e.g., coal, petroleum, cement, sand, and gravel). Foreign cargo data reported in tables 10.1–10.4 exclude domestic cargo that moves between US ports by ship and barge or between US mainland ports and Puerto Rico.

Major Marine Terminals

The major marine terminals in the Port of Baltimore are both public and private operations. Together, these terminals provide a variety of handling facilities from general cargo to those specializing in handling metals, bulk cargoes, containers, and automobiles. Baltimore's inland location and large nearby market area, when combined with the state's impressive highway and railroad networks, enhance its locational assets. The port is highly integrated with other modes of transportation. The Norfolk Southern and CSX Railroads have facilities at several of the terminals. Interstate 95, the "Main Street of the East Coast," is in close proximity. Although the Maryland Port Administration has been proactive in its efforts to keep the Port of Baltimore modern and competitive, it faces serious challenges beyond the port itself. Issues related to the dredging of shipping channels leading into Baltimore as well as regional and national competition from other ports must be constantly addressed.

Six public marine terminals reside in the Port of Baltimore:

1. Dundalk Marine Terminal is the largest and most versatile terminal, with thirteen berths and direct rail access. It handles containers; automobiles; steel; farm, break-bulk, construction, and forest products; and "roll-on/roll-off" (RO/RO) equipment. (RO/RO includes farm and construction equipment, but not automobiles.) Baltimore is the leading RO/RO port in the United States, and it is also the East Coast hub for Wallenius Wilhelmsen, the world's largest RO/RO carrier, which operates out of the Dundalk Marine Terminal. It has direct rail access, allowing the Norfolk Southern Railroad to deliver dozens of pieces of farm and construction equipment at once. Sev-

eral automobile processors maintain operations at Dundalk; Ports America runs a private container operation there. The port's proximity and good rail and highway connections to the Midwest's major farm and construction equipment manufacturers have all helped it become the leading US port for the export of combines, tractors, hay balers, and for the import of excavators and backhoes.

2. Fairfield Auto Terminal was developed by Toyota Motor Company and opened in 1988. This premier 50-acre (20-hectare) facility helped make Baltimore a leading vehicle importer and exporter. Located only 3.5 miles (5.6 kilometers) from I-95, its processing facilities include a car wash, body shop, and accessory installation shop. By 2011, the entire terminal was leased by Mercedes Benz.
3. Intermodal Container Transfer is a facility that moves cargo between bulkhead (ships) and railhead. Adjacent to the Seagirt Marine Terminal, this 70-acre (28-hectare) operation allows cargo to catch a train to almost anywhere in the United States. CSX Intermodal operates the on-dock rail yard, which is tied into a national intermodal system.
4. North Locust Point, an older terminal of great historical importance to Baltimore, has adapted to societal and economic change over the past century. It has received immigrants, been a cargo pier for the Baltimore and Ohio Railroad (B&O), and handled many other forms of cargoes. Today, this multiuse 90-acre (36-hectare) terminal has adapted to handle forest products, break-bulk, liquid bulk, RO/RO, and containerized cargo. It is also home to several latex (rubber) importers. In 2002, Archer Daniels Midland shifted its grain exports to Virginia, ending a 130-year shipping tradition that played an important role in the economic development of Maryland.
5. South Locust Point can handle any type of general cargo and is especially adept at handling break-bulk, RO/RO, and forest product cargoes. In 1988, it doubled in size to 80 acres (32 hectares). South Locust Point Terminal can handle the needs of medium-sized steamship lines, multipurpose vessels, and any cargo that needs to hit the road immediately. Interstate 95 runs rights past the front gate of this terminal, which also has direct access to CSX rail lines.
6. Seagirt Marine Terminal is a 275-acre (112-hectare) high-tech facility with state-of-the-art cargo handling equipment. Opened in 1990, it now specializes in containers and can handle 150,000 containers annually. The railhead at the Intermodal Container Transfer Facility is only 1,000 feet (305 meters) away.

The Port of Baltimore also has five private terminals:

1. Baltimore Metal and Commodities specializes in handling metals, soft commodities, and project cargo shipped as break-bulk. It is located adjacent to Fort McHenry.

2. Canton Marine Terminal—a noncontainerized facility handling bulk, break-bulk, and RO/RO cargo—is situated on the Patapsco River near the Seagirt Marine Terminal.
3. Masonville Auto Facility is a high-tech automobile processing facility that opened in September 2000. It is owned by the Maryland Port Administration and leased to ATC Logistics. This state-of-the-art facility includes an advanced auto tracking system, options installation shop, car wash, and office. The tracking system assigns a bar code for each car so that auto dealers can directly access the facility's inventory. Masonville covers 50 acres (20 hectares) with a 94,000-square-foot (8,460-square-meter) building for processing automobiles.
4. Ruckert Marine Terminal specializes in handling metals, ores, fertilizer, and other dry bulk, break-bulk, and primary materials. It covers 100 acres (41 hectares).
5. Sparrows Point Terminal is a bulk and break-bulk loading and unloading facility. It is located adjacent to the Mittal Sparrows Point steel mill and is serviced by both the Norfolk Southern and CSX Railroads.

Cruise Business

A surprise in the story of the Port of Baltimore is its rise as a departure port for cruise ships. In 2010, Baltimore was the fifth-largest cruise port on the East Coast by number of passengers on departing cruises. The port's new cruise ship terminal opened in 2006, and by 2010 the port had a record year for cruises, with even more cruises booked for 2011.

Dredging Shipping Channels

Nothing can be shipped in or out of the port unless the shipping channels are maintained. The Maryland Port Administration is a strong advocate of dredging and deepening shipping channels leading into the Port of Baltimore, which has a main channel that is 50 feet (15.2 meters) deep all the way from Cape Henry at the mouth of the Chesapeake Bay. The state of Maryland has a twenty-year dredging plan to keep the main channel open to ensure safe passage for vessels, but vessels are much larger than they were just thirty years ago. If a 50-foot channel is not maintained, the economic impact on the port would be devastating. But a major issue surrounding dredging is the placement of spoil material.

Off the northwest end of Kent Island in the middle of the Chesapeake Bay lies the muddy bottom of Kent Island Deep, 78 feet deep (23.8 meters). This 4-mile (6.44-kilometer) stretch has become highly controversial. Maryland plans to use the Kent Island Deep to dump mud and silt dredged from the shipping channels, eventually filling it to a uniform depth of 45 feet (13.72 meters). Opponents fear that algae growth in shallow water will kill shellfish, but the Maryland Port Administration disagrees and cites the danger of ships running aground and business fleeing the Port of Baltimore.

The state also has plans to deepen the port's secondary channel coming from the Chesapeake and Delaware Canal (C&D) in the northern bay. Additional pressure has been put on the Maryland Port Administration by major container shipping lines that use the port. They would like to see the C&D and the channel from the north deepened. Losing even a single shipping line would be a severe blow to the competitiveness of the port, especially in the lucrative container shipping business. Opponents of deepening the C&D cite several reasons for their position. Even if it were deepened, the new large container ships could not use the C&D. The US Army Corps of Engineers has consistently failed to prove that any economic benefits will derive from spending more than $100 million on this project, which will not likely add jobs nor additional shipping to the port (Gilchrest 2000). When the C&D was deepened in 1975, some residents along the canal saw their wells go dry or become contaminated. Shoreline erosion worsened, and silting in Elk River accelerated. Chesapeake City, which is bisected by the canal, had its water supply cut, resulting in the need for a second water and sewerage plant on the other side of the canal. In 1975, the US Army Corps of Engineers and the Port Administration promised that the deepening the canal would increase business at the port, but it has not. Instead, traffic through the C&D has declined. This is the dilemma: while dredging may harm the environment, failing to dredge may harm the economic health of the port and the Maryland economy at large.

The Kent Island Deep (renamed Site 104 by planners) is an economically cheap solution. Dumping at Site 104 would not require the usual building of dikes and drainage or the grading typical of other sites (Little 1999). Barges could simply be towed to the site and the material dumped. The prospect of Site 104 becoming a dumping ground drew strong opposition from several environmental groups, as such activity would bury crabs, maybe some Atlantic croaker, and invertebrates. (The site is probably too deep to sustain populations of oysters, clams, and plants.) The US Army Corps of Engineers has estimated that mud and silt dumped at the surface could drift to other areas, where they could affect crabs, clams, and oyster beds. Nutrient loading is also a concern; the addition of nitrogen and other nutrients could cause the growth of algae, which eventually die, decompose, deplete oxygen, and create an anaerobic environment in which organisms cannot survive. Other federal agencies, such as the US Environmental Protection Agency and the US Fish and Wildlife Service, have questioned the assertion by the state of Maryland and the US Army Corps of Engineers that dumping would not hurt the environment. In the summer of 2000, after intense lobbying by citizens and environmental groups, the governor ruled out dumping at Site 104.

Most shipping comes to the Port of Baltimore from the south through a set of channels that are dredged regularly to keep them at a depth of 50 feet (15.2 meters), which is as deep as any East Coast port. Although a vessel coming from the north through the C&D can save about five hours of travel time, the C&D and the channels leading to Baltimore are only 35 feet (10.67 meters) deep. Most container ships have a maximum draft of 39.5 feet (12.12 meters), which is the depth of the Panama Canal. Since the 1980s, Maryland has been seeking federal permission to deepen this approach to 40 feet (12.19 meters). The federal government, which pays 75 percent of the dredging costs, will not approve the project unless it feels that economic benefits will exceed the expense. There are other limitations to enhancing the approach from the north. Some ships cannot use the C&D because of the low height of several bridges over the canal. Although a ship can save time and fuel by using the canal, it is expensive in other ways. It can cost a large container ship around $9,500 in pilot fees (pilots from both Delaware and Maryland are required) to have the ship guided through the canal. Pilot fees up the Chesapeake Bay from the south cost much less. Because pilot fees through the C&D often exceed savings in fuel costs, many ships opt for the southern route.

What are the alternatives? Even without Site 104, dredging to keep the 50-foot channel open can continue with dumping at the two current sites at Hart-Miller and Poplar Islands. By some estimates, the two sites will be filled by 2015 at the earliest. Another alternative is the 72,000 acres (28,800 hectares) of federal land at Aberdeen Proving Grounds along the bay shoreline of Harford and Baltimore Counties. But this site has serious drawbacks. Much of it is wetland habitat. And some of the site has been used as a dumping ground for up to thirty million rounds of unexploded ordinance that would present a risk to workers.

Another artificial island could be created in the Chesapeake Bay similar to the current main 11,000-acre (4,400-hectare) dumping site at Hart-Miller Island, which sits at the mouth of Back River and is made entirely of mud dredged from shipping channels. While many environmentalists favor artificial islands as opposed to bottom dumping, others see the creation of artificial islands as a stopgap measure. The continuing flow of silt down the Susquehanna River into the upper bay is a natural process that will not stop. In the long run, artificial islands could fill much of the upper bay.

The federal government maintains several fill sites inland along the C&D. The containment walls could be raised and dredging hauled there until around 2025. The major drawback with this alternative is that the mud would have to be hauled 45 miles (72.42 kilometers) from the main shipping channel, and then over land. As a last resort, there is always the Atlantic Ocean. But there has not been any major scientific analysis of the affect of dumping dredged material on the continental shelf, and the ocean's location far away and the cost of movement would be prohibitive.

The carrying of silt into the Chesapeake Bay by numerous tributaries is an age-old natural process. Maintaining artificial shipping channels to support the economic activities centered on the Port of Baltimore will certainly have costs, both environmental and economic. This dilemma will continue to demand attention and require creative and insightful long-range regional planning. The Maryland Port Administration, various federal agencies, environmental groups, and the citizens of Maryland will continue to seek a balance among these various interests and costs. Both the Chesapeake Bay ecosystem and the tremendous economic engine centered on the Port of Baltimore are critically important to Marylanders.

Future Prospects for the Port of Baltimore

During the 1980s and early 1990s, the Port of Baltimore faced serious challenges from rail deregulation, labor conflicts, loss of shipping lines, and a shrinking number of cargo vessels worldwide (*Baltimore Sun* 1999a). In the early 1980s, Baltimore had a reputation among the shipping lines of the world as the "port that didn't work in the rain." At the first sign of a sprinkle, longshoremen walked off the docks. The port also had high labor wages and strict rules, which put the Port of Baltimore in a weak position that competitor ports effectively exploited. By the late 1990s, Baltimore had begun to redefine and establish itself once again as an important East Coast maritime center. The gloom and pessimism of the 1980s gave way to cautious optimism, and the labor situation turned around. Today, longshoremen have not only agreed to flexible hours, but they have also agreed to work in the rain. The port has targeted specialty, or "niche," cargoes as well as turned once again to its geographical advantage of being relatively close to the population centers of the Midwest. The 50-foot (15.24-meter) shipping channel, a more efficient workforce, and modern facilities have all helped to keep the Port of Baltimore competitive.

Until the summer of 1999, Baltimore was hindered by the fact that the CSX Railroad was the only game in town. CSX, and indeed railroading in the United States, started in Baltimore. But by 1999, CSX had been gradually moving its operations out of the city. The Maryland Port Administration realized that CSX favored the larger railhead at Norfolk down the bay. During the summer of 1999, a new railway came to Baltimore, the Norfolk Southern Railroad. Eager to win business, the Norfolk Southern has rivaled CSX by offering more attractive services and rates. Another development affecting ports worldwide is the jumbo container ship, which has greatly affected economies of scale.

Not all ports are able to accommodate these new ships, which are nearly three and a half football fields in length and carry 6,000 truck-size aluminum containers (a 50% increase over the largest container ships of the late 1990s). Baltimore is one of the few East Coast ports with a channel deep enough to handle jumbo container ships.

The state of Maryland has invested heavily in the Port of Baltimore. Today, Baltimore has among the best and most modern terminals on the East Coast. Unlike many of its competitors, Baltimore still has space for expansion. Even so, there have been setbacks. In the spring of 1999, the Port of Baltimore failed to win a bid to become the East Coast center for the mammoth Maersk Line. Instead, Maersk Line decided to go with the New York–New Jersey Hudson River location. CSX, owner of 50 percent of the shipping line, pushed hard for the New York–New Jersey site. Although this decision was a severe blow to the Port of Baltimore, in the long run the attention given to Baltimore may work to its advantage. The bid for Maersk Sea-Land brought together many interests in Baltimore, including longshoremen, the state government, terminal operators, the business community, and others. Baltimore had been seriously considered, and many shipping companies liked what they saw. Since 1999, Baltimore has attracted more automobile and other RO/RO cargo. Paper and other forest products now flow through the port in greater volume. Paper and forest products are not usually time-sensitive cargo; therefore Baltimore's more inland geographic location and the time it takes to cruise up the bay are not critical hindrances.

The container business is important to the port because it is among the most economically lucrative types of shipping. It creates more jobs and spin-off businesses than most other forms of cargo (*Baltimore Sun* 1999b). The decision to center the Port of Baltimore's operations on containerization was made in the 1950s. Critics of this new shipping method said the trend would not last, but they were proven wrong. During the late twentieth century, container shipping revolutionized ocean shipping and helped Baltimore maintain its position as a major US general cargo port. In 2009, Baltimore ranked thirteenth among US ports for container business, and business has been growing since. Today the Port of Baltimore's largest customer is the Taiwanese container shipping company Evergreen.

The Port of Baltimore remains an important economic engine not only for Maryland but for the wider surrounding region. The state has made heavy investments in the port and has worked closely with labor unions, businesses, and local governments to keep Baltimore competitive. The geographical assets of the port are vital to its success. The port serves not just Maryland companies and individuals, but also various importers and exporters in numerous states, especially in the mid-Atlantic and midwestern regions. The Port of Baltimore is a critical link to places across the entire nation. It entered the twenty-first century on a strong footing, but its continued revival is not a sure thing. Competition among East Coast ports will continue to be ruthless and will require constant vigilance to keep the Port of Baltimore viable.

Air Transportation

Three major international airports, 141 public and private regional and smaller airports, and fifty-nine heliports serve the state of Maryland (Maryland Department of Economic and Community Development 2003). The leading commercial air transport facility in the state is Baltimore–Washington International Thurgood Marshall Airport (BWI), located in Anne Arundel County between Baltimore and Washington, DC, just east of the Baltimore–Washington Parkway. The other two major international airports in the region are Reagan National Airport (formerly Washington National) and Dulles International Airport, both in nearby Virginia. Since the early 1960s, all three airports have competed for the region's market, which stretches into Delaware, New Jersey, Pennsylvania, Virginia, and Washington, DC.

After World War II, when aviation technology had proven that flying boats were no longer necessary for crossing the oceans by air, officials selected an inland location for a new airport at the site of Friendship Church in Anne Arundel County. In June 1950, President Harry S. Truman dedicated the new Friendship International Airport, owned at that time by the City of Baltimore. This modern airport was the first in the country with the new pier concept of design, and until 1962 it was the only airport in the region capable of handling large passenger jets. A pier design includes several narrow buildings with aircraft parked on both sides. One end of the pier connects to a ticketing and baggage claim area. Although piers offer high aircraft capacity in a simple design, they often result in a long distance from the check-in counter to the gate. During the 1960s, competitive pressure from Washington National and Dulles International Airports eroded Friendship's traffic. In 1972, the state of Maryland purchased the Friendship International Airport from Baltimore for $36 million and in 1973 changed its name to Baltimore–Washington International Airport to reflect its more regional scope. Since that time, there has been a constant proactive program of improvements for BWI. The name changed once again in 2006 to Baltimore–Washington International Thurgood Marshall Airport to honor this Maryland native and former Supreme Court justice. The airport currently covers over 3,596 acres (1,439 hectares) and has five piers (four domestic and one international). It is operated by the Maryland Department of Transportation's Aviation Administration.

In 1980, the country's first intercity air-rail passenger station was opened at BWI, just six minutes from the terminal along Amtrak's Washington–New York–Boston line. BWI is now also connected directly with downtown Baltimore by a light-rail line. By 2010, BWI was served by forty airlines, including twenty-five commercial airlines plus charter, commuter, and air cargo airlines. On a daily basis BWI handles 694 scheduled arriving and departing domestic flights, eighteen international flights, and ten cargo flights. The major airlines serving BWI with passenger service include American, America West, Continental, Delta, Northwest, Piedmont, Southwest, United, and US Airways. In 2010, Southwest Airlines had the largest passenger market share at BWI (55.3%).

Air Canada, Air Tran, Bahamas Air, British Airways, and US Airways provide international service at BWI. Among major airports in the United States, BWI ranks twenty-third for passengers and thirtieth for cargo. The primary cargo commodities shipped out of BWI are mostly high-value Maryland products: machine parts, electrical and electronic machines, aircraft and spacecraft components, seafood, chemicals, and pharmaceutical and biological products. BWI experienced significant growth in the number of passengers during the 1990s. In 2010, BWI handled 21.94 million passengers, putting BWI just behind Dulles International Airport. The top five airports of domestic origin and destination to and from BWI in 2010 were, in order, Boston, Orlando, Atlanta, Fort Lauderdale, and Tampa. BWI's growth can be attributed mainly to domestic passenger travel. It does not have the major international direct and nonstop destinations that Dulles and Philadelphia have, and the slow decline in international travel through BWI is troubling (Hancock 2006).

The investment and growth at BWI has been beneficial to the economy of Maryland. The Maryland Aviation Administration estimates that BWI generated 99,913 direct and indirect jobs in 2010, with over 10,000 full- and part-time employees at the airport. That same year BWI generated $5.1 billion in business revenues for central Maryland. The state's investment in BWI is critical because the spatial interactions enabled by modern and efficient transportation keep Maryland's people and businesses connected to other places essential to its modern economy.

The remaining smaller airports and heliports are spread throughout all of the counties of Maryland. Commuter lines serve some of these airports; some are private company airstrips, and some heliports serve hospitals and businesses. The largest of these airports are in Cumberland, Baltimore County, Easton, Frederick, Hagerstown, and Salisbury.

Highway and Rail Transportation

Although Maryland is one of the smallest states in the country, it is still 324 highway miles (521.4 kilometers) from the far western city of Oakland in Garrett County to Crisfield in Somerset County on the Eastern Shore (fig. 10.2). An extensive network of over 29,000 miles (46,670 kilometers) of highways and about 1,000 miles (1,609 kilometers) of rail routes crisscross the state. The highways consist of 4,376 miles (7,409 kilometers) of municipal roads, 19,973 miles (32,143 ki-

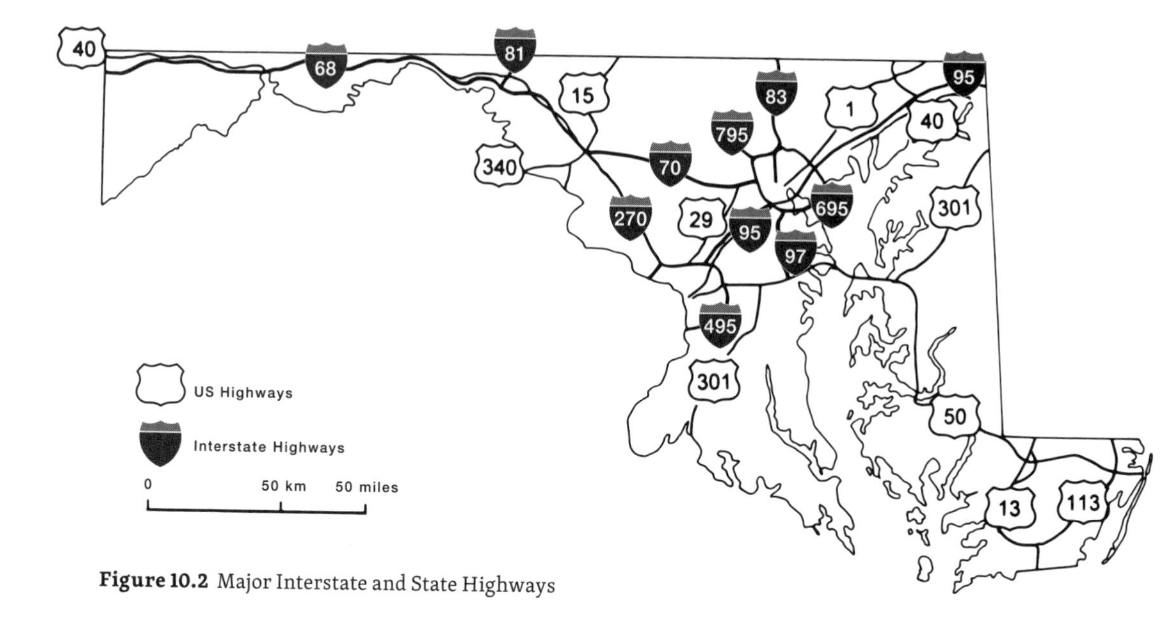

Figure 10.2 Major Interstate and State Highways

lometers) of county roads, and 5,230 miles (8,417 kilometers) of state roads (Maryland Department of Economic and Community Development 2003). Most parts of the state are easily accessible via multilane, limited-access highways.

The state's highways are well connected and integrated into the US Interstate System. Many cities in the eastern half of the United States are reachable by truck from Maryland in one to two days. Atlanta, Boston, Buffalo, New York, and Pittsburgh can all be reached overnight, and Chicago and St. Louis are a two-day drive away. About 37 percent of the US industrial market and 31 percent of the nation's population are within overnight delivery distance by truck. Some 350 common and contract-carrier trucking firms operate in Maryland.

Two major public intercity bus systems centered on Baltimore and Washington, DC, as well as by several smaller operators serve over 150 communities in Maryland. The Maryland Transit Authority (MTA) operates the 15.4-mile (24.64-kilometer) light-rail system in Baltimore. The original light-rail line runs from Hunt Valley (north of Baltimore City in Baltimore County) south to Pennsylvania Railroad Station, and then via Oriole Park at Camden Yards to the Cromwell Station/Glen Burnie area in Anne Arundel County. The expansion line extended service from Amtrak's Pennsylvania Station to BWI. The Metro Subway line runs from Owings Mills, northwest of Baltimore in Baltimore County, to central Baltimore, ending at Johns Hopkins University Hospital. Although the light-rail system is rather limited in its geographic extent, it is coordinated with MTA bus service.

There are places in the Baltimore region that the public transport system still does not serve well, which could be an impediment to continued growth and prosperity. Some communities have large labor pools, while others have plenty of available jobs. The two need to be connected. Although economic growth and prosperity surround West Baltimore, for example, the region is not a full participant in the local economy. It is essentially landlocked because Maryland's transportation services do not connect its citizens to where the jobs are (Ambrose 2000).

The Washington Metropolitan Area Transit Authority (WMATA) operates a much more extensive rail system. WMATA was created by an interstate compact in 1967 in order to plan, develop, build, finance, and operate a balanced regional transportation system in the Washington, DC, metropolitan area. The District of Columbia, as well as the states of Maryland and Virginia, support and run WMATA. The bus and light-rail systems extend into Montgomery and Prince George's Counties in Maryland as well as Arlington, Fairfax, and Loudoun Counties and the cities of Alexandria, Fairfax, and Falls Church in Virginia. The rail system opened in 1976 and currently includes eighty-three stations and 103 miles (166 kilometers) of rail line. The integrated Metrorail and Metrobus systems of WMATA serve an area of 1,500 square miles (3,885 square kilometers) and 3.4 million people.

THE RED LINE

Seeking to improve the light-rail system in Baltimore, planners have proposed several additional routes. The proposed Red Line would be an east–west route beginning at Interstate 70 near the Social Social Security Administration at Woodlawn in Baltimore County and run through West Baltimore with an intermodal stop at the West Baltimore MARC station. Then it would continue downtown, intersecting with the existing light-rail line, and then to East Baltimore with stops in the gentrifying areas of Canton, Fells Point, and Patterson Park. Different groups, such as the Maryland Transit Administration's Citizens' Advisory Committee, have proposed alternative routes, continuing a vigorous debate.

If built, the Red Line will run through West Baltimore along Route 40. A portion of Route 40 was built as a divided highway when there was a plan to extend Interstate 70 through West Baltimore to downtown. The existing highway section was built with a wide median to accommodate a light-rail line. Baltimore City officials have said that the Red Line will run underground through downtown from Martin Luther King Jr. Boulevard eastward to at least President Street. But there are some disagreements over the exact route and where the line will be above and below ground. Backing for a new light-rail line exceeds support for an alternate dedicated bus-line proposal by a two-to-one margin. Baltimoreans were still severely divided over Red Line alternatives as of 2011, when severe national, state, and city economic crises had put the project on hold.

Maryland has an important place in the history of US railroading. In 1827, a group of forward looking Baltimore merchants, led by Philip E. Thomas and George Brown, petitioned the Maryland General Assembly to charter the Baltimore and Ohio Railroad. Construction on this first railroad in the United States began on July 4, 1828. By May 1830, the B&O had reached Ellicott Mills in Howard County. It reached Cumberland in 1842 and Wheeling, Virginia (now West Virginia, founded in 1863), in 1853. With the birth of the B&O, the US railroad industry was born, and Baltimore was secured in its dominance of Maryland commerce (Chappelle et al. 1986). At first, horses pulled the rail cars. Soon thereafter railways saw the rapid development of steam locomotives, led in large part by the successful run of Peter Cooper's *Tom Thumb* in August 1830. By 1857, the B&O reached St. Louis, Missouri; the railroad proved instrumental in moving troops and supplies for the Union during the Civil War. By the early 1900s, the B&O connected Chicago, New York, and Philadelphia, but hard times lay ahead. By the 1950s, the B&O was mainly a freight carrier facing severe financial problems. In 1965, it merged with the Chesapeake and Ohio Railroad, and in 1973, a merger with the Western Maryland Railroad led to the formation of the Chessie System. In 1980, the Chessie System merged with the Seaboard System Railroad to create CSX.

Today the rail freight system in Maryland remains focused on Baltimore. Specialized facilities in the city include piggyback, container, rail-to-truck transfer, automobile terminals, and coal, grain, and ore loading and storage. Hagerstown and Cumberland are important secondary rail centers. Rail traffic is heaviest on the lines into western Maryland and along the urban corridor both north and south of Baltimore (fig. 10.3). A number of short-line, light-density rail routes throughout the state currently operate with state and federal subsidies, including Chesapeake Railroad, Eastern Shore Railroad, Maryland and Delaware Railroad, Maryland Midland Railway, and Winchester and Western Railroad. Today the major rail carriers serving Maryland include Amtrak, CSX, and Norfolk Southern.

Rail commuter service from Washington, DC, to Baltimore connects to service continuing north. Nearly one-fifth of the passenger traffic between New York City and the Washington–Baltimore region is by rail. Both Washington and Baltimore are along the federally designated high-speed corridor running between Newport News, Virginia, and Boston. The new high-speed Acela Express trains operated by Amtrak run nonstop from Washington to New York City, covering the distance in less than two and a half hours. As commuter air travel in the northeast corridor continues to take longer owing to flight delays, security lines, airport congestion, and longer travel times to and from airports, rail passenger trains are looking more attractive to travelers.

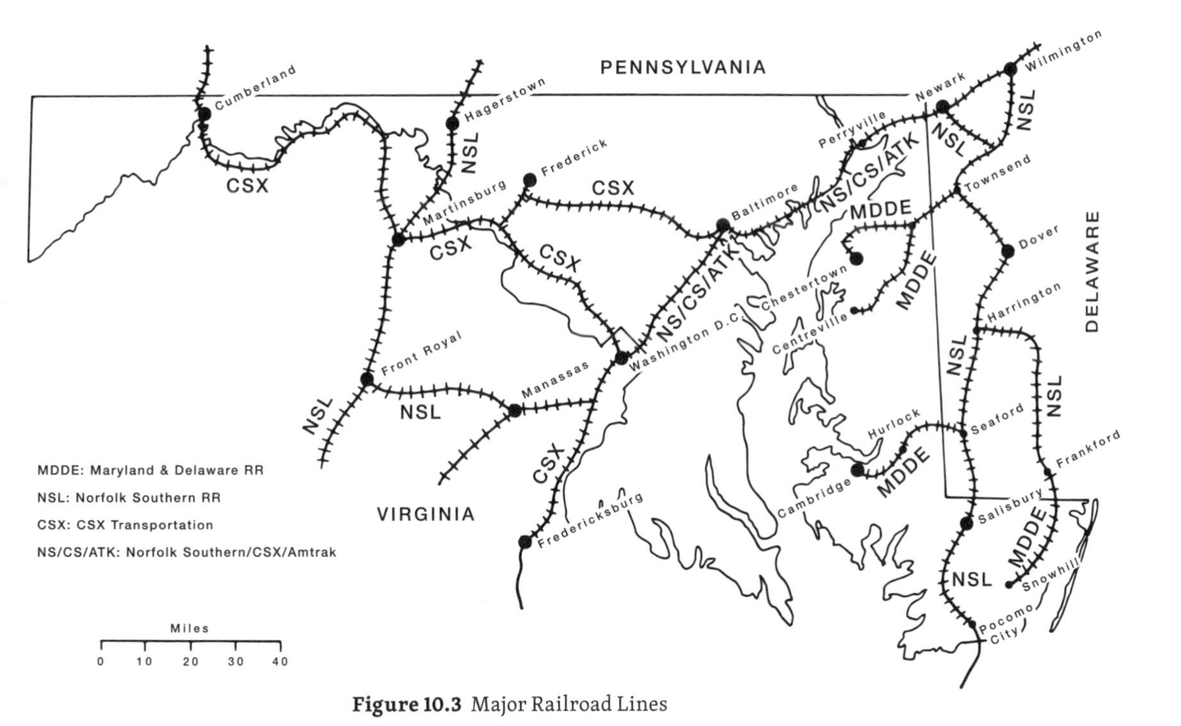

Figure 10.3 Major Railroad Lines

Communications

Maryland is near the top of the national mass-media communications hierarchy, with two major national centers in Baltimore and Washington, DC, as well as several smaller, more local regional centers. Major centers—such as New York City and Los Angeles—are at the top of the national communications hierarchy. Baltimore falls into the group of second-tier centers along with cities such as Boston, Chicago, Philadelphia, and San Francisco. Because of its national and international prominence as a political-governmental center, Washington has a special status near the top of the communications hierarchy.

Numerous newspapers and broadcasting centers reside within Maryland. Twelve daily state newspapers and two in Washington, DC, serve the people of Maryland. Outside of Baltimore and Washington, the only daily newspapers are in the regional urban centers of Annapolis, Cambridge, Cumberland, Easton, Elkton, Frederick, Hagerstown, Rockville, Salisbury, and Westminster. Over fifty weekly newspapers are published in Baltimore, Washington, and the regional urban centers in addition to a group of smaller urban centers across the state. Lower-order local weekly newspapers are far more dispersed and ubiquitous than higher-order daily newspapers. Higher-order urban functions generally require a much larger population (and therefore market) for support, but even those newspapers in the largest cities of Maryland and the nation face severe competition from various electronic news services. Newspapers are in the process of adapting to this new environment, with some publishing news stories online. The newspapers that have been slow to adapt are disappearing.

Marylanders have a choice of radio stations (fifty-four AM and eighty-two FM stations), plus a number of radio stations broadcasting from Washington, DC. There are fifteen television stations in Maryland: eight are commercial and seven are state-operated public stations, with the Maryland Public Broadcasting Commission being the licensee. Washington is home to seven commercial stations and one public station. Maryland and Washington television stations also serve populations and markets in adjoining areas of Delaware, Pennsylvania, Virginia, and West Virginia. During the 2000 national presidential election, for example, all of the national and regional polls correctly predicted that Maryland would vote Democrat and support presidential candidate Al Gore with its electoral votes. But numerous television and radio advertisements were broadcast from Salisbury, some supporting Gore and some supporting the Republican candidate, George W. Bush. Why was the Bush campaign spending money in state that would surely be lost? The answer lies in the geographic location of Salisbury and the market area of its radio and television stations. The Bush advertisements targeted voters in nearby Delaware (which is served by broadcasting stations in Maryland and Philadelphia), whose votes were critical in a close election. Because purchasing airtime from smaller Maryland regional stations is far less expensive than from major Philadelphia stations, the decision was made to beam the political advertisements into Delaware from Salisbury.

Cable television subscription systems augment the range of commercial channels in Maryland. Hagerstown covers much of western Maryland and nearby Pennsylvania and West Virginia. Salisbury's cable system covers the Eastern Shore. The spatial diffusion of cable television in Maryland has followed the typical reverse-hierarchical pattern that has occurred elsewhere in the nation. Cable television started in nonmetropolitan western Pennsylvania in order to bring television to places that could not receive television airwave signals. In more recent decades, cable television has specialized and worked itself up the urban hierarchy to larger cities.

For marketing and rating purposes, Nielsen Media Research divides the United States into 210 designated market areas (DMAs), or regions within which the population can receive the same television and radio stations. DMAs are identified by the largest cities, usually located near the center of regions. These market areas are used to compile ratings for television (Nielsen Media Research) and radio (Arbitron). Market regions often overlap (i.e., people residing on the border may be able to receive content from either market). In Maryland, there are four

DMAs: Baltimore, Pittsburgh, Salisbury, and Washington, DC. Garrett County in far western Maryland falls within the Pittsburgh DMA. Marylanders in that region access Pittsburgh television and radio stations as well as newspapers. The Washington DMA extends from Allegany County in the west to St. Mary's County in southern Maryland. The Baltimore DMA extends from Carroll County east to Cecil County, and south to Dorchester County on the Eastern Shore. The smallest DMA, Salisbury, covers Somerset, Worcester, and Wicomico Counties.

As in the rest of the nation, the Internet now connects Marylanders to individuals and organizations worldwide. The Internet has changed the way many people communicate, access information, shop for goods and services, and redefine their geographic activity space. Because Maryland is a prosperous state within a generally wealthy country, its people are well connected, as state and local governments as well as businesses have invested heavily in modern communications technologies. Although the digital divide is still present in Maryland, it is much narrower than in other regions of the country and the wider world. The people of Maryland have a broad base of mass-communications media available. In an increasingly globalized information age, Marylanders are well served by modern communications systems from the international down to the local scale.

11 Governing Maryland

Maryland is organized into twenty-three counties plus Baltimore City, which for most purposes is legally considered the equivalent of a county. Four counties are in western Maryland, eight in the central part of the state (including Baltimore City), three in southern Maryland, and nine on the Eastern Shore. Most of the counties are between 250 and 500 square miles (648 and 1,296 square kilometers); Baltimore, Dorchester, Frederick, and Garrett are larger, and Calvert and Baltimore City are smaller.

Counties are generally the most important form of government in Maryland. They have gradually assumed responsibility for basic services—such as schools, police, fire protection, and public works—that elsewhere are often the responsibility of municipalities. As in many states to the south of Maryland, the county level of government is the backbone of local government. To the north, at the other end of the political organizational scale in New England, towns reign supreme. New England towns have their own local governments, hold town meetings, and render the county level of government much less influential than in Maryland. Although the county form of government dominates in Maryland, there are still some municipal governments.

At the core of the structure of Maryland's government is local responsibility for functions and services, under broad policy guidance from the state government. Economic and social changes have been more easily accommodated because of the great amount of flexibility in this system. Over the past three centuries, Marylanders have created and sustained a system of local government that is so responsive it may well prove to be among those state systems most capable of appropriately and quickly addressing the severe challenges of urban development, decay, and sprawl (Spencer 1965).

Although some argue that Maryland state and local governments are so responsive and effective because of the state's relatively small geographic size, such reasoning would seemingly lend itself to a more compact and homogeneous land and society, which is not the case. Although geographically small when compared to many other states, Maryland still presents great diversity in both its physical geography and social structure. The social and economic issues on the Eastern

Shore differ from those in the rapidly growing Washington-Baltimore urban corridor. Environmental issues in far western Maryland differ from those in southern Maryland. Citizens in Garrett County in the far west can at times feel distant from what is going on at the state capitol in Annapolis. Maryland's accomplishments in local governance have been both difficult to achieve and a struggle to sustain.

The counties and municipalities that form Maryland's basic units of government have a high degree of structural uniformity. The absence of clutter and multiple local governments that so often compound and clog the local processes in other states is a main strength of the Maryland system. It is also flexible enough to accommodate the great variety of physical, social, cultural, and economic environments found in different parts of the state. Each local government in Maryland reflects in its operations and relations both the prominent features of its own environment and the general requirements and limitations of constitutional and legislative provisions (Spencer 1965). In short, Maryland has a system of government that is capable of being geographically expressive of its diversity.

A good example of flexibility in local government is Wye Island in Queen Anne's County. In 1974, nationally prominent developer, James Rouse (a native of Easton in nearby Talbot County), announced plans to buy most of Wye Island and to develop it in a carefully planned manner. After a series of exhaustive socioeconomic and physical geographical studies, the Rouse Company unveiled its development plan at a public session in Centerville. The Queen Anne's County commissioners were not supportive. The local government refused to rezone Wye Island below the minimum 5-acre (2.20-hectare) plot, and shortly afterward the Rouse Company withdrew its proposal, which had cost nearly $900,000 to prepare (Gibbons 1977). The people and the county government had been heard.

The county unit originally began as a broad geographic area within which judicial and record-keeping functions were performed locally in the county seat, which was usually located near the geographic center of the county. The original guiding principle was that the county seat should be no more than one day's ride by horseback from any point in the county (Spencer 1965). In 1683, Maryland's legislature passed the "town act," systematizing the establishment of city and town municipalities. Municipal governments held the responsibility of providing services such as public health, construction and maintenance of public thoroughfares, and some fire and police protection. County and incorporated municipalities serve largely different—yet partially overlapping—populations. Municipal governments often serve densely settled populations in small areas, whereas county governments serve populations that are both spatially clustered as well as dispersed over larger geographic areas. Municipalities and counties have traditionally divided responsibilities in a functionally complementary manner.

Municipal Government

A number of incorporated cities and towns in Maryland have self-governing powers, making them distinct from unincorporated places. The municipal citizens of incorporated places participate in a level of government that is closest to the people. To be successful, local self-government depends on a high level of awareness and commitment by a participating citizenry.

Today there are only 157 municipal governments in all of Maryland. Of the twenty-three counties, only Baltimore and Howard have no incorporated cities or towns. Excluding Baltimore City, the populations of Maryland's municipalities range from about 40,000 to 50,000 people. Over 700,000 people (about 12% of Maryland's population) live in incorporated cities and towns other than Baltimore City (Maryland Municipal League 2000). Although the median population of these municipalities is 1,300, the population of each of the largest six municipalities (Baltimore City, Bowie, Frederick, Gaithersburg, Hagerstown, and Rockville) exceeds the population of one-third of Maryland's counties.

Each municipality has its own charter that outlines those powers and areas of responsibility its citizens have agreed to assume. All municipalities, regardless of size, are treated equally under state law. The powers available to them when drafting their charter are set forth in the Maryland State Constitution (Article XI-E) and the Annotated Code of Maryland (Section 23A). Maryland state law recognizes county and municipal governments as coequal local entities, meaning that county government cannot impose itself over a municipality (Maryland Municipal League 2000). The combination of functions and powers claimed by municipalities in their charters varies widely and is not closely related to population size. Some smaller, more rural municipal governments are responsible for a wide range of services, including police, roads, refuse, water, sewer, planning, and zoning. Even so, some of the larger municipalities have opted to provide fewer services and to depend on county, state, or regional governments as providers instead. Although municipal governments are autonomous from their county's government, for example, they may depend on the county to provide police, water, and sewer services.

Baltimore City is a chartered municipality in a class of its own. Originally an incorporated municipality within Baltimore County (Baltimore County once had an incorporated municipality, although it no longer does), Baltimore City in 1851 parted ways with Baltimore County. Baltimore City has been a separate incorporated independent unit ever since. In 1915, Maryland voters ratified Article XI-A of the Maryland Constitution, allowing counties to adopt charter home rule. There are three forms of county government in Maryland: county commissioner, charter home rule, and code home rule. Both charter and code home rule offer broader powers to local governments to exercise self-governance than the traditional county commissioner form. The trend throughout Maryland is for counties to move toward a home rule form of government (Washington County, Maryland 2006).

Baltimore City adopted a charter and began exercising charter home rule in 1918 (Maryland Municipal League 2000). Its charter established joint governance with powers shared by the mayor and city council. The city also has an important board of estimates, which is responsible for drawing up the annual budget and running the city on a day-to-day basis. Members of the board of estimates include the mayor, president of the city council, comptroller, public works director, and city solicitor.

In 1915, changes to the state constitution uniformly gave the right of charter home rule only to counties (including Baltimore City). It did not, however, uniformly give the right to local municipalities to establish independent charter government. It was not until November 1954 that Maryland voters ratified Article XI-E of the state constitution, giving municipalities the right to opt for charter home rule. This was the first major change in the legal status of Maryland's municipal governments since the general assembly granted the first municipal charter in 1683 (Maryland Municipal League 2000). Prior to 1954, it was possible to create a new municipality, but each case was handled separately and required that the Maryland General Assembly enact an individual piece of legislation. The 1954 article regularized the process for creating new municipalities.

The advantages of grassroots municipal government usually highlight more direct control by the people of their own destiny. Local government is made up of citizens who are part of the community, are more sensitive to local needs, and have the power to propose changes in the structure of government and to amend the charter as local needs evolve. Citizens have more direct access to their elected officials. Yet there are some disadvantages to local government, including higher local taxes to pay for services. Without committed citizen volunteers, local government usually does not work well. The economies of scale, which can enable county governments to reduce costs, are often not possible at the more local level.

The geographical pattern of incorporated cities and towns in Maryland is uneven, but several major characteristics emerge. There is a large cluster of municipalities in Montgomery and Prince George's Counties in metropolitan Washington, DC. In the stretch of counties running from Harford south to St. Mary's, municipalities are sparse and are completely absent in Baltimore and Howard Counties. Central and western Maryland have clusters of municipalities, as does Dorchester County on the Eastern Shore.

Geographical overlap into adjacent counties of several municipalities underscores the fact that incorporated cities and towns are

BORDER-STRADDLING TOWNS

Two municipalities straddle the border between Maryland and Delaware. Marydel, incorporated in 1929, overlaps from Caroline County in Maryland into Kent County in Delaware. The Maryland part of Marydel is incorporated, but the Delaware part is not. Delmar also sits astride the border between Wicomico County, Maryland, and Delaware. Incorporated in 1899, Delmar's motto is "the little town too big for one state." Each town elects its own mayor and council, raises its own revenue, and passes its own ordinances. The revenues of the two incorporated Delmars are then put into a common pool with a single town manager. The police station, waste treatment plant, elementary school and town hall are in Delmar, Maryland. The library and volunteer fire department are in Delmar, Delaware. Middle and high school students from Delmar, Maryland, may go to school in Delmar, Delaware, or Salisbury, Maryland. Both Delmar and Marydel lie astride the famous Mason-Dixon Line surveyed and marked in 1768 by surveyors Mason and Dixon.

autonomous from county government. Such overlaps include Millington (incorporated in 1890) in Queen Anne's and Kent Counties, Mount Airy (incorporated in 1894) in Frederick and Carroll Counties, Queen Anne (incorporated in 1953) in Queen Anne's and Talbot Counties, and Templeville (incorporated in 1865) in Caroline and Queen Anne's Counties.

Changing Political Regions

An anonymous writer once called Maryland "America in miniature." Although there is some truth in this description, there is also much in Maryland's character that is unique. Despite its small size, Maryland contains a rich and contrasting variety of physical and cultural landscapes: wealth and poverty, cosmopolitan urban centers and rural areas, plains and mountains, centrality and isolation, manufacturing and farming—and the list does not end here.

It is perhaps in its politics that heavily Democrat Maryland is not typical of the entire country. When was the last time you heard of Maryland being used as a representative state in a national political poll? Maryland also has a distinct culture flavored by its horse breeding and racing industry; lifestyle associated with the Chesapeake Bay; racial and ethnic diversity in the urban core from Baltimore to Washington; and population of diplomats, federal workers, and institutions related to the close proximity of the nation's capital. Longtime US Senator Paul Sarbanes of Maryland once stated that if states had souls, Maryland's would be the Chesapeake Bay. Maryland, a state with a long history and rich traditions, entered the twenty-first century pulsating with exciting changes from the upper Eastern Shore ruled by its "squirearchy," across the fox-and-hounds country of the Piedmont, to the coal mines and new commercial developments of Garrett County. H. L. Mencken once wrote an essay describing Maryland as the most "average" of states. In doing so, this distinguished and celebrated Baltimore writer recognized the presence in Maryland of the elemental character of the nation. Much has changed since the times of Mencken, however; Maryland has become so much more than just average. Perhaps it is finally time to discard the "America in miniature" description of Maryland.

The political landscape of Maryland is certainly not static, even though—when considered at the national level—Maryland is routinely described as a blue state that always votes Democrat in national elections (table 11.1).

At the intrastate geographical scale, Maryland is not so politically homogeneous. In 2008, the *Baltimore Sun* printed an election guide leading up to the presidential primary (*Baltimore Sun* 2008a). For the purpose of dissecting and analyzing its political landscape, the *Baltimore Sun* divided Maryland into eight regions: the Eastern Shore, urban corridor (subdivided into six regions), and western Maryland.

The Eastern Shore

Caroline, Cecil, Dorchester, Kent, Queen Anne's, Somerset, Talbot, Worcester, and Wicomico Counties constitute the Eastern Shore region. Outside some historic African American enclaves and pockets of wealth along the rivers and bay waterfront (especially on the upper Eastern Shore north of the Choptank River), the Eastern Shore is a relatively homogeneous, conservative region. Republicans do well here in local and statewide elections. Although there has been an influx of new residents from the Western Shore (since the opening of the Bay Bridge in 1952), as well as an increasing number of commuters across the Bay Bridge, the region has remained predominantly conservative and Republican. A good number of the people who have purchased summer residences in Wicomico and Worcester Counties near Ocean City still vote in their home counties on the Western Shore. Many natives of the Eastern Shore are quick to state their distinct identity. One old saying goes, "We don't give a damn for the whole state of Maryland; we're from the Eastern Shore" (Meyer 2003). Tobacco, fishing, and agricultural patterns going back to the seventeenth century still affect the culture and politics of the region.

North of the Choptank River, wealthy retirees and businessmen who own valuable waterfront estates have a strong influence. On the lower Eastern Shore, south of the Choptank, there are more working-class watermen, farmers, and some residents who are suspicious of government. The lower Eastern Shore is home to many Southern Democrats who have not voted for a Democratic presidential candidate since that

Table 11.1. Maryland's presidential elections, 1912–2012

Year	*Democrat*	*Republican*	*Year*	*Democrat*	*Republican*
1912 P	Champ Clark	Theodore Roosevelt	1964 P	Daniel Brewster	no contest
1912 G	**Woodrow Wilson**	William Howard Taft	1964 G	**Lyndon Johnson**	Barry Goldwater
1916 P	no contest	no contest	1968 P	no contest	no contest
1916 G	**Woodrow Wilson**	Charles E. Hughes	1968 G	Hubert Humphrey	**Richard M. Nixon**
1920 P	no contest	Leonard Wood	1972 P	George Wallace	Richard M. Nixon
1920 G	James M. Cox	**Warren Harding**	1972 G	George McGovern	**Richard M. Nixon**
1924 P	no contest	Calvin Coolidge	1976 P	Edmund G. Brown	Gerald Ford
1924 G	John Davis	**Calvin Coolidge**	1976 G	**Jimmy Carter**	Gerald Ford
1928 P	no contest	Herbert Hoover	1980 P	Jimmy Carter	Ronald Reagan
1928 G	Al Smith	**Herbert Hoover**	1980 G	Jimmy Carter	**Ronald Reagan**
1932 P	no contest	Herbert Hoover	1984 P	Walter Mondale	Ronald Reagan
1932 G	**Franklin Delano Roosevelt**	Herbert Hoover	1984 G	Walter Mondale	**Ronald Reagan**
1936 P	Franklin Delano Roosevelt	no contest	1988 P	Michael Dukakis	George H. W. Bush
1936 G	**Franklin Delano Roosevelt**	Alf Landon	1988 G	Michael Dukakis	**George H. W. Bush**
1940 P	no contest	Thomas Dewey	1992 P	Paul Tsongas	George H. W. Bush
1940 G	Franklin Delano Roosevelt	Wendell Wilkie	1992 G	**Bill Clinton**	George H. W. Bush
1944 P	no contest	unpledged	1996 P	Bill Clinton	Robert Dole
1944 G	**Franklin Delano Roosevelt**	Thomas Dewey	1996 G	**Bill Clinton**	Robert Dole
1948 P	no contest	no contest	2000 P	Al Gore	George W. Bush
1948 G	**Harry S. Truman**	Thomas Dewey	2000 G	Al Gore	**George W. Bush**
1952 P	Estes Kefauver	no contest	2004 P	John Kerry	George W. Bush
1952 G	Adlai Stevenson	**Dwight Eisenhower**	2004 G	John Kerry	**George W. Bush**
1956 P	Estes Kefauver	Dwight Eisenhower	2008 P	Barack Obama	John McCain
1956 G	Adlai Stevenson	**Dwight Eisenhower**	2008 G	**Barack Obama**	John McCain
1960 P	John F. Kennedy	no contest	2012 P	Barack Obama	Mitt Romney
1960 G	**John F. Kennedy**	Richard M. Nixon	2012 G	**Barack Obama**	Mitt Romney

Source: compiled from various sources
Note: Winners of general elections are shown in boldface. G indicates the general election candidate for each party; P indicates the Maryland primary winner for each party.

party took a strong civil rights stand starting in the late 1960s. Although seven of the nine Eastern Shore counties have more registered Democrats than Republicans, there has been a consistent pattern of Republican victories in the region. In Somerset and Dorchester Counties, a significant number of African American enclaves exist, but their votes for Democrats are not enough to change the overall pattern of a Republican stronghold.

The Urban Corridor

Baltimore City

Baltimore City routinely turns out large numbers of voters for the Democrat Party. By far the largest city in the state, Baltimore's local and state politics are dominated by the urban social issues of poverty, crime, and education. Behind Montgomery and Prince George's Counties, Baltimore City is the third-largest stronghold of the Democratic base in Maryland. Baltimore has always had a diverse population; it has been home to a significant number of immigrants going back to the colonial period. Following the Civil War, Baltimore became one of the major centers of African Americans in the United States. The African American population of Baltimore soared again after World Wars I and II (Pietilia 2010). Soon thereafter, Baltimore developed as the heavy industrial center of Maryland. Although much of Baltimore's heavy industry is now gone, influences of the strong ties between the Democrat Party and organized labor remain.

Baltimore Suburbs

The bellwether counties of Anne Arundel, Baltimore, Carroll, Harford, and Howard—although now with fewer voters than the DC region—still have a strong influence on Maryland politics. Anne Arundel, Carroll, and Harford Counties form the core of the Republican base in Maryland. Anne Arundel County has become a true political battleground. The northern part of the county borders liberal Democratic Baltimore City and contains some older suburbs, while southern Anne Arundel County is more rural, conservative, and Republican. Fast growing, affluent Howard County has steadily become more Democrat in recent years and has supported Democratic presidential candidates in the past few elections. When James Rouse, the famous developer of the planned town of Columbia in Howard County, got his project underway in 1967, he was destined to change a once-rural Republican county into a community of people, jobs, and commerce that is now a strongly Democrat county. It is difficult to guess in hindsight whether Rouse realized that his planned utopia would change the political balance of Howard County. Baltimore County is a different story. Historically dominated by conservative Democrats who readily vote across party lines, Baltimore County is today considered a true swing county, and its political offices are often up for grabs. The county contains many of the older Baltimore suburbs; nationally, older suburbs tend to vote Democrat and are usually on the left of the political spectrum. Baltimore County planners have established the Urban-Rural Demarcation Line, where rural areas outside the line, which are predominantly Republican, are preserved and high-density, older suburbs along with Baltimore City, which are predominantly Democrat, sit inside the line.

Frederick County

Frederick Country is no longer a true western Maryland county when it comes to politics, economics, and demographics. Frederick connects to Washington, DC, by I-270 as well as by commuter rail, and as a result the county has become heavily populated with an influx of newcomers from the Washington area. Though still predominantly Republican, Democrats are steadily gaining ground in Frederick, which has become a political battleground. A noticeable shift has taken place in Frederick County away from its historical alliance with conservative western Maryland and toward the progressive liberal ideas of Montgomery County (*Baltimore Sun* 2008a). The extension of Metro lines and the MARC train from Washington farther out into Montgomery County and toward Frederick County has extended the residential reach of more progressive liberal commuters.

Montgomery County

Montgomery County is the wealthiest and most liberal county in Maryland. Among all counties in the United States, Montgomery is one of the five wealthiest in terms of median family income. A thriving high-tech economy, numerous consulting firms, a population of diplomats, and a large cadre of federal workers all form the political base in this progressive and classical liberal county. People from Montgomery County usually support progressive social causes, including abolition of the death penalty, legalizing same-sex marriage, and gun control. Montgomery County integrated all of its schools only four years after *Brown v. the Board of Education* ruling by the US Supreme Court in 1954, an early indicator of its progressive liberalism. As a group, people from Montgomery County tend to support greater levels of government involvement in the economy, environment, and health care, which is a position that goes against the grain of more conservative parts of Maryland and the United States in general. Yet Montgomery County—the most populous jurisdiction in the state—is a significant force in Maryland politics. The population boom in the county began after World War II with the expansive suburban growth around Washington, DC, which was attracting many people from the outside the area with its many federal jobs. Today, the federal government employs over one-quarter of the county's workforce. With this social and economic profile, it should come as no surprise that Montgomery County is a Democratic political powerhouse in the state.

Prince George's County

Prince George's County is one of the wealthiest majority-black counties in the United States. It is home for a large African American population (65%), most of whom vote Democrat (Lang and Lang 2009). Within Maryland, it is considered one of the most reliably Democratic counties. Unlike Montgomery County, however, Prince George's County is not as progressively liberal. Religion still plays a major role in the county's public life. People here are also more socially conservative than their neighbors in Montgomery County. This county now has the second-largest population in Maryland, and it is a solid and reliable Democrat stronghold. People in Prince George's County are highly dependent on federal employment.

Southern Maryland

The dominating theme in Calvert, Charles, and St. Mary's Counties in recent years has been growth. Just as Frederick to the northwest has become a bedroom community of Washington, DC, so, too, have the counties of southern Maryland to the southeast. Although this region has long been more favorable to Republicans, it is becoming a growing power center for Democrats in the state. In the 2000 presidential election, 49 percent of Charles County residents voted Democrat (that percentage increased to over 62% in the 2008 election). Population growth in southern Maryland has been largely from new residents commuting to Washington, bringing with them their progressive liberal values. Calvert and St. Mary's Counties still usually vote Republican, but with an increasing number of Democratic voters.

Western Maryland

The western Maryland region includes Allegany, Garrett, and Washington Counties. Outside its largest cities of Hagerstown and Cumberland, this far western region of Maryland is overwhelmingly rural, and the people who live here are often described as self-reliant, mountain-country conservatives. Western Maryland's climatic zone is colder than other parts of the state, and its mountains differentiate it topographically from the rest of Maryland. The rugged terrain of the Appalachian Highlands has been a barrier to easy east–west movement, in earlier times limiting the flow of ideas, technology, and values. These physical geographical differences are accompanied by a noticeable difference in attitude among western Marylanders, which is described as "generally socially and economically conservative, but with more of a libertarian streak than is found in other parts of the state" (*Baltimore Sun* 2008a). Over the years, western Maryland has developed a culture that is unique from the rest of Maryland. The geographic pattern that evolved is sometimes called a core-periphery relationship (Hanna 1995). The core-periphery relationship has been present since the beginning of construction on the National Road in 1802. Outsider wealthy industrialists, such as John Garrett, president of the Baltimore and Ohio Railroad, led what many consider an exploitation of the region. Company towns were built and profits were realized from monopolies on the provision of basic needs. Once the railroad was completed, the timber removed, and the coal mined, companies departed, usually leaving behind a despoiled environment. As a result, the population felt alienated and isolated (Lesh 2010; Parks and Wiseman 1990).

It is not uncommon for western Marylanders to express a feeling of detachment from the rest of the state. "The rest of the state thinks Maryland ends somewhere down around Frederick. Maryland don't want us, West Virginia won't have us, and we don't want Pennsylvania" (Meyer 2003). The region's history of corporate exploitation helps to explain its social and economic conservatism with strong antigovernment and antipolitics sentiments. The majority of residents in these three far western counties are staunch Republicans. Feeling ignored by the political power brokers in Annapolis and Baltimore, they associate with the Appalachian culture of western Pennsylvania and West Virginia. During the Civil War, western Marylanders strongly supported the Union, and it has been a solid Republican region since (Brugger 1988). Garrett County has never voted for a Democratic presidential candidate. Western Marylanders as a group generally have a lower standard of living than the rest of the state. They are more rural, have less education, and are primarily white. In Maryland, as in the rest of the nation, these characteristics tend to be qualities of a Republican region (Lesh 2010).

Maryland's Political Landscape

One analyst has likened Maryland's political landscape to a soft-shell crab sandwich. The two halves of the bun are the Republican areas of western Maryland and the Eastern Shore, and in between is the lump crab meat that is the dense area of Democrats from metropolitan Baltimore to the Washington, DC, suburbs (Lesh 2010). The regional patterns described above support this caricature of political Maryland. The four largest jurisdictions in Maryland—Baltimore, Montgomery, and Prince George's Counties along with Baltimore City account for about 57 percent of the total population of the state—all voted Democrat in the 2008 presidential election. In that election, the number of Democratic votes from just these four jurisdictions exceeded the total number of Republican votes statewide.

Klinkner (2004) has defined a "landslide county" as one that gave 60 percent or more of its votes to a particular party. Using his definition, Montgomery and Prince George's Counties as well as Baltimore City were landslides for the Democrats. Democrats traditionally have stronger support in urban areas, while Republicans can usually count on rural areas for strong support. Democrats tend to fare better than Republicans in jurisdictions with higher education levels; in the 2008 presidential election, the counties with the highest numbers of college graduates—Montgomery and Howard—voted Democratic. Anne Arundel and Frederick Counties, with over 30 percent college graduates, voted Republican. Yet Democrats won in Montgomery and Howard

Counties in a landslide, while Republicans won by less than 2 percent in Anne Arundel and Frederick Counties. Anne Arundel and Frederick Counties are battleground counties, as noted above. Interestingly, Frederick County's increase in Democratic votes has coincided with an increase in its percentage of college graduates (Lesh 2010). At the other end of the spectrum, Maryland has six counties with populations where 15 percent or less have graduated from college: Allegany, Caroline, Dorchester, Garrett, Somerset, and Washington. All are either in western Maryland or on the Eastern Shore (in the buns of the soft-shell crab sandwich), and all voted Republican in the 2008 election. In fact, Allegany, Dorchester, and Garrett were Republican landslide counties. One must be careful not to draw the conclusion that Democrats are better educated than Republicans, however. Many residents of Baltimore City and the surrounding older suburbs (a Democratic stronghold) have low educational achievements. Educational achievement is but one socioeconomic piece of Maryland's electoral puzzle.

In Maryland, as in the nation, racial diversity is another factor related to political affiliation and voting. Seven counties in Maryland have white populations greater than 85 percent (Garrett County is 98% white)—Allegany, Carroll, Cecil, Frederick, Garrett, Queen Anne's, and Washington—and all seven voted Republican in 2008. Among them, Allegany, Carroll, Garrett, and Queen Anne's were Republican landslides. The jurisdictions with the lowest percentages of white population (Baltimore City, Charles County, Montgomery County, Prince George's County, and Somerset County) were Democratic landslides, except for Somerset, where Republicans won by less than 3 percent.

In the 2008 presidential election, the states that voted for the Republican candidate had higher percentages of nuclear families, which are generally defined as a family group consisting of a pair of adults and their children. This family structure contrasts with blended families, single-parent families, and other nontraditional households. Nationally, the percentage of families that can be considered nuclear families is 21.4 percent. The pattern in Maryland closely resembles the national pattern. The counties with the highest percentages of nuclear families—Calvert, Carroll, Frederick, Howard, and St. Mary's—all voted Republican in 2008 except for Howard. Among the jurisdictions with low percentages of nuclear families are Baltimore City, Baltimore County, Kent County, and Prince George's County. All voted Democrat in 2008. Kent County on the Eastern Shore went Democratic in 2008, but by a margin of fewer than fifty votes. The relationship between nuclear families and voting patterns is noticeable, but not extremely strong. Maryland does have some Democratic counties with high percentages of nuclear families. Nationally, California and New Jersey rank fourth and fifth by percentage of nuclear families, but both are Democratic strongholds. Still, the impact of family structure on voting patterns is important to consider as another piece of the puzzle.

The percentage of the population that is foreign born is another social characteristic that sometimes relates to voting patterns. About 12.3 percent of Maryland's population is foreign born, almost the same as the 12.5 percentage for the nation. Only three jurisdictions fall above the state level: Howard, Montgomery, and Prince George's Counties. All voted Democratic in 2008. Falling just below the state average of foreign born, but still with high percentages, are Baltimore City and Frederick. Of all these jurisdictions with high foreign-born percentages, only the battleground of Frederick voted Republican. Those counties with the lowest percentages of foreign-born voters all voted Republican. In Maryland, foreign-born populations are a strong indicator of party alliance. Republicans generally do well in areas with the lowest percentages of foreign born (Lesh 2010).

Nationally, Democrats tend to live in areas with high population densities that are served by mass transit, more so than Republicans. In Maryland, 7.2 percentage of its people are mass-transit commuters, led by Baltimore City as well as Baltimore, Howard, Montgomery, and Prince George's Counties. All of these jurisdictions went Democrat in the 2008 election. The counties with the lowest percentages of mass-transit commuters (Allegany, Caroline, Cecil, Garrett, and Kent) all voted Republican except for Kent. This pattern is generally reflective of the strength of Democratic strongholds in high-density urban areas, and Republican strongholds in low-density rural areas.

Maryland is usually glossed over at the national level as just another strong Democratic blue state. Yet the political landscape of the state when viewed at a more local geographical scale reveals a competitive and active puzzle that in many ways reflects relationships found nationwide.

12 Urban Geography

Knowing the size, number, and spacing of settlements on the landscape is an important basis for understanding many human activities. The locations of poultry processing plants on the Eastern Shore and Orioles Park at Camden Yard in Baltimore, for example, relate strongly to their proximity to urban markets. Cities are directly important to the surrounding territory (hinterland) they serve and influence. The extent of these tributary areas varies according to the size and functions of cities. As an example, Oakland in Garrett County is a regional urban center providing important functions for far western Maryland. Oakland is in turn nested within the hinterland of the much larger urban center of Cumberland in Allegany County. Frederick is a larger regional center with a greater tributary area and more functions, and it in turn falls within the greater Washington, DC, metropolitan area. Baltimore and Washington are even larger urban centers with functions that serve the entire state and beyond with their national and global connections. The urban system resembles Russian wooden dolls where multiple small dolls are nested within a larger doll.

From cities radiate public policies, economic decisions, and cultural trends such as music, language, and clothing fashion. Cities contain banks, colleges and universities, libraries, museums, sports teams, commercial establishments, and all the elements that, when added together, direct the development of the landscape. The rural areas of Maryland complement their local urban centers, and the entire urban system of Maryland is focused on the port city of Baltimore. It is in the cities of Maryland that the geographies of production and consumption interlink. Major decision-making institutions are usually located in cities. Traditionally, we connect to the outside world through our cities.

Interurban Relationships

For many years, geographers have studied the structures of regions and the sizes of urban service centers. One general finding is that as the size category (usually measured by population or number of urban functions) increases, the number of settlements decreases, meaning that as we group urban places in Maryland, we should expect to find fewer large cities but many smaller settlements. This pattern is known as the *rank-size relationship*. The number, size, and distribution of settlements on the landscape form a hierarchy of cities. Developed regions such as Maryland have a rather well-balanced hierarchy, with the number of settlements decreasing in a regular manner as the settlement size hierarchy ascends. Less developed regions do not usually display this balanced hierarchical pattern. In those regions there are often many smaller settlements, only a few intermediate size settlements, and then a large primate city at the top of the hierarchy.

It is also common for urbanologists to speak of the *order* of a settlement. In the United States, New York City is a first-order, or alpha, city. Chicago, Los Angeles, and Philadelphia are second-order cities. Washington, DC, and Baltimore, when ranked separately, are third-order centers, but when their metropolitan areas are combined, they form the fourth-largest combined statistical area (CSA) in the nation. Because of its special governmental function as the nation's capital, Washington has an importance and service area that exceed its population size.

The settlements in an urban hierarchy are interrelated in a functional manner. Large centers such as Baltimore and Washington, DC, link directly to other large centers in the nation and to the world by communications and transportation. Strong information flows radiate from these high-order centers. New trends in fashion, music, and technology usually appear first in high-order centers and then diffuse down the hierarchy to towns and rural areas.

Maryland's Settlement Hierarchy

In the settlement hierarchy for Maryland, the only urban place with a population over 90,000 is Baltimore. At the bottom of the hierarchy reside 157 rural settlements. A population size of 2,500 is used to differentiate between urban and rural places.

The smallest settlements in Maryland are *roadsides*, which consist of a few residences and perhaps a service station, motel, or small general or variety store. One example of a roadside, Port Tobacco Village in Charles County, is listed in the 2010 US Census of Population as having fifteen residents. The next-largest rural settlement is the *hamlet*. Many years ago, geographer Glenn Trewartha once described the unincorporated hamlet as "the first hint of thickening in the settlement plasma; it is neither purely rural nor purely urban, but neuter in gender, a sexless creation midway between the more determinate town and country" (Trewartha 1943). There are a large number of hamlets across Maryland, and the common functional factor for each is usually a rural post office. The *village* has more functions than a hamlet. In addition to the functions found in hamlets, villages typically have an automobile dealership, hardware store, drugstore, church, and elementary school. The largest rural settlement, the *town*, offers additional functions, usually including a bank, high school, and professional services (doctor, lawyer, and dentist).

Smaller urban places service a more local, rural hinterland. A number of intermediate-size cities are regional business service centers. Others are transportation centers, special function centers (e.g., serving government administration or military installations), or suburbs of Maryland's highest-order center, the *metropolis*. Maryland's large metropolitan milieu is a complex system of industrial parks, high-tech centers, residential areas, cultural amenities, and shopping malls, all alive with vitality but also throbbing with traffic jams, pollution, crime, and decay. This is the dilemma of Maryland's urban centers—they simultaneously represent society's hopes, accomplishments, and creativity and also reflect its despair in the crime, poverty, and breakdown of many values that have defined Maryland culture over the centuries.

Urban Patterns

Maryland's urban system has a pattern with three distinct characteristics: a dominating concentration stretching from Baltimore to Washington, DC; a number of smaller cities around the state serving as regional centers; and many smaller towns and villages in the interstices serving local areas.

The large Washington–Baltimore metropolitan complex is a tangle of industrial plants, transportation routes, townhouses, office buildings, education facilities, malls, affluence, poverty, and other attributes of contemporary urbanism. The increase in sprawl from these large cities has been rapid. The urban corridor has many types of economic activities ranging from Baltimore's shipyards to the many new high-tech industries in Montgomery County. As recently as 1990, a well-defined rural hinterland surrounded the smaller regional centers in Maryland. But today many of these areas are also experiencing their own sprawl and suburbanization. Suburbs now surround the regional center of Frederick, which today is considered part of the Washington metropolitan area. Salisbury continues to sprawl across Wicomico County. In places such as Frostburg, Hancock, Oakland, and Westernport, however, the urban–rural interface is still somewhat more pronounced.

Population Distribution and Urbanization

As in the entire United States, the population of Maryland has steadily become more urbanized. The US Census of Population of 1920 first reported the population of the United States as being majority urban, at 51.2 percent. Maryland nearly reached majority urban in 1900, when its population reached 49.8 percent urban. The 1910 census officially made Maryland a majority urban state (50.8% urban). By the 2010 census, Maryland had become about 95 percent urban.

Population density is a rather simple value found by dividing the population of an areal unit by its total land area. Although useful for generalizations, simple population density has some serious shortcomings. Population density assumes an equal spread of people across the land surface of the geographic unit. The overall US population density in 2010 was 83.8 persons per square mile. New Jersey had the highest state density with 1,175.3 persons per square mile. Alaska had the lowest with 1.2 persons per square mile. But even within New Jersey there are extensive nonurban areas with very low population densities, and Alaska has some areas with much higher densities than the overall state figure. Density tells only part of the story. For a more refined picture of population, one must more specifically describe the distribution within a state.

Population Concentration Index

The population concentration index (CI) determines how much of a state's population needs to shift among the counties to give each county the same population density. The higher the CI, the more concentrated the population (i.e., the more people need to shift to create an equal density in all counties; Marcus 2006).

The CI can be calculated as:

$$CI = \frac{\sum_{i=1}^{n} (\text{abs}\,(\%POP_i - \%AREA_i))}{2},$$

where $\%POP_i$ is the population of county *i* as a percentage of the total Maryland population; $\%AREA_i$ is the land area of county *i* as a percentage of the total Maryland land area; and $\sum_{i=1}^{n}$ is the sum of the absolute values of these differences. This value, when divided by 2, equals CI, which may be multiplied by 100 to find the percentage.

For all twenty-three counties plus Baltimore City, the CI in 1900 equaled 42.5% (fig. 12.1), meaning that 42.5% of Maryland's population would have to shift to create an equal density across the state. The CI peaked at 56% in 1970, and declined thereafter to 47.5% in 2005.

For all twenty-three counties but excluding Baltimore City, the CI in 1900 was 21.5%. It peaked at 45% in 1970, declining to 42% in 2005. The significantly lower CI in the data set excluding Baltimore City reflects its large, dominating population size and the strong role it plays in the distribution of residents throughout Maryland.

For the United States as a whole, the population has gradually become less concentrated since 1900. Because entire states are the units of measurement at the national level, this pattern reflects the general southwestward spread of the US population during the twentieth century. The movement of the geographic center of population west and southwestward since the first US Census in 1790 also reflects this spatial demographic process. What was once a national population concentrated on the East Coast is now a population still dominated by

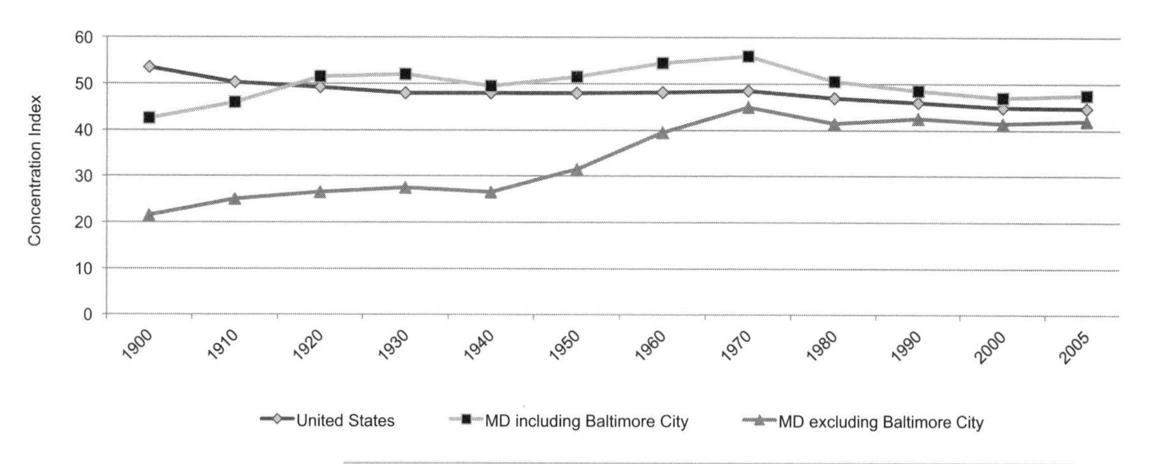

Figure 12.1 Population Concentration Index

megalopolis but with other significant centers of concentration to the west.

The population CI for Maryland presents a pattern that differs from the national pattern. Counties are the units of measurement at the state level, although smaller units within counties could also be used. As the areal units of measurement get smaller (more local) and more numerous, the data more closely reflect individual people, a process that "unmasks" data aggregated in the much larger spatial units and demonstrates the critical role that geographical scale plays in calculating the CI.

Considering all twenty-three counties and Baltimore City, the CI shows a steady increase in the concentration of Maryland's population from 1900 to 1970. Thereafter the CI begins to decline. The increasing CI during the first period describes an increasingly urbanized Maryland. The dip during the 1930s occurred during the Great Depression, when more people stayed in rural areas, and some urbanites even moved to the rural areas of Maryland. The lack of jobs in the cities and towns of Maryland during the Great Depression temporarily slowed down the migration from rural to urban as well as the concentration of the state's population. After 1940, people began to move to the urban centers of Maryland once again. The onset of World War II brought with it many urban jobs, especially in the defense-related industries of Baltimore, and in the defense, service, and administrative sectors in the Washington, DC area.

The decline of the CI after 1970 reflects accelerated suburbanization and even the movement of people to rural areas within commuting distances of urban centers. The construction of the US Interstate Highway System played a strong facilitating role in this spatial reorganization of Maryland's landscape. Its effects on the landscape kicked in noticeably during the 1970s. The third trend line in figure 12.1 shows the CI excluding Baltimore City. Because Baltimore City is so much larger than the next tier of cities in Maryland, it is useful to see how the population concentrated in the rest of the state. Although the CI is lower, the trend line also increases from 1900 to 1970, with a decline and leveling-off after 1970. The slight decline in urban Maryland outside Baltimore City during the 1930s is also discernable.

What we can conclude from this analysis is that Maryland has a population that continues to urbanize. Given a fixed land area and a growing population, the population density continues to increase. The concentration of people in Maryland in the large urban centers peaked in 1970 and has since decentralized to smaller urban centers, suburbs, and exurbs. Maryland remains a highly urbanized state within the huge megalopolitan conurbation on the East Coast of the United States.

Planned Towns

The building of new, planned communities is an old concept that came to the United States from England. Two of the largest such communities in Maryland are Columbia, the most recent new town in Maryland, and preceding it Greenbelt in Prince George's County.

Greenbelt

In 1935, President Franklin Delano Roosevelt created the Resettlement Administration. Its Suburban Resettlement Division was responsible for the planning and development of three new towns: Greenbelt, Maryland; Greenhills, Ohio (5 miles or 8 kilometers north of Cincinnati); and Greendale, Wisconsin (7 miles or 11.3 kilometers from Milwaukee). The planners sited Greenbelt on a 3,370-acre (1,365-hectare) tract of land in Prince George's County, just 13 miles (21 kilometers) from downtown Washington, DC. The land was second-growth woodland and worn out tobacco farmland.

The plan for Greenbelt can be largely credited to Ruxford Tugwell, head of the Resettlement Administration. Intended as low-cost housing for low-income families, Greenbelt incorporated elements of Ebenezer Howard's English "garden city" as well as features from another new town, Radburn, New Jersey. Intended to be a garden city surrounded by belts of green space, Greenbelt was built on a natural ridge and combined a community center in close proximity to an elementary school and recreation facilities. The town was designed for walking with minimal interference from automobile traffic. Plentiful sidewalks and several underpasses under the main roads were also built.

> The income of the new residents ranged from 800–2000 dollars. Roughly sixty-three percent of the original residents were Protestants, thirty percent Catholic, and seven percent Jewish. No blacks were allowed, nor were women to work outside of the home. Greenbelt, as a unit, valued family and recreation. Out of necessity, it also seems to have valued economy. (Greenbelt Museum 2013)

In 1937, Greenbelt was incorporated as a town, and the first hand-picked residents moved in. The religious mix was intended to reflect the mix in the nation as a whole. Although built to accommodate low-income families, Greenbelt was not a racially integrated community. By 1940, Greenbelt had slowly grown to 2,831 residents. With heavy employment needs in the capital's defense agencies during World War II, a housing shortage developed. In 1941, builders erected about 1,000 new frame row houses in Greenbelt. These houses were not well planned; they faced streets, had narrow setbacks, and lacked interior parks and playgrounds. Between 1952 and 1954, the federal government sold Greenbelt to a housing cooperative and developers. A tract of 1,362 acres (552 hectares) was deeded to the US National Park Service and was developed as Greenbelt Regional Park. Little subsequent residential construction took place until after 1960.

In 1954, the Baltimore–Washington Parkway was completed, including a four-way exit at Greenbelt that generated heavy traffic on Greenbelt Road connecting into US Route 1 at College Park, along which sprang strip commercial establishments and a regional shopping center. In 1964, the Capital Beltway (Interstate 495) was completed with two exits for Greenbelt. With this new accessibility, Greenbelt became a prime area for development. Starting in 1962, developers built many new garden apartments, and by 1967 the population had doubled. The original Greenbelt idea disappeared in the new areas being built around the original center. Greenbelt today is no longer the low-income town originally planned. Major employers nearby include the NASA Goddard Space Flight Center, National Security Agency, and University of Maryland.

The green space sold to developers in 1954 has since been developed, and major roads now trisect and fragment the town. Acres were split into minimum-size lots with minimum houses built on them, and wooded acres were bulldozed into submission. Yet historic Old Greenbelt remains remarkably intact. The original planned community of residences still sits in a parklike setting near the old town center. Strong community efforts for historical preservation are largely credited for preserving the historic heart of the town (Shprentz 1999). Old Greenbelt is the core around which the modern community has been built, and it has matured into an attractive suburb of Washington, DC.

Columbia

The second major planned community in the Washington–Baltimore corridor is Columbia in Howard County. After having observed with dismay the unplanned development in the Washington–Baltimore corridor, developer James Rouse decided to provide a better alternative. Born and raised in Easton, Talbot County, Rouse wanted to create a community with the neighborly life he knew there. In the early

GREENWAYS

One of the most exciting topics in recreation planning since the early 1990s has been the development of greenways. These protected corridors of open space allow for various approaches to both park planning and land conservation. Greenways in Maryland come in many shapes and sizes, each with its own unique character. A greenway can be a protected creek bed, a forested corridor, a ridgeline, a stream valley park, or a converted railroad or utility right-of-way (Maryland Department of Natural Resources 2000).

An example of a converted railroad right-of-way is the popular 20-mile (32-kilometer) Northern Central Rail Trail (NCRT) running from the Ashland Road area in Hunt Valley, Baltimore County, northward into Pennsylvania. Opened in 1984 along the 152-year-old Northern Central Railroad right-of-way, the NCRT lies within the boundaries of Gunpowder Falls State Park. The NCRT has historical significance. In 1861, president-elect Abraham Lincoln traveled its route on his way to Washington, DC. During the Civil War, it was an important rail link, carrying troops, food, and supplies to the Union Army. In April 1865, the body of the assassinated President Lincoln was carried along the NCRT on its way from Washington to Columbus, Ohio. The 13-mile (21-kilometer) Baltimore and Annapolis Trail in Anne Arundel County is another converted rail right-of-way in the heavily populated central Maryland region.

The Maryland Greenways Commission was established in March of 1990 with the responsibility of coordinating the efforts of several public agencies and programs to create a statewide system of greenways in every region of the state. By 2000, over 900 miles (1,448 kilometers) of greenway corridors existed in Maryland. Other corridors are in the process of being established, with state officials identifying about 1,000 miles (1,609 kilometers) of potential greenways. The Maryland greenway network includes long-distance trails such as the 184-mile (296-kilometer) Chesapeake and Ohio (C&O) Canal National Historic Park, an unpaved trail along the former towpath of the C&O Canal. The trail runs from Georgetown in Washington to Cumberland in Allegany County, where it ends at the Western Maryland Railroad Station, now a visitor's center. The National Park Service owns and operates the C&O Canal path. Another long-distance greenway in Maryland is a section of the Appalachian Trail, which runs from Maine to Georgia.

The greenways of Maryland reside in its cities, suburbs, and rural areas. Although most of the state's greenways are public land, many greenways also incorporate some private land. The Department of Natural Resources has designed some of Maryland's greenways to be pristine corridors for the protection and movement of plants and animals, and these areas are not usually open for human recreational use or access.

Greenways can be established in a number of ways. Land can be donated, purchased with public funds, or made available with a right-of-way procurement. Program Open Space (POS) provides state funding to acquire greenways as well as other recreation and open-space needs, such as parkland, forests, and wildlife habitat. POS provides 100 percent funding to buy land needed for state, county, and city parks. POS also provides up to 75 percent of the development costs of these projects. Maryland's real estate transfer tax funds this program; one-half of 1 percent of the purchase price of a home or land is placed into a fund for POS. Since being established in 1969, POS has funded the purchase of land for more than 5,000 county and municipal parks and conservation areas. The law stipulates that Maryland's counties should use their local share of the transfer tax revenue to set aside 30 acres (12.14 hectares) of land for parks and recreation for every 1,000 people (Wheeler 2006). Not all counties have met this goal, as the program has not always allowed Maryland to keep pace with development and population growth. The fund has been diverted at times to help balance the state budget. When property values increase, so does the cost of acquiring open space. The greatest needs for parks and other open-space areas are generally in the most developed areas, where land prices tend to be the highest. Maryland's population grows every year, but the amount of land remains the same. About 20 percent of Maryland is developed, and about 19 percent is protected with conservation easements or publicly owned open space. POS has generally been good for business, as hiking and biking greenway trails generate jobs and income. Home values are generally higher around parks and protected spaces. New businesses, especially in the service sector, prefer to locate in places with parks and high-quality environments. Tourism is big business in Maryland, and quality open-space and recreational facilities attract tourists. State funding for greenways also comes from the Maryland Department of Transportation's Enhancement Program, which provides a 50/50 match with local funds. Access to agricultural land is possible through the Agricultural Land Preservation Foundation, whereby farmers sell development rights on their land by establishing an agricultural preservation district, and then sell the easement back to the foundation (Maryland Department of Natural Resources 2000).

Maryland will continue to develop its greenway network well into the future, but it will require the cooperation of public agencies and private interests as well as the many jurisdictions of the state. Many of the existing greenways already cross several jurisdictions. Regional greenways include the Lower Susquehanna Heritage Greenway, Patapsco Regional Greenway, Patuxent Regional Greenway, and Pocomoke River Greenway. Future plans for interstate greenways include the Allegheny Highlands Trail, Marshyhope Ponds Greenway, Nanticoke River Greenway, Potomac River Greenway, and Tri-State Greenway. All of these projects presently involve interjurisdictional groups that are working with federal, state, and local agencies in a cooperative effort.

Maryland will also be part of two national greenway trails that are still in the planning stage. The East Coast Greenway (ECG) is a developing trail system that will eventually span nearly 3,000 miles (4,827.9 kilometers) as it winds its way between Canada and Key West, linking all the major cities of the Eastern Seaboard. Over 25 percent of the route is already on safe, traffic-free paths. The ECG will be the nation's first long-distance, interurban, multiuser transportation trail network. It will be a city-to-city trail system connecting existing and planned trails with new corridors that will incorporate canal towpaths, parkway corridors, abandoned railroads, waterfronts, and parks. The first part of the ECG runs from Boston to Washington, DC. The American Discovery Trail (ADT), another long-distance trail project, is also currently being planned. It will run east-west across the United States and incorporate trails, parks, wilderness areas, and historic-cultural routes to form a continuous trail. Various established trails and greenways across Maryland will be part of the ADT, including the Anacostia Stream Valley Trail, C&O Canal, and Fort Circle Trail.

Although Maryland has been a pioneering and proactive state when it comes to establishing green infrastructure, it is no reason to become complacent. Before Europeans colonized Maryland in the early seventeenth century, the land was about 95 percent forested with the other 5 percent as tidal marsh. By 1993, both of those figures had reduced by half. Suburbanization and development on large house lots consume excessive amounts of land, much of it formerly rural. This development often occurs in a scattered geographic pattern that fragments the landscape—forests are divided and parcels isolated by roads, houses, and malls. Wildlife habitat and migration corridors are lost, and normal ecosystem functions such as absorption of nutrients, recharging of water supplies, and replenishment of soil are disturbed or destroyed (Weber 2003). At least 180 plant and thirty-five animal species have disappeared from Maryland, and another 310 plant and 165 animal species are endangered. It was estimated in a 2003 study by the Maryland Department of Natural Resources (Weber 2003) that today Maryland has only 2 million acres (809,389 hectares) of ecologically significant land that has not been consumed by human development. Of these 2 million acres of green infrastructure, about 70 percent is unprotected. Maryland's green infrastructure is under severe pressure from development. Without a vigorous effort by both the public and private sectors, the remaining vulnerable green space will be further reduced and fragmented.

1960s, Rouse built the Village of Cross Keys—a mixture of high-rise and garden apartments with a village square surrounded by upscale stores—in Northwest Baltimore. It was a step toward the conceptual development of Rouse's goal for Columbia—a planned community designed to enhance the lives of the people in it and to earn a profit at the same time (Dozer 1976).

The Rouse Company planners found that in the 1960s on the Eastern Seaboard, there were just two areas bracketed by the dual pressures of two major metropolitan centers less than 50 miles (80.5 kilometers) apart: the corridors of Providence-Boston and Washington-Baltimore. Because no sizeable tract of land was available between Providence and Boston, Rouse turned to the open land of Howard County in Maryland. Quietly buying land piece by piece (149 parcels), Rouse acquired 15,200 acres (6,156 hectares) by 1964 (Moryadas 1968). Rouse purchased the parcels in secrecy so as not to alert local landowners or to cause the price of land to skyrocket. Rouse borrowed much of the money from Connecticut General Life Insurance Company to finance the project, paying an average price of $1,500 per acre. Eventually, the Rouse Company owned 10 percent of Howard County. In October 1963, Rouse informed Howard County commissioners that he owned a good part of the county and wished to build a new town on it. He argued that it would be impossible for Howard County to remain rural forever and that it was a good alternative to the uncoordinated growth experienced in other parts of Maryland. Throughout 1964, Rouse traveled around the county presenting his plans to residents (Pickett 1964). Finally, in August 1965, the Howard County commissioners voted to pass the New Town District Zoning Ordinance. One year later, construction on Columbia began.

Community Research Development, the developers of Columbia, held a series of seminars with experts in transportation as well as behavioral and social scientists to formulate their community concept. The group agreed that the nucleus of the residential community would be the neighborhood, consisting of 500 to 600 families grouped around a lower school, a convenience store, nursery, tot lots, and other neighborhood facilities. Five or six neighborhoods would constitute a village of 2,500 to 3,500 families. The village center would feature a library, higher-order shopping establishments, and junior and senior high schools. A cluster of villages would constitute a town. The town center would be the focal point of the area with a concentration of urban functions, including office buildings, department stores, theaters, a hospital, shopping center, and other units.

In 1970, Columbia had only 8,800 residents. It grew to 53,000 by 1980 and 76,000 by 1990. By the mid-1970s, some 20,000 people had jobs in Columbia and 700 businesses had located there. By 2006, Columbia had about 95,000 residents, about 35 percent of the population of Howard County. Columbia is the largest unincorporated place in Maryland (only Howard and Baltimore Counties have no incorporated places). Businesses continue to flock to Columbia and surrounding areas in Howard County. Since the late 1970s, numerous industrial parks have been built. By 2006, there were fifteen office, industrial, and research parks in Columbia. Although many Columbians commute to Baltimore or Washington, DC, more are now employed in Columbia and at nearby federal agencies such as Fort Meade and the National Security Agency.

The median household income of Columbians in 2010 was $93,801, significantly above the Maryland statewide median household income of $70,050, which is among the highest median household income in the United States, statistically at the top of the heap alongside New Jersey and Connecticut. Over 90 percent of Columbians have a high school diploma or higher degree, 59 percent have bachelor's degrees, and 29 percent have graduate degrees. Columbia has a racial mix of residents, too: white (non-Hispanic), 64.4 percent; black, 21.5 percent; Hispanic, 4.1 percent; two or more races, 2.8 percent; Asian Indian, 2 percent; Chinese, 1.9 percent; Korean, 1.8 percent; American Indian, 0.8 percent; and others, 2.1 percent. Columbia has an innovative school system with higher education available at Howard Community College in addition to branch campuses of Antioch College, Johns Hopkins University, and Loyola College, among others. Although Columbia is an attractive place to live, it is not utopia. There are growing problems of crime, integration, and self-government. Relative to other places in metropolitan Baltimore, however, Columbia still stands out as a good place to live.

The Washington-Baltimore Metropolis

Throughout most of the history of the United States, the neighboring cities of Baltimore and Washington, DC, were distinct entities separated by rural open space. But both cities slowly sprawled into their adjacent countryside, inevitably coalescing into one metropolitan area.

After the 1990 census, the cities of Washington and Baltimore and their suburbs were designated as the Washington-Baltimore consolidated metropolitan statistical area (CMSA). Geographical areas given the CMSA federal designation by the US Office of Management and Budget have a high degree of economic and social integration with overlapping labor markets. The Washington-Baltimore CMSA includes the core cities of Washington and Baltimore, and also Northern Virginia, central Maryland, one county on Maryland's Eastern Shore, and two counties in the eastern panhandle of West Virginia. After the 2000 census, this region was designated as the Washington-Baltimore-Northern Virginia combined statistical area (CSA). The population of this entire CSA reached 8,570,632 in 2010, making it the fourth-largest CSA in the United States. In addition to the Washington-Baltimore-Northern Virginia CSA, there are also the Cumberland metropolitan statistical area (MSA); Hagerstown-Martinsburg MSA; part of the Philadelphia-Camden-Vineland CSA (which includes Cecil County, Maryland); Salisbury MSA; and the micropolitan areas of Easton, Cambridge, and Ocean Pines. Garrett County is the only county in Maryland not defined as being part of any metropolitan area.

Baltimore City

Baltimore figures prominently in Maryland's historical and contemporary geography. While its importance to the economic and cultural life of Maryland cannot be denied, Baltimore's shortcomings need to be more clearly defined, understood, and addressed. Creative leadership from politicians, businesspeople, educators, and all Marylanders must be forthcoming if Baltimore is to—at a minimum—mitigate the severity of its ills. Although Baltimore has much in common with other North American cities (commercial activities, large size, various zones of activity, and functions), it has its own rhythm, pace of life, physical site, and climate, all of which give it a unique character. Baltimore is a dynamic urban organism whose morphology has changed greatly over the past fifty years. Some years ago, geographer Sherry Olson described this city at the individual human scale:

> the Baltimore economy not only produces steel for the North Atlantic community, and electric drills for the nation, distributes cars and groceries to six states, and fries chicken for Baltimoreans, but as a byproduct it produces Baltimore society—the class structures of consumption and the geographical structures of distribution. It produces in some measure a "Baltimore person," with a sense of competition and vulnerability, and an awareness of the limits on his ability to control his own personal life. (Olson 1976)

Areal Growth

Baltimore has grown as the metropolis of Maryland to become one of the largest cities in the United States. Originally part of Baltimore County, city and port grew rapidly in the early nineteenth century, and by 1830 Baltimore was the third-largest city in the country. In 1851, Baltimore City separated from Baltimore County to become an independent jurisdiction. Today it is an independent, incorporated city with a legal status equivalent in most ways to each of the twenty-three counties of Maryland.

The size and shape of Baltimore City in 1851 was significantly different from its much larger contemporary areal extent, and its growth did not occur without conflict. The transformation of Maryland from a predominantly rural-based to urban society has been a fundamental and central historical and geographical theme since the eighteenth century. This process resulted in sharp conflicts between jurisdictions, most notably between Baltimore City and Baltimore County. As the city began to grow rapidly in the early nineteenth century, a suburban ring developed on its three landward borders with the rest of Baltimore County. Anxious to expand its municipal tax base and political power in Annapolis, Baltimore City sought to acquire suburban land, resulting in several annexations right up until 1918.

Baltimore Town started out as a 60-acre (24-hectare) tract of land in 1730. In 1745, it added 10 acres (4 hectares) with the acquisition of the adjoining village of Jonestown, and the Maryland General Assembly approved twelve more annexations between 1745 and 1783. Because the annexed parcels were small and did not shift political power, their incorporation into the city was not controversial. By the late eighteenth century, Baltimore City's taxes exceeded those in the rest of the county, so further annexations (by consent) ceased for a time. Although still considered part of Baltimore County, Baltimore City was an independent municipality. By 1797, Baltimore City had a mayor and city council; Baltimore County commissioners had almost no taxing power within the city.

By 1816, the "precincts" (suburban areas adjoining the city) had about 12,000 residents who accounted for one-third of Baltimore County's population and over 40 percent of the county's total real estate value (Coyle 1909). Baltimore City attempted to capture this wealth, land, and population through a petition to Annapolis to annex over 13 square miles (33.67 square kilometers). Like Baltimore City at the time, the precincts were heavily Republican. Baltimore County Federalists could easily capture all four of the county's state delegate seats if they could rid themselves of the precincts, so Baltimore County supported the annexation of 1817, which did not alter the political balance because Baltimore City was limited to two delegates regardless of its population size (Arnold 1978).

Once again, the population in the suburbs immediately surrounding the city—known as "the Belt"—began to grow. Conflict between city and county became so intense that in 1851 Baltimore City separated from Baltimore County, whose seat of government moved from the city to the small village of Towsontown, 7 miles (11.27 kilometers) north

(Arnold 1978). A few years later, Baltimore County became temporarily immune from further annexations because the Maryland State Constitution of 1864 did not allow the transfer of land from one county to another without the consent of the residents of the area to be shifted. Baltimore County assumed that Baltimore City was equivalent to a county under this law.

Public services in the Belt were notoriously poor. Fire and police protection were inadequate, as were water and sewer lines (if they existed at all). Schools in the Belt were so bad that many Baltimore County residents paid a fee to send their children to Baltimore City schools. Residents of the Belt—mostly from the heavily populated Eastern District that included Highlandtown and Canton—defeated an annexation attempt by the city in 1874, but the city's 1888 annexation bid was successful. This time, each district of the Belt (Eastern, Northern, and Western) voted separately. Only the East voted not to join the city. Baltimore City had gained another 7.5 square miles (19.43 square kilometers) and about 38,000 people from Baltimore County. In the 1890s, newly electrified streetcars ran from Baltimore City out into the county. In the early twentieth century, roads were improved to accommodate automobiles. A new suburban belt grew to the north and west of the city, and to the east, the populations of the industrial areas of Highlandtown and Canton grew rapidly. The nearly 35,000 people in this district lived without sewers. Police and fire protection remained poor, and the schools were a health and fire hazard (Arnold 1978). The people of the Eastern District asked Baltimore County for a bond issue in 1914 to improve its infrastructure and services, but up-county farmers and more prosperous middle-class suburbanites defeated the bond. After considerable further political conflict between the city and county, the 1918 annexation passed. Baltimore extended its boundaries to the east, north, and west. The area of Baltimore City increased from 32.05 to 91.03 square miles (83 to 236 square kilometers). Highlandtown was now a neighborhood of Baltimore City, even though for many years its residents referred to the rest of Baltimore City as "West Highlandtown." Although the Great Depression and World War II nearly halted further suburban growth, it rapidly resumed in 1945. Owing to new laws and the new political balance between the city and county, no additional annexations have occurred since.

The suburban population has continued to grow and spill over into the four counties adjoining the city, as well as into Carroll County and even Queen Anne's County on the Eastern Shore. The clear urban-rural separation of the eighteenth century has long disappeared. The fingers of urbanization now reach out from Baltimore City across what is today called the Baltimore–Towson MSA, embracing 2,710,489 people in Baltimore City and six counties.

Neighborhoods and Diversity

During the 1980s and 1990s, residents were fleeing Baltimore City in record numbers. Baltimore City's first response was to try to achieve fiscal stability rather than to focus on the reasons people were leaving (e.g., deteriorating infrastructure and quality of life). But in order to address to these challenges, Baltimore City needs fiscal stability concomitant with long-term investments in its future (O'Malley 2000). Today, Baltimore City's most critical quality-of-life issues are crime, drug addiction, vacant houses, and substandard public schools. The city's success in addressing these issues depends on a long-term commitment to a comprehensive social and economic development strategy that integrates various initiatives.

Baltimore City has lost almost one-third of its population since 1950, even though the city does attract new residents and some areas are growing. The phenomenal commercial, residential, and retail development in Canton in recent years is one example of a Baltimore neighborhood that is courting new residents. Historic buildings in neighborhoods such as Arcadia, Ashburton, Marble Hill, and Roland Park draw new residents, too. The revitalization of Fells Point and Federal Hill provides yet more proof that neighborhoods can turn the tide and attract people. Many of Baltimore's problems are concentrated in the inner core of the city. In 2010, it was estimated that Baltimore City had about 14,000 vacant and uninhabitable properties and 4,000 vacant but habitable properties, many of which are in poor physical condition. Most of these are row houses clustered in the inner city, an area with a weak housing market and where the social fabrics are severely distressed. When all of these factors are combined, they lead to a weakened community confidence, areas unattractive to new residents, and disinvestment (O'Malley 2000).

The basic geographical community unit of Baltimore City is the neighborhood, and there are about 131 of them in the city. Using US Census data, Johns Hopkins University maintains the online Historical Census for Baltimore City, which gives detailed social and economic information by neighborhood for the years 1970, 1980, 1990, and 2000 (http://webapps.jhu.edu/census/). These data document the wide range of diversity across the neighborhoods of Baltimore City for the period 1970–2000 (e.g., Guilford in north-central Baltimore is 89% white and only 6% black, while Mondawmin in West Baltimore is 99% black). Homeland in north-central Baltimore, just west of York Road, has a median family income of $103,570. Just to the south, along York Road and Greenmount at North Avenue, the neighborhood of Waverly has a median family income of only $17,650. In Canton in Southeast Baltimore, 11.24 percent of the people live below the federal poverty line. Cedarcroft (nested within Chinquapin) in north-central Baltimore has

0 percent below the poverty level. Although Baltimore neighborhoods have continued to change since 2000, Johns Hopkins University data reveal macro geographical patterns that are still useful.

Social and economic contrasts reflected in these statistics are visible on the Baltimore landscape. There is a definite pattern of racial diversity and concentration in Baltimore (fig. 12.2). From the concentration of African Americans in the city core, a major prong stretches northwest along Liberty Heights Avenue into Baltimore County. Another major prong stretches east along Pulaski Highway, and a third extends north between York Road on the west and Belair Road on the east. The western border of this last prong is sharp and shows up as the division between North Baltimore-Guilford-Homeland on the west and Chinquapin Park-Belvedere on the east, generally coinciding with York Road and Greenmount Avenue. To the west of the divide, the population is heavily white and prosperous. Just across the road to the east, the population is predominantly African American with lower incomes.

Geographically, Baltimore has little racial diversity east of Greenmount Avenue along North Avenue. Another area of low diversity stretches to the northwest along Liberty Heights Avenue. The pattern

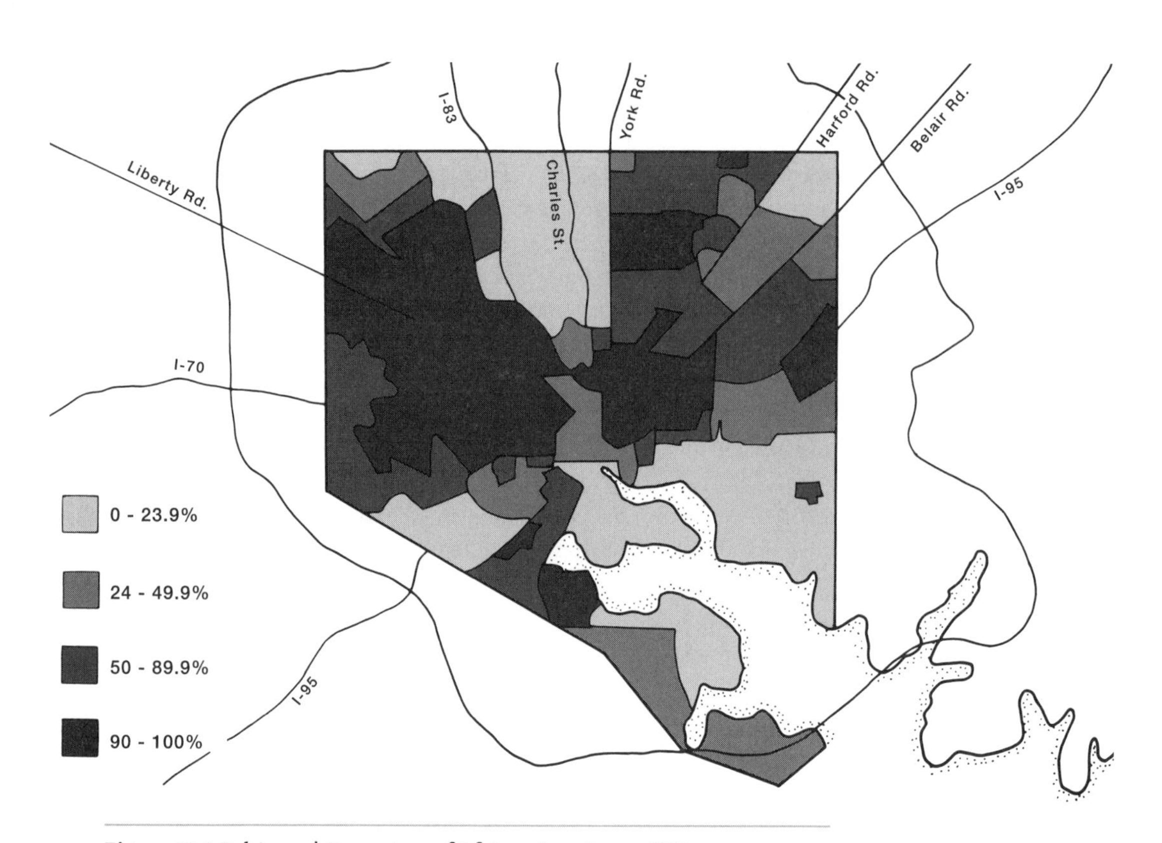

Figure 12.2 Baltimore's Percentage of African Americans, 2010

of low diversity corresponds closely to the geographical pattern of African Americans in figure 12.2.

Percentages of families living below the federal poverty level follow a predictable pattern. The high percentages in the city core and stretching northwest correspond with the dense African American population. The low poverty rates along the northwest boundary correspond to a population that is predominantly white and more prosperous. Low poverty rates stretching to the northeast correspond with racially diverse middle-class neighborhoods. The geographical pattern mirrors previous patterns for other socioeconomic variables. Better maternal and child health (which is determined by three variables—available prenatal care for mothers, birth delivery at full term, and birth weight) in the northwest is expected. The poorer maternal and child health in the east-central core and in the Cherry Hill neighborhood in the south-central city correspond to a poor, heavily African American population, but with some racial diversity.

All of these spatial patterns reinforce each other, and the emerging geographical pattern of prosperous and underprivileged areas is distinct. Collecting socioeconomic data and mapping them are a major part of what various state and city planning agencies do. Although the process is time consuming and costly, the collection and interpretation of these kinds of data are critical. Yet it is not the end product. The real challenge is to use this information to find solutions, formulate effective policies, and then fund and effectively implement those policies.

Ground Rent

One economic issue that is centered on Baltimore but has wider social implications is ground rent. Parts of Maryland—especially in Baltimore City—still cling to an arcane system of property ownership where the land is owned by one entity and the buildings by another. The origins of ground rent in Maryland go back to 1632, when King Charles I approved the Maryland land grant to Cecilius Calvert, the First Lord Baltimore. Calvert's agents in Maryland collected rent from colonists who were allowed to build on his lands. Calvert, in turn, paid an annual land rent to the king in the form of two Indian arrowheads and one-fifth of any gold and silver found in Maryland. Since no silver and no significant gold were found, the rent remained largely symbolic. After the country won its independence from England, the Maryland legislature continued the practice of ground rent by empowering landholders with the right to collect land rent in the form of cash or trade. Because cash was scarce, rent could be paid in kind, often with tobacco or labor. Although some landowners accepted a token rent, the overriding principle of ground rent survived.

As Baltimore was growing rapidly and taking its place among the large East Coast cities as an industrial center during the nineteenth century, the demand for housing increased dramatically. Developers responded by building miles of red brick row houses with their hallmark white marble stairs. To reduce building costs and to make these homes affordable to working-class families, developers leased the land on which houses were built. Minimal ground rents were paid to the ground-rent holder, usually the developer. After World War II, when many GIs returned home to Baltimore, ground rent continued as a means of providing affordable housing to more people. Although opposition to ground rent increased in the 1940s, the practice was mainly seen as a positive tool. An attempt to introduce legislation in Annapolis in 1959 to phase out ground rents failed. Ground rents spread into surrounding counties, and today they are still found in parts of Anne Arundel and Baltimore Counties.

Ground rents are usually ninety-nine-year leases that can be renewed. Payments are usually semiannual and in most cases range from $50 to $150; the amount is stated in the original property lease. The law does not allow the ground rent or the redemption rate to be changed. Maryland law requires ground-rent owners to redeem (sell) the ground rent to the owners of the real estate (property) if requested. The redemption amount is determined by taking the annual ground rent and multiplying it by a redemption rate; the rate depends on the date the lease was created.

By the early twenty-first century, ground rents in Baltimore City had led to many problems. Because current ground rents in some cases date to the nineteenth century, real estate owners must often identify and search for the ground-rent owner, which is not always easy because Baltimore's system of record keeping has been poor. Real estate sellers who cannot easily find the ground-rent owner can go to the recorder of deeds in Baltimore City, where the information must be looked up manually. Real estate owners who ignore a ground rent that is indicated on the original lease do so at their own peril. If a homeowner cannot locate the ground-rent owner, the safe approach is to establish an escrow account with a bank. By Maryland law, the time limit for collecting back-ground rent is three years. In recent years, some individuals and groups were acquiring ground-rent ownerships of numerous properties in Baltimore City. When a real estate owner does not pay ground-rent payments, for any reason, the law is on the side of the ground-rent owner. Between 2000 and 2006, there were about 4,000 lawsuits filed in Baltimore City by ground-rent owners seeking either possession of homes or thousands of dollars in fees. In over 500 cases, judges decided in favor of ground-rent holders. As property values increased, it became increasingly attractive for unscrupulous ground-rent holders to attempt to seize homes and sell them for a profit, the homeowner in turn losing everything. In 2006, the number of ground-rent lawsuits increased by 73 percent, mainly in gentrifying neighborhoods such as Patterson Park and Washington Village (which is sometimes called "Pigtown") with increasing property values (*Baltimore Sun* 2006).

The exact number of Baltimore City properties with ground rent is not known, but is estimated anywhere from 74,000 to 120,000. Areas with a high percentage of homes with ground rents correspond closely with previously viewed patterns of high poverty, low incomes, and African American concentrations. The stark contrast along York Road running north-south in the north-central part of the city shows that ground rents are prevalent on the east side (an area of lower incomes) and are largely absent on the more affluent west side. Other areas with a high percentage of properties with ground rents include Cherry Hill in the south, a northeast prong southeast of Pulaski Highway, a prong stretching northwest along Reisterstown Road, and heavy concentrations just east and west of the downtown area.

Many of those who lose their homes do so because they are unable to pay the attorney's fee on top of overdue ground rent. In early 2007, legislation was introduced in Annapolis to protect real estate owners by allowing ground-rent owners to place a lien on the property but not seize it. Maryland Senate Bill 106/House Bill 172 amended the Real Property Articles that prohibited the creation of any new ground rents. Additional legislation from 2007 prohibits ground-rent leaseholders from collecting more than three years of back-ground rent. Ground leaseholders now have three years to officially register their ground-rent leases or face forfeiture of future collection of the leases. The law also ensures that leaseholders and lessees have each other's contact information, so collection problems should not result from a failure to receive payment notices. Leaseholders now have the burden to give timely notice to the lessee of an outstanding debt or to provide an option for the lessee to buy out the lease and own the property. By mid-2007, leaseholders challenged the constitutionality (both on a national level and in Maryland) of the laws prohibiting new ground rents and requiring leaseholders to comply with Maryland's registry requirement.

Washington, DC

Although not part of Maryland, Washington, DC, directly influences much of the state, especially in surrounding Charles, Montgomery, and Prince George's Counties. No social geography of Maryland would be complete without recognizing the importance of the relative location of this capital city sited on land that was once part of Maryland.

Despite the importance of the United States in world affairs, its capital is of modest size when compared to other capital cities such as London, Paris, and Tokyo. But land area and population size are not the best indicators of Washington's importance. While other larger capitals often developed as multipurpose urban places, Washington was built as—and largely remains—a single-purpose city, functioning principally as the seat of the federal government. Although employment in a wide variety of other activities (research firms, national headquarters of corporations and nonprofit organizations, interest groups, tourism, etc.) has exceeded government employment, many of these activities support the city's national and international administrative and decision-making roles. The complexity of organization and administration centered on this city can be understood only in the regional context of its entire metropolitan area extending into Maryland, Virginia, and West Virginia.

Between 1774 and 1789, the Continental Congress met in eight different towns and cities, eventually determining that the new republic needed a central seat of government. Through compromise, the Continental Congress decided to locate the new national capital on a central and accessible estuary that was navigable for seagoing vessels. The Northern states favored the Delaware River estuary, the Southern states the Potomac. The southerly location eventually prevailed, and in 1790 George Washington finalized the specific location (Harper and Ahnert 1968). The site he selected along the Potomac River lay close to the geographical center of the thirteen states' area and population. George Washington expected the new capital to become the commercial center and leading seaport of the country, but the growth of nearby Baltimore outpaced Washington, and the capital became part of the commercial hinterland of Baltimore.

The 10- by 10-square-mile (16- by 16-square-kilometer) area for the capital transferred to the federal government in 1791 by Virginia and Maryland was designated as the District of Columbia, including the ports of Alexandria on the Virginia side of the Potomac River and Georgetown on the Maryland side. George Washington commissioned Pierre L'Enfant, a French military engineer, to design the city plan. L'Enfant's plan resembled his earlier one for Versailles, France: a rectangular grid superimposed with broad, diagonal-running avenues, many converging on the Capitol and White House. Other significant focal points in the form of circles were included in the plan. Today, the city is divided into quadrants, with the Capitol as the central focus.

In 1800, the government moved from Philadelphia to its new seat, which in 1794 had acquired the name Washington City (Harper and Ahnert 1968). Most of the development subsequently took place on the Maryland side of the Potomac, and in 1846 the Virginia side of the District of Columbia returned by plebiscite to the state of Virginia. Today, the City of Washington and the District of Columbia are coextensive, being governed by the same municipal government and considered the same entity. The city has a mayor and a full municipal government, but until 1871, when Georgetown finally joined the city, there were multiple jurisdictions within the District of Columbia. The US Congress is the final authority in the city, and it has the authority to overturn any measures passed by the municipal government, which means that residents of Washington have a unique status within the nation, lacking representation within the federal government (i.e., residents of Washington do not have elected representation in the US Senate or House of Representatives).

Until 1920, most of the population growth in what is now the metropolitan area took place within the District of Columbia itself. Since then, the growth rate of the suburbs has been higher than the central city. By 1953, the population of the suburban area, stretching beyond the boundaries of the District, had surpassed that of the central city. Little land remains for housing within Washington, which has over 60 percent of its land in nontaxable public and semipublic ownership (including streets and parks), while the average for other US cities is 34 percent (Harper and Ahnert 1968). The bulk of civilian employment in the federal government is in the executive branch, with the legislative and judicial branches employing less than 10 percent of the total. In addition to the administrative functions of its many departments and agencies, the federal government maintains extensive research operations. The more than fifty major research facilities in the metropolitan area include the Beltsville Agricultural Research Center, Central Intelligence Agency, Defense Intelligence Agency, NASA Goddard Space Flight Center, National Bureau of Standards, National Geospatial-Intelligence Agency, National Institutes of Health, National Weather Bureau, Naval Medical Center, Naval Observatory, Naval Oceanographic Office, Naval Research Laboratory, Smithsonian Institution, US Geological Survey, and Walter Reed Army Medical Center.

Many private nonprofit organizations and businesses specializing in research and consulting find it advantageous to be located in or near Washington, DC. Geographical proximity to government personnel and the vast information resources of government libraries and archives remain strong attractive factors. Several major universities enhance the research environment, including American University, Catholic University, Gallaudet University (the first school for the advanced education of the deaf, established in 1857), George Mason University, Georgetown University, George Washington University, Howard University, and the University of Maryland. Numerous other research universities maintain a significant presence in the city, such as Johns Hopkins University's Paul H. Nitze School of Advanced International Studies. The nation's capital is second only to New York among US metropolitan areas in the total number of scientists. Nearly all of the nation's 500 largest corporations have some presence in the Washington area. Another sizeable sector of the city's economy is tourism and conventions. Tourism generates over $10 billion annually in direct spending and generates about 260,000 jobs. The millions of visitors who come to the nation's capital every year include tourists, conventioneers, politicians, diplomats, government officials, military personnel, scientists, educators, students, lobbyists, businesspeople, petitioners, and others. Washington is indeed one of the major crossroads of the world.

TOURISM

Although tourists come to Maryland from all over the United States and overseas, most visitors live within the region. The top source of domestic out-of-state visitors to Maryland is its immediate neighbor to the north, Pennsylvania, followed by Virginia. This should not be surprising, owing to the geographic proximity and the significant population sizes of these neighboring states. Delaware accounted for only 2.3 percent of visitors to Maryland in 2001, owing to its small size and being overshadowed by the larger states in the region. New York and New Jersey rank third and fourth, respectively. Florida ranks fifth among places of origin for visitors. Located along Interstate 95, the major north–south highway on the East Coast, Maryland receives a large number of visitors from Florida traveling north and south, especially going northward during the summer and holiday seasons. The significant number of visitors from California (seventh), and Ohio (eighth) and various other states (totaling 15.6%) reflects the drawing power of sites of national significance such as Annapolis, Antietam Battlefield, Fort McHenry, and especially the National Capital in Washington, DC.

The central region of Maryland receives the most visitors, with Baltimore City being the most visited Maryland site. Its many attractions include Fort McHenry; the Inner Harbor with its complex of shops, the National Aquarium, Maryland Science Center, and Hard Rock Café; the Baltimore Orioles, Ravens, and other sporting events; and of course shopping. The Maryland Office of Tourism and Development has documented the most popular activities of persons visiting Maryland, and shopping leads the list (Maryland Department of Economic and Community Development 2003). The Eastern Shore received the second-highest number of visitors in 2001. After Baltimore City, Ocean City is the leading attraction for tourists in Maryland. People come to Ocean City for the beach, amusements, and festive atmosphere of this longtime resort town on the Atlantic Coast. Summer travel across the Bay Bridge and along Route 50 to Ocean City can be a trying experience during the frequent heavy traffic periods, however. Maryland's colonial state capital and the US Naval Academy in Annapolis are major attractions for visitors. Ranking third for total visitors, Annapolis continues to attract tourists from across the United States and overseas to its beautiful bayside harbor, restaurants, and many shops.

Being close to Washington, DC, is also beneficial to the commercial tourist trade of Annapolis, as it is for much of the rest of the state. People who come from afar to visit the nation's capital usually visit sites in Maryland as well. Indeed, there is a geographic agglomeration of major cultural, historic, and recreational attractions between Norfolk, Virginia, and Philadelphia that has become a regular part of the grand American family automobile tour.

Urban Sprawl

There is a growing concern in all parts of the United States that runaway urbanization and its decaying cities, suburban strip malls, extensive residential subdivisions, traffic jams, and worsening pollution are destroying the nation's forests, farmlands, and open space. Although this perception of the changing landscape is continually gaining support, it is not unquestioned. Both proponents and critics of policies to slow or halt urban sprawl present powerful arguments. The sides have vigorously engaged in this debate, and the outcome will have far-reaching effects on how the landscape will evolve.

The Process of Sprawl

A global shift in human settlement patterns has occurred in recent years. In 2007, for the first time, more people worldwide lived in urban than rural areas, which may be "the single greatest change our society has undergone since its transition from a nomadic hunter-gatherer lifestyle to a sedentary agricultural one" (Withgott and Brennan 2007). Since the beginnings of urbanization several thousand years ago, urban centers have depended on an outside hinterland to supply them with food, fiber, labor, and other resources. That is still the case today.

Nationally, as in Maryland, many cities by the last quarter of the twentieth century had more working age people than they had jobs for, which led to rising unemployment, poverty, crime, and despair. Starting in the 1950s, when the lack of employment was increasingly crippling cities, many affluent urban dwellers moved to the suburbs, attracted by lower population density, more open space, lower crime rates, better schools, lower-priced homes, and lower real estate taxes. State and local zoning laws, federal housing policies, and the construction of the US Interstate Highway System all worked to facilitate the exodus to the suburbs (Withgott and Brennan 2007). As the suburbs attracted more people from the cities, those who could not afford to move were left behind, exacerbating the decline of the central cities and creating a vicious cycle of migration, urban decline, and suburban sprawl over the farmlands and open spaces of the country that increased in intensity as time passed. Early suburbs adjacent to central cities eventually became crowded. These older inner suburbs then began to suffer the conditions of the central cities. New successive rings of suburbs grew, ranging farther and farther from central cities. Suburbanite commuters were traveling longer distances to work (often in the central city) and spending more of their day stuck in traffic.

The phenomenon of suburban growth and spread came to be called *sprawl*. Many of the definitions of sprawl were judgmental, describing its effects with negative connotations. Defining sprawl as "the spread of low-density urban or suburban development outward from an urban center" is one nonjudgmental definition with widespread support (Withgott and Brennan 2007). Regardless of how it is defined, urban sprawl results in the allocation of more physical space per person than in cities (i.e., lower densities outside the cities). Based on this process, some researchers further define sprawl as the physical spread of development at a rate greater than the rate of population growth. From 1950 to 1990, for example, the population of fifty-eight major US metropolitan areas grew by 80 percent, but the land they covered increased by 305 percent (Withgott and Brennan 2007).

Managing Sprawl

More Marylanders are feeling the effects of sprawl on their daily lives. Commuting times are getting longer, pollution is worsening, open space is disappearing at an alarming rate, and the pressures from growth are enormous. Beginning in 1997, Maryland attempted to manage its growth through a whole series of policies and programs that became known as *smart growth*. In 1998, the director of a land-use institute in Michigan called the Maryland Smart Growth program "the most promising new tool for managing growth in a generation" (Goodman 1999). Even so, smart growth has fallen short of fulfilling its promise.

Smart Growth

A basic goal of smart growth is to build up, not out. Economic investment focuses on existing urban centers and multistory, multiuse buildings over one-story single-family homes spread over large areas. Related to smart growth is a movement among planners and architects called *new urbanism*, a method of planning based on designing neighborhoods with walkable scales, where homes, businesses, schools, public buildings, and other amenities are located close together. The goal is to create neighborhoods in which most needs can be met without using a car. Table 12.1 summarizes ten principles of smart growth.

Three main factors shape smart growth and antisprawl programs in Maryland (Cohen 2002).

1. Widespread public desire to preserve the health of the Chesapeake Bay.
2. Strong resistance to state intervention in local land-use planning.
3. Political tension between urbanized and less populated jurisdictions.

Maryland faced the inability of state and local planning regulatory practices to control suburban sprawl between 1990 and 2000, when Baltimore City lost over 84,000 residents, but the population of the Baltimore region outside the city grew by nearly 256,000. As a result, Maryland leaders were thinking about mechanisms to curtail sprawl before public officials in most other states had even begun to define the problem. No major book written on sprawl in the United States is without a section on Maryland, a pioneering state in growth management. In 1997, Governor Paris Glendening created the Office of Smart Growth to oversee the state's efforts, and the Maryland legislature passed laws and programs to promote smart growth.

Although Maryland's policies and programs brought to light the issue of smart growth, the reality is that implementation has been slow and the outcomes thus far disappointing (Cynkar 2007a). "To bring about

Table 12.1. Ten principles of smart growth

1.	Mix land uses
2.	Take advantage of compact building design
3.	Create a range of housing opportunities and choices
4.	Create walkable neighborhoods
5.	Foster distinctive, attractive communities with a strong sense of place
6.	Preserve open space, farmland, natural beauty, and critical environmental areas
7.	Strengthen and direct development toward existing communities
8.	Provide a variety of transportation choices
9.	Make development decisions predictable, fair, and cost effective
10.	Encourage community and stakeholder collaboration in development decisions

Source: US Environmental Protection Agency, 2005

something other than conventional suburban development, you're really turning a supertanker. Glendening's Office of Smart Growth and the policies and procedures he put in place have just started to make a dent in solving sprawl in Maryland" (Flint 2006). Smart growth implementation in Maryland has encountered significant resistance from some counties. In Maryland, county and municipal governments approve new development. The state gives incentives to limit growth—for example, by providing funds for water and sewer service, highways, schools, and economic development—but only in areas approved by the state for growth under smart growth programs. Counties may still opt to approve development outside these areas, but without state assistance.

Overall, decisions at the local level still induce sprawl. Marylanders continue to chew up farmland, fragment agricultural land, and move farther from urban cores (Cynkar 2007b). Maryland is sandwiched between Baltimore and Washington, two major metropolitan areas that have grown together, and is strongly influenced by Philadelphia. Real estate in these urban areas is expensive, so people seek lower land prices in formerly rural areas farther away from cities. It is not in the densely populated counties (Baltimore City as well as Baltimore, Montgomery, and Prince George's Counties) that population is growing fastest, but in the formerly rural counties such as Calvert, Cecil, Charles, and Washington.

Conclusion Maryland Transformed

Assessing Maryland's future is not simply an exercise in extrapolating the past. Several factors are critical to the state's success in the future, including demographic shifts and population growth, environmental challenges, technological innovation, and global economic influences. These factors will act synergistically to present the people of Maryland with both challenges and new opportunities as they strive for sustainable development. The future state will be even more deeply integrated with the global system, as will the rest of the nation. Changes taking place in many other parts of the world are already affecting Maryland residents in Chevy Chase, Elkton, Frederick, Oakland, Salisbury, and Westminster.

As the Maryland landscape changes over time, patterns from the past will continue to influence its land and people. Today the processes of population growth, landscape transformation, and environmental change are occurring at an increasingly rapid rate. No longer is it adequate to watch trends unfold over a relatively long period of time, reacting only when problems approach the crisis stage. With so many rapid changes bringing myriad challenges, Maryland is experiencing "time compression," running on fast forward, and its public and private decision makers need to be more proactive. No sooner does one new landscape emerge than it transforms into another, the parchment continually and rapidly being erased and redrawn. Indeed, Maryland is a palimpsest—its changes over time visible below the surface—with the new being created over the old, but with vestiges of the old still seen on the landscape.

The introduction to this book discussed some of the major challenges to Maryland's quality of life (table I.1). Among these challenges is environmental degradation, especially air and water pollution and the serious condition of the Chesapeake Bay. Another set of challenges focuses on major transformations in the state's economy. Not only do many residents make a living in jobs that did not exist just two decades ago, they must also adapt rapidly to meet present and future economic changes. The last group of challenges, population growth, is fundamental to all other challenges facing Maryland. The growth rate of the Maryland population will soon lead to one million more people in the state. These additional residents, and with them their geographical distribution and characteristics, will continue to be the primary factor affecting both the future natural environment and economy of the state.

In the late 1990s, Maryland's state government began to address the serious issues of population growth and urban sprawl. The results of these efforts have fallen far short of expectations. But even if fully implemented and embraced by jurisdictions statewide, smart growth measures are not designed to halt growth; instead, they seek to redirect and manage growth and its impact on the landscape, and to limit growth to specific geographic areas of Maryland. The multipronged efforts of smart growth seek to preserve farmland and open space, withhold state funding for development that exacerbates urban sprawl, and revive older settlements by making them more appealing. Despite these measures, Maryland's population is projected to grow from its present 5.5 million to about 6.3 million by 2025. Considering natural increase, net interstate migration, and net international migration, Maryland added on average 130 people per day during the 1990s; that growth continues today. Although recognized as a national leader in smart growth planning, Maryland must take smart growth to the next level if it is to sustain its environment, infrastructure, governmental services, economy, and the "Maryland way of life." Estimates project that during the first quarter of the twenty-first century, Maryland will see as much land developed as during its first 366 years as a state (Bouvier and Stein 2000).

More residents of Maryland are coming to realize that well-intentioned smart growth policies are not enough to sustain the state's high quality of life. In a survey sponsored by the Maryland Higher Education Commission, "Solutions for Maryland's Future," Maryland voters cited education (specifically primary and secondary), economic issues, and crime and safety as the most important issues facing Maryland's leaders (Maryland Higher Education Commission 2006a). Higher education, health care, and the environment were among the second tier of challenges.

Workers today compete for jobs not only with other Maryland residents and those from other states, but also with highly educated people from all over the world. Maryland students need an education that helps them to survive in a world where ideas, money, services, and people are constantly in motion, freed from the constraints of state and national boundaries (Kamdar 2007). Numerous international surveys indicate that Americans are not competing well in global higher education. To thrive in the global human capital economy, Maryland must make additional investments in education, especially in higher education. Maryland's increasingly diverse population is a critical component of the state's future. Black, Hispanic, Asian, American Indian, and other minority children are gradually replacing the white non-Hispanic population. These minority groups represent an ever-growing part of Maryland's future workforce. The majority of Maryland's minority children come from families with incomes that are lower than those of white non-Hispanics; they will require better access to higher education and increased support through need- and merit-based scholarships and programs if they are to play a successful role in sustaining Maryland's quality of life.

By 2025, Maryland is projected to have nearly one million new residents, 500,000 new homes, and 600,000 new jobs. Where will Marylanders live, work, and play? Between 1973 and 2010, developed lands in Maryland increased by 154 percent while population grew by only 39 percent. Today, over 50 percent of all developed land in Maryland holds only 15 percent of the state's houses. Since 1973, over 1 million acres (400,000 hectares) of agricultural and forestlands have been developed. If development trends do not change by 2025, another 400,000 acres (160,000 hectares) of farms and forest will be lost.

The landscape and socioeconomic structures of the state are transforming rapidly. How Maryland residents make a living is changing. Minority populations are growing, and with them social values are evolving, too. Can all of these factors be brought together and focused on a path to sustainable growth for Maryland? The future of Maryland's landscape and lifestyle is in the hands of its political, educational, business, and nonprofit organizational leaders as well as the public.

The story of Maryland is a story of a special place. From its western mountains to central urban centers to the Chesapeake Bay and Atlantic beaches, Maryland is still a tremendous place to live, work, and play. Its rich cultural, historical, and geographical heritages are undeniable, but in this age of affluence, a gap is growing between affluent and poorer Marylanders. Can Maryland's leaders help all its citizens, including the poor, realize their human potential? Will Maryland residents—natives and newcomers alike—make the commitments, sacrifices, and investments necessary to sustain this place called Maryland?

REFERENCES

AgrAbility Project. "Broiler Chicken Production in the US." *AgrAbility Quarterly* 5, no. 2 (March 2005): 3. http://fyi.uwex.edu/agrability/files/2010/02/AgrAbility-Quarterly-March-2005.pdf.

Alexander, John, and Ley Gibson. *Economic Geography*, 2nd ed. Englewood Cliffs, NJ: Prentice Hall, 1979.

Alexander, Robert. "Baltimore Row Houses of the Early Nineteenth Century." *American Studies* 16 (1975): 65–76.

Alford, John. "The Chesapeake Bay Oyster Fishery." *Annals of the Association of American Geographers* 65 (1975): 229–32.

Alliance for the Chesapeake Bay. "Striped Bass Reproduction Up in Maryland." *Bay Journal* (November 1999): 8.

———. "Food for Thought: NRI Illustrates Rate That Sprawl Is Eating Landscape." *Bay Journal* (April 2000): 19.

Ambrose, Eileen. "Gaps in Transportation Hurt State, Experts Say." *Baltimore Sun*, November 10, 2000, C1.

American Medical Association. *Physician Characteristics and Distribution in the US*. Chicago: American Medical Association, 2009.

Arnett, Earl, Robert Brugger, and Edward Papenfuse. *A New Guide to the Old Line State*. Baltimore: Johns Hopkins University Press, 1999.

Arnold, Joseph. "Suburban Growth and Municipal Annexation in Baltimore, 1745–1918." *Maryland Historical Magazine* 73, no. 2 (Summer 1978): 109–28.

Association of American Medical Colleges. *2011 State Physician Workforce Data Book*. Washington, DC: 2011. www.aamc.org/download/263512/data.

Atkinson, Robert, Luke Stewart, Scott Andes, and Stephen Ezell. *Worse Than the Great Depression: What the Experts Are Missing about American Manufacturing Decline*. Washington, DC: Innovation Technology & Innovation Foundation, 2012. http://www2.itif.org/2012-american-manufacturing-decline.pdf.

Baker, Henry S. "Maryland Tobacco: Certain Aspects of a 300 Year Old Enterprise." PhD diss., Rutgers University, 1957.

Baltimore Sun. "A Bright Future for Baltimore's Port." November 14, 1999a, C2.

———. "Port Seeks Business for Future." November 28, 1999b, D4.

———. "Baltimoreans Going to Court for Not Paying Rent." December 6, 2006.

———. "Election 2008: US in Miniature." February 11, 2008a, 6E.

———. "In a Rich State, Some Are Still Left Behind." March 9, 2008b, 19A.

———. "Controlling Growth." March 13, 2008c, 16A.

Bernard, Richard M. "A Portrait of Baltimore in 1800: Economic and Occupational Patterns in an Early American City." *Maryland Historical Magazine* 69 (1974): 341–60.

Blankenship, Karl. "Maryland Begins Ariakensis Oyster Research in Three Rivers." *Bay Journal* (June 2004): 1. http://www.bayjournal.com/article/maryland_begins_ariakensis_oyster_research_in_3_rivers.

Blood, Pearl. *The Geography of Maryland*. Boston: Allyn and Bacon, 1961.

Bode, Carl. *Maryland: A History*. New York: W. W. Norton, 1978.

Bouvier, Leon, and Sharon McCloe Stein. *Maryland's Population in 2050: Is Smart Growth Enough?* Washington, DC: Negative Population Growth, 2000.

Brooke, Richard, and Henry Baker. "A Thermometrical Account of the Weather for One Year, Beginning September 1753." *Philosophical Transactions of the Royal Society of London* 51 (1759): 58–69.

Brugger, Robert. *Maryland: A Middle Temperament 1634–1980*. Baltimore: Johns Hopkins University Press, 1988.

Cahill, Julie. "The Crab: A Big Catch or Just a Trap?" *Cross Sections* 13, no. 2 (1996): 24–28.

Capps, Randolph, and Karina Fortuny. "The Integration of Immigrants and Their Families in Maryland: The Contributions of Immigrant Workers to the Economy." Washington, DC: Urban Institute, 2008.

Caudill, Henry M. *Night Comes to the Cumberland*. Boston: Little, Brown, 1962.

Chappelle, Suzanne, Jean H. Baker, Dean R. Esslinger, Whitman H. Ridgway, Constance B. Schulz, and Gregory A. Stiverson. *Maryland: A History of Its People*. Baltimore: Johns Hopkins University Press, 1986.

Chesapeake Bay Foundation. "American Shad: The Forgotten Fishery." Annapolis, MD: Chesapeake Bay Foundation, 2010. http://www.cbf.org/about-the-bay/more-than-just-the-bay/creatures-of-the-chesapeake/american-shad.

Clark, Marsha. "Coal Mining in Western Maryland: A Closer Look." *Metro News* 6, no. 10 (February 1978): 2.

Clemons, Josh. "Supreme Court Rules for Virginia in Potomac Conflict." University, MS: National Sea Grant Law Center, 2003. http://nsglc.olemiss.edu/SandBar/SandBar2/2.4supreme.htm.

Clemons, Paul. "From Tobacco to Grain: Economic Development of Maryland's Eastern Shore, 1660–1720." PhD diss., University of Wisconsin, 1974.

Coddington, James, and David Derr. *An Economic Study of Land Utilization in the Tobacco Area of Southern Maryland*. College Park: University of Maryland, 1939.

Cohen, James R. "Maryland's Smart Growth: Using Incentives to Combat Sprawl." In *Urban Sprawl: Causes, Consequences and Policy Responses*, edited by G. Squires, 293–320. Washington, DC: Urban Institute, 2002.

Committee on Non-Native Oysters in the Chesapeake Bay. *Non-Native Oysters in the Chesapeake Bay*. Washington, DC: National Research Council, National Academies Press, 2003.

Coyle, Wilbur F., ed. *Records of the City of Baltimore: Eastern and Western Precinct Commissioners, 1810–1817*. Baltimore: Baltimore City, 1909.

Cynkar, Amy. "Sprawl Nation." *Urbanite*, January 2007a, 59–61.

———. "The State of Sprawl." *Urbanite*, February 2007b, 63–65.

Dance, Scott. "Severstal to Sell Sparrows Point Steel Mill to Renco Group." *Baltimore Business Journal*, March 2, 2011, 1.

Dozer, Donald M. *Portrait of the Free State: A History of Maryland*. Cambridge, MD: Tidewater Press, 1976.

Earle, Carville, and Ronald Hoffman. "Staple Crops and Urban Development in the Eighteenth-Century South." *Perspectives in American History* 10 (1976): 7–78.

Edgar, J. *Coal Processing and Pollution Control*. Houston: Gulf Publishing, 1983.

Edwards, Jonathan. "Maryland's Metallic Mineral Heritage." *Maryland Conservationist*, August 1967, 1.

Fincham, Michael. "The Battle over Blue Crabs: Capping the Last Great Fishery." *Maryland Marine Notes* 12, no. 1–2 (1994a): 4.

———. "The Blue Crab in Winter." *Maryland Marine Notes* 12, no. 1–2 (1994b): 5.

Flint, Anthony. *This Land: The Battle over Sprawl and the Future of America*. Baltimore: Johns Hopkins University Press, 2006.

Franklin, John, and Alan Doelp. *Shocktrauma*. New York: St. Martin's Press, 1980.

Garrett County Development Corporation. *Feasibility Study for a Coal Preparation Facility*. Oakland: Garrett County, Maryland, 1978.

Gibbons, Boyd. *Wye Island: The True Story of an American Community's Struggle to Preserve Its Way of Life*. Baltimore: Johns Hopkins University Press, 1977.

Giddens, Paul H. "Land Policies and Administration in Colonial Maryland, 1753–1769." *Maryland Historical Magazine* 28 (1933): 142–71.

Gilchrest, Wayne. "Deepening C&D Canal Not the Best Solution." *Bay Journal*, September 2000, 19.

Goldstein, Mark. "Income Inequality Continues to Grow in Maryland." Baltimore: Maryland Department of Planning, 2002. www.mdp.state.md.us/msdc/income_inequality/incomeinequality_1980_2000.pdf.

———. "Maryland's Changing Demographics." Paper presented at the Maryland State Department of Planning Leadership Challenge XV, June 13, 2006. http://planning.maryland.gov/msdc/Affiliate_meeting/2011/SDC_Affiliates_Meeting_Nov10_2011.pdf.

Goodman, Peter S. "An Unsavory By-Product: Runoff and Pollution." *Washington Post*, August 1, 1999, A1.

Gottmann, Jean. *Megalopolis: The Urbanized Northeast Seaboard of the U.S.* New York: Twentieth Century Fund, 1961.

Greenbelt Museum. "History." Greenbelt, MD: Greenbelt Museum, 2013. http://greenbeltmuseum.org/history/.

Hancock, Jay. "BWI Menu for Foreign Trips Offers Cold, Water." *Baltimore Sun*, November 12, 2006.

Hanna, Stephen P. "Finding a Place in the World Economy: Core-Periphery Relations, the Nation-State and the Underdevelopment of Garrett County, Maryland." *Political Geography* 14, no. 5 (1995): 451–72.

Harper, Robert, and Frank Ahnert. *Introduction to Metropolitan Washington*. Washington, DC: Association of American Geographers, 1968.

Henry, Christine. "Bethlehem Celebrates New Mill amid Worries." *Baltimore Sun*, September 22, 2000, 1D.

Hoover, Edgar. *An Introduction to Regional Economics*. New York: Alfred Knopf, 1975.

Horton, Tom. "Government, Business Seek to Fight Pollution." *Baltimore Sun*, September 8, 2000, 2B.

Hungerford, Edward. *The Story of The Baltimore & Ohio Railroad: 1827–1927.* 2 vols. New York: Arno Press, 1928.
International Trade Administration. "Maryland: Expanding Exports and Creating Jobs through Trade Agreements." Washington, DC: US Department of Commerce, January 2014. http://www.trade.gov/mas/ian/statereports/states/md.pdf.
Kamdar, Mira. *Planet India.* New York: Scribner, 2007.
Kenny, Hamill. *The Origin and Meaning of the Indian Place Names of Maryland.* Baltimore: Waverly Press, 1961.
Klinkner, Phillip A. "Red and Blue Scare: The Continuing Diversity of the American Electoral Landscape." *Forum* 2, no. 2 (2004): 1–12.
Koontz, Michael, Laura Niang, Timothy Sletten, Mark Stunder, Radhika Narayanan, and Kent Barnes. "Maryland Hazard Analysis." Baltimore: Maryland Emergency Management Agency, 2000.
Kuff, Karen. "Gold in Maryland." Baltimore: Maryland Geological Survey, 1987. http://www.mgs.md.gov/esic/freeseries.html.
Lang, Robert, and Jared Lang. "The 2008 Presidential Race: A Geographic Analysis of Four Virginias." Blacksburg, VA: Metropolitan Institute of Virginia Tech, 2009.
Leffler, Merrill. "Oyster Farming vs. Oyster Hunting." College Park: Maryland Sea Grant, University of Maryland, 2000. http://mdk12.org/instruction/curriculum/hsa/biology/oysters/history/.
———. "A New Oyster for the Bay?" *Chesapeake Quarterly* 1, no. 3 (2002): 2–9.
Lerner, Neal. "In the World of the Watermen, Window on a Work Culture: Riding the Wake of a Chesapeake Bay Crabber." *Christian Science Monitor,* June 2000, 11.
Lesh, Jonathan. "America's Political Landscape: A Focus on Maryland." Towson, MD: Towson University Department of Geography, 2010.
Little, Robert. "Bottom of the Bay Stirs Controversy." *Baltimore Sun,* November 23, 1999, A1.
LoLordo, Ann. "Sun Journal: Earth Lovers Battle Green Aliens." *Baltimore Sun,* June 11, 2000, 2A.
Luckenbach, Mark. "Crassostrea Ariakensis: Panacea or Pandora?" College Park: University of Maryland Center for Environmental Science, March 2004. http://ian.umces.edu/pdfs/ian_newsletter_8.pdf.
Main, Gloria. "Maryland and the Chesapeake Economy, 1670–1720." In *Law, Society, and Politics in Early Maryland,* edited by L. G. Carr Aubrey Land and Edward Papenfuse, 134–55. Baltimore: Johns Hopkins University Press, 1977.
Mann, Cindy, Andy Scheider, and Sara Thom. "The State Tobacco Settlements." Washington, DC: Center on Budget and Policy Priorities, 1999.
Marcus, Morton J. "Density: How Concentrated Is Our Population?" *Indianan Business Review* 81 (Summer 2006): 1–5. http://www.ibrc.indiana.edu/ibr/2006/summer/summer06.pdf.
Maryland Bureau of Mines. "Annual Report of the Maryland Bureau of Mines." Westernport: Maryland Department of Environment, Land Management Administration, Mining Program, 1994–2009.
Maryland Business Journal. "Oyster Industry Is Alive and Well." *Maryland Business Journal* (August-September 1980): 20.
Maryland Department of Business and Economic Development. *MaryLand*

of Opportunities: Rankings. Baltimore: Maryland Department of Business and Economic Development, 2013. http://www.choosemaryland.org/factsstats/pages/rankings.aspx.

Maryland Department of Economic and Community Development. *Maryland Statistical Abstract*. Towson, MD: Regional Economic Studies Institute, Towson University, 2003.

Maryland Department of Natural Resources. *A Review of the Maryland Coal Industry*. Columbia, MD: Versar, 1994.

———. "Introduction to Maryland's Greenways." Annapolis: Maryland Department of Natural Resources, 2000. www.dnr.state.md.us/greenways/intro.html.

Maryland Department of Transportation. "The World Port of Baltimore." Baltimore: Maryland Department of Transportation, Maryland Port Authority, 1978.

Maryland Geological Survey. "Maryland Prehistory." Baltimore: Maryland Geological Survey, 1976. (Revised by Dennis Curry, 1980.)

Maryland Grape Growers Association. "The 2010 Maryland Vineyard Survey." Damascus: Maryland Grape Growers Association, 2013. http://www.marylandgrapes.org/vineyards/2010VineyardSurvey.pdf.

Maryland Higher Education Commission. "Solutions for Maryland's Future." Baltimore: Maryland Higher Education Commission, 2006a. www.mhec.state.md.us/higherEd/SolMDFuture.asp.

———. *Trend Book*. Baltimore: Maryland Higher Education Commission, 2006b.

Maryland Municipal League. *Maryland's 157: The Incorporated Cities and Towns*. Annapolis: Maryland Municipal League, 2000.

Maryland Port Administration. "Foreign Commerce Statistical Report 2009." Baltimore: Maryland Port Administration, 2009.

———. "Port of Baltimore: Press Release." Baltimore: Maryland Port Administration, April 23, 2012.

Maryland Wine and Grape Advisory Committee. *Maryland Wine: The Next Vintage*. Annapolis: Maryland Department of Agriculture, 2005.

McCann-Murray, Sherry. *A Geologic Walking Tour of Building Stones of Downtown Baltimore, Maryland*. Educational Series No. 10. Baltimore: Maryland Geological Survey, 2001.

McKnight, Tom. *Regional Geography of the United States and Canada*. Upper Saddle River, NJ: Prentice Hall, 1997.

Mertz, Tawna. "Can Bay's Oysters Make a Comeback?" *Bay Journal*, September 6, 1999, 1.

Meyer, Eugene. *Maryland: Lost and Found Again*. Centreville, MD: Tidewater Press, 2003.

Miller, Fred. "Maryland Soils." University of Maryland Cooperative Extension Services Bulletin 212. College Park: University of Maryland, 1967.

Mitchell, Robert. "The Formation of Early American Culture Regions: An Interpretation." In *European Settlement and Development in North America: Essays on Geographical Change in Honor and Memory of Andrew Hill Clark*, edited by James Gibson, 66–90. Toronto: University of Toronto, 1978.

Mitchell, Robert, and Edward Muller. "Interpreting Maryland's Past: Praxis and Desiderata." In *Geographical Perspectives on Maryland's Past*, 1–50. College Park: University of Maryland Geography Department, 1979.

Mountford, Kent. "Lewis Eugene Cronin, 1917–1998: A Gentleman and a Scholar." *Bay Journal*, March 1, 1999, 2.

Nalewajko, Paul. "Impacts of the Poultry Industry for the Delmarva Region." San Francisco: San Francisco State University, 2004. http://www.slideserve.com/kemp/impacts-of-the-poultry-industry-for-the-delmarva-region-by-paul-nalewajko.

National Park Service. "Assateague Island National Seashore North End Restoration Project." Washington, DC: National Park Service, 2004. http://www.nps.gov/asis/naturescience/upload/ProjectIntroduction.pdf.

Olson, Sherry. *Baltimore*. Cambridge, MA: Ballinger, 1976.

O'Malley, Martin. "Economic Growth Strategy for Baltimore City: Building on Strengths." Baltimore: Baltimore City, 2000. http://www.baltoworkforce.com/documents/reports_econ_growth_strat.pdf.

Papenfuse, Edward. "Economic Analysis and Loyalist Strategies during the American Revolution: Robert Alexander's Remarks on the Economy of the Peninsula or Eastern Shore of Maryland." *Maryland Historical Magazine* 68 (1973): 173–95.

Parks, A. Franklin, and John B. Wiseman. *Maryland: Unity in Diversity.* Dubuque, Iowa: Kendall Hunt, 1990.

Petzrick, Paul. "An Overview of the Western Maryland Coal Combustion By-Products/Acid Mine Drainage Initiative." Annapolis: Maryland Department of Natural Resources, Maryland Power Plant Research Program, 1999. http://www.techtransfer.osmre.gov/nttmainsite/library/proceed/ccb1996/back.pdf.

Pickett, Edward G. "Rouse Describes New City's Plans." *Baltimore Sun,* November 12, 1964, 56.

Pietilia, Antero. *Not in My Neighborhood*. Chicago: Ivan R. Dee, 2010.

Porter, Frank W. *Maryland Indians, Yesterday and Today.* Baltimore: Maryland Historical Society, 1983.

Raitz, Karl. "The National Road: Life into Landscape." *Humanities* 15, no. 2 (1994): 20–24.

Reed, J. C., Jr., and J. C. Reed. "Gold Veins near Great Falls, Maryland." US Geological Survey Bulletin 1286. Reston, VA: US Geological Survey, 1969.

Reutter, Mark. "Shadow of Steel's Lost Empire." *Baltimore Sun,* October 19, 2001.

———. *Making Steel.* Champaign: University of Illinois Press, 2004.

Richards, William, and Paul Ticco. "The Suminoe Oyster, Crassostrea Ariakensis, in the Chesapeake Bay." Virginia Sea Grant Program Grant No. NA96RG0025. Charlottesville: University of Virginia, 2002.

Schmidt, Martin. *Maryland's Geology.* Centreville, MD: Tidewater, 1993.

Sewell, Jane Elliot. *Medicine in Maryland: The Practice and the Profession 1799–1999.* Baltimore: Johns Hopkins University Press, 1999.

Shelsby, Ted. "Shore Tomato Canneries Dwindle with Crop." *Baltimore Sun,* August 24, 1980, K7.

Shelsby, Ted, and Heather Dewar. "Scientists Identify New Clam Parasite." *Baltimore Sun,* June 2, 2000, B2.

Shprentz, Deborah Sheiman. "Greenbelt, Maryland: Preservation of a Historic Planned Community." *Cultural Resource Management Bulletin* 22, no. 8 (1999): 53–56.

Simon, Anne. *The Thin Edge.* New York: Morrow-Avon, 1985.

Skaggs, David Curtis. *Roots of Maryland Democracy, 1753–1776*. Westport, CT: Greenwood Press, 1973.

Spencer, Jean E. *Contemporary Local Government in Maryland*. College Park: University of Maryland Bureau of Government and Research, 1965.

Thompson, Derek, ed. *An Atlas of Maryland*. College Park: University of Maryland Department of Geography, 1977.

Thoreen, Tim. "The Office of Surface Mining: Abandoned Land Mine Program." *Restoration and Reclamation Review* 3, no. 2 (Spring 1998): 1–7. http://conservancy.umn.edu/bitstream/11299/58963/1/3.2.Thoreen.pdf.

Trewartha, Glenn. "The Unincorporated Hamlet." *Annals of the Association of American Geographers* 33 (1943): 32.

US Bureau of Economic Analysis. *Regional Data: GDP and Personal Income*. Washington, DC: US Department of Commerce, Bureau of Economic Analysis, 2012. http://bea.gov/iTable/index_regional.cfm.

US Census Bureau. *Statistical Abstract of the U.S., State Rankings*. Washington, DC: US Department of Commerce, Census Bureau, 2010.

———. *Income*. Washington, DC: US Department of Commerce, Census Bureau, 2013. http://www.census.gov/hhes/www/income/index.html.

US Department of Agriculture. "2002 Census Volume 1, Chapter 1: State Level Data—Maryland." In *Census of Agriculture*. Washington, DC: US Department of Agriculture, 2002. http://www.agcensus.usda.gov/Publications/2002/Volume_1,_Chapter_1_State_Level/Maryland/.

Van Ness, James. "Economic Development, Social and Cultural Changes: 1800–1850." In *Maryland: A History, 1632,1974*, edited by Richard Walsh and William Fox, 188–190. Baltimore: Maryland Historical Society, 1974.

Vokes, Harold E. *Geography and Geology of Maryland*. Revised by Jonathan Edwards Jr. Baltimore: Maryland Geological Survey, 1968.

Warner, William. *Beautiful Swimmers*. New York: Penguin, 1976.

Washington County, Maryland. *The Task Force on Home Rule: Report to the County Commissioners for Washington County, Maryland*. Hagerstown: Washington County, Maryland, February 2006, 6.

Weaver, Kenneth, James Croffoth, and Jonathan Edwards. *Coal Reserves in Maryland: Potential for Future Development*. Annapolis: Maryland Geological Survey, 1976.

Weber, Ted. "Maryland's Green Infrastructure Assessment: A Comprehensive Strategy for Land Conservation and Restoration." Annapolis: Maryland Department of Natural Resources, Watershed Services Unit, 2003. http://dnrweb.dnr.state.md.us/download/bays/gia_doc.pdf.

Wheeler, Timothy. "Program Might Snuff Out Tobacco as Major Maryland Crop." *Baltimore Sun*, July 19, 2000, 1A.

———. "Md. Open-Space Program Lags behind Development." *Baltimore Sun*, August 7, 2006, 1A.

Withgott, Jay, and Scott Brennan. "Cities, Forests, and Parks: Land Use and Resource Management." In *Essential Environment: The Science behind the Stories*, chap. 9. San Francisco: Pearson/Benjamin Cummings, 2007.

Zimmerman, Tim. "How to Revive the Chesapeake Bay: Filter It with Billions and Billions of Oysters." *US News and World Report*, December 29, 1997, 63.

INDEX

Page numbers in *italics* refer to figures and tables.